10 0607408 4

KU-279-685

COUNTERING TERRORISM AND WMD

3 OCT 11 .00   HALL 06

Despite recent attempts to to international terrorism, including the US strikes against Afghanistan and Iraq, President Bush's declaration of a "War on Terror", this book argues that there is a fundamental need to consider how to consolidate and build efforts against terrorism through the creation of an informal global counter-terrorism network.

Drawing from their experience with Terrorism Early Warning groups and centers established in the USA, the editors explain why the need for such a network is so pressing and elaborate how it could be formed. Bringing together the opinions of experts from clinical medicine and public health, economics, public policy, and law enforcement, along with those from the military and intelligence communities, this volume strives to give some coherence and direction to the process of creating a global counter-terrorism network. It first identifies the nature of a global counter-terrorism network, then shows how such a global network could be created, and finally provides some guidelines for gauging its future effectiveness.

This book will be of great interest to researchers and professionals in the fields of terrorism and security studies, national security and intelligence studies.

Peter Katona is Associate Professor of Clinical Medicine at the David Geffen School of Medicine, UCLA. He is a member of the Los Angeles County Bioterrorism Advisory Committee for Public Health Preparedness and Response and the Biopreparedness Work Group Committee of the Infectious Diseases Society of America.

Michael D. Intriligator is Professor of Economics, Political Science, and Public Policy at UCLA where he is also Co-Director of the Jacob Marschak Interdisciplinary Colloquium on Mathematics in the Behavioral Sciences.

John P. Sullivan is a practitioner and researcher specializing in conflict studies, terrorism, intelligence, and urban operations. He is a member of the Los Angeles Sheriff's Department where he currently serves as a lieutenant. He is co-founder of the Los Angeles Terrorism Early Warning (TEW) Group and coordinates many of its activities.

CASS SERIES: POLITICAL VIOLENCE
Series Editors: Paul Wilkinson and David Rapoport

TERRORISM VERSUS DEMOCRACY: THE LIBERAL
STATE RESPONSE
Paul Wilkinson

AVIATION TERRORISM AND SECURITY
Paul Wilkinson and Brian M. Jenkins (eds)

COUNTER-TERRORIST LAW AND EMERGENCY POWERS IN
THE UNITED KINGDOM, 1922–2000
Laura K. Donohue

THE DEMOCRATIC EXPERIENCE AND POLITICAL VIOLENCE
David C. Rapoport and Leonard Weinberg (eds)

INSIDE TERRORIST ORGANIZATIONS
David C. Rapoport (ed.)

THE FUTURE OF TERRORISM
Max Taylor and John Horgan (eds)

THE IRA, 1968–2000: AN ANALYSIS OF A SECRET ARMY
J. Bowyer Bell

MILLENNIAL VIOLENCE: PAST, PRESENT AND FUTURE
Jeffrey Kaplan (ed.)

COUNTER-TERRORIST LAW AND EMERGENCY POWERS IN
THE UNITED KINGDOM, 1922–2000
Laura K. Donohue

RIGHT-WING EXTREMISM IN THE TWENTY-FIRST CENTURY
Peter H. Merkl and Leonard Weinberg (eds)

COUNTERING TERRORISM AND WMD

Creating a global counter-terrorism network

Edited by
Peter Katona, Michael D. Intriligator
and John P. Sullivan

University of Nottingham
Hallward Library

Routledge
Taylor & Francis Group

LONDON AND NEW YORK

1006074084

First published 2006
by Routledge
2 Park Square, Milton Park, Abingdon, Oxon OX14 4RN

Simultaneously published in the USA and Canada
by Routledge
270 Madison Ave, New York, NY 10016

*Routledge is an imprint of the Taylor & Francis Group,
an informa business*

© 2006 Peter Katona, Michael D. Intriligator and John P. Sullivan
selection and editorial matter; individual chapters © the contributors

Typeset in Times by
RefineCatch Limited, Bungay, Suffolk
Printed and bound in Great Britain by
Antony Rowe Ltd, Chippenham, Wiltshire

All rights reserved. No part of this book may be reprinted or
reproduced or utilized in any form or by any electronic,
mechanical, or other means, now known or hereafter
invented, including photocopying and recording, or in any
information storage or retrieval system, without permission in
writing from the publishers.

British Library Cataloguing in Publication Data
A catalogue record for this book is available from the British Library

Library of Congress Cataloging in Publication Data
Katona, Peter, Dr.
Countering terrorism and WMD : creating a global counter-terrorism
network / Peter Katona, Michael Intriligator and John Sullivan.
p. cm.—(Political violence series)
Includes bibliographical references and index.
1. Terrorism. 2. Terrorism—Prevention—International
cooperation. 3. Weapons of mass destruction. I. Intriligator,
Michael D. II. Sullivan, John P., 1959– III. Title. IV. Political
violence.
HV6431.K316 2006
363.325′17—dc22

ISBN 10: 0–415–38498–2 (hbk)
ISBN 10: 0–415–38499–0 (pbk)
ISBN 10: 0–203–08746–1 (ebk)

ISBN 13: 978–0–415–38498–8 (hbk)
ISBN 13: 978–0–415–38499–5 (pbk)
ISBN 13: 978–0–203–08746–6 (ebk)

CONTENTS

CONTENTS

CONTENTS

ILLUSTRATIONS

Figure

Tables

CONTRIBUTORS

Arabinda Acharya is Manager, Strategic Projects at the International Centre for Political Violence and Terrorism Research in the Institute of Defence and Strategic Studies, Nanyang Technological University, Singapore. His area of research includes conflict, political violence and human security. His area of specialization is South and Southeast Asia. His most recent publications include *Conflict and Terrorism in Southern Thailand* (coauthored with Rohan Gunaratna and Sabrina Chua, 2006).

Philip Bobbitt, A.W. Walker Centennial Chair in Law, University of Texas at Austin, is one of the nation's leading constitutional theorists. His interests include not only constitutional law but also international security and the history of strategy. He has published six books: *Constitutional Interpretation* (1991), *Democracy and Deterrence* (1987), *U.S. Nuclear Strategy* (with Freedman and Treverton) (1989), *Constitutional Fate* (1982), *Tragic Choices* (with Calabresi) (1978) and most recently *The Shield of Achilles: War, Peace and the Course of History* (2002). He is a member of the American Law Institute, The Council on Foreign Relations, the Pacific Council on International Policy, and the International Institute for Strategic Studies, and has served as Associate Counsel to the President, Counselor on International Law at the State Department, Legal Counsel to the Senate Iran-Contra Committee, and Director for Intelligence, Senior Director for Critical Infrastructure and Senior Director for Strategic Planning at the National Security Council. He is a former trustee of Princeton University and a former member of the Oxford University Modern History Faculty and the War Studies Department of King's College London.

Anneli Botha, Senior Researcher on Terrorism, Institute for Security Studies (ISS), South Africa, holds a Master's degree in Political Studies (Strategic Studies and African Politics) from the Rand-Afrikaans University, and is a former captain in the South African Police Service: Crime Intelligence.

Robert J. Bunker is a consultant to the Counter-OPFOR Program, National Law Enforcement and Corrections Technology Center-West, a program of the National Institute of Justice. He is also a member of the Los Angeles Terrorism Early Warning group. He holds degrees in political science, government, behavioral science, anthropology-geography, social science and history, and has counter-terrorism operational experience and training. He has taught in unconventional warfare, national security and counter-terrorism programs at the university and responder training levels. His most recent publications are the edited works *Non-State Threats and Future Wars* (2003) and *Networks, Terrorism and Global Insurgency* (2005).

Lindsay Clutterbuck, is a Research Leader with RAND Europe, specializing in terrorism and security-related issues. He recently retired from the Metropolitan Police Service in London where he served for over twenty-one years in a variety of counter-terrorism roles in the Specialist Operations Department at New Scotland Yard. For his Ph.D, he researched the origins and evolution of terrorism and counter-terrorism on the British mainland. He is a Senior Associate Research Fellow at Kings College London. He has contributed chapters to two books on contemporary counter-terrorism issues: "The Changing Face of International Terrorism in the UK," in *The Evolving Threat: International Terrorism in the post 9–11 Era* (ed. A. Pargeter, forthcoming, 2006) and "Law Enforcement," in *Attacking Terrorism: Elements of a Grand Strategy* (ed. A.K. Cronin and J.E. Ludes (2002). In addition, he has written two historically based articles on terrorism and counter-terrorism in the academic journal *Terrorism and Political Violence*: "The Progenitors of Terrorism: Russian Revolutionaries or Extreme Irish Republicans" (2004) and "Countering Republican Terrorism in Britain: Its Origins as a Police Function" (2006).

Barry Desker is Director, Institute of Defense and Strategic Studies (IDSS), Singapore, former Ambassador to Indonesia and a former CEO of the Trade and Development Board.

Michael D. Intriligator received his SB (Bachelor of Science) degree in Economics from the Massachusetts Institute of Technology in 1959, his MA in Economics from Yale University in 1960 and his Ph.D. degree from the Massachusetts Institute of Technology in 1963. He is Professor of Economics at the University of California, Los Angeles (UCLA), where he is also Professor of Political Science, Professor of Policy Studies in the School of Public Policy and Social Research, and Co-Director of the Jacob Marschak Interdisciplinary Colloquium on Mathematics in the Behavioral Sciences. He is also a Senior Fellow of the Milken Institute in Santa Monica, California and the Gorbachev Foundation of North America in Boston, Massachusetts. He is a Fellow of the Econometric

Society, a foreign member of the Russian Academy of Science, an AAAS Fellow of the American Association for the Advancement of Science, and an elected member of the Council on Foreign Relations (New York) and the International Institute for Strategic Studies (London). He is the author of more than 200 journal articles and other publications in the areas of economic theory and mathematical economics, econometrics, health economics, reform of the Russian economy, and strategy and arms control. He is the author of *Mathematical Optimization and Economic Theory* (1971, also translated into Spanish and Russian) and of *Econometric Models, Techniques, and Applications* (1978, also translated into Greek and Spanish; second edition, with Ronald G. Bodkin and Cheng Hsiao, 1996). He is coeditor, with Kenneth J. Arrow, of *The Handbook of Mathematical Economics* (1981, 1982, 1985); coeditor, with Zvi Griliches, of *The Handbook of Econometrics* (1982, 1983, 1986); and coeditor, with Kenneth J. Arrow, of the book series "Handbooks in Economics" published by Elsevier.

Brian M. Jenkins is a senior advisor to the president of the RAND Corporation and one of the world's leading authorities on terrorism. He founded the RAND Corporation's terrorism research program 30 years ago, has written frequently on terrorism, and has served as an advisor to the federal government and the private sector on the subject. He is a former Army captain who served with Special Forces in Vietnam, and also a former deputy chairman of Kroll Associates. He served as a captain in the Green Berets in the Dominican Republic and later in Vietnam (1966–1970). In 1996, he was appointed by President Clinton to be a member of the White House Commission on Aviation Safety and Security. He has served as an advisor to the National Commission on Terrorism (1999–2000) and in 2000 was appointed as a member of the US Comptroller General's Advisory Board. He is also a special advisor to the International Chamber of Commerce (ICC) and a member of the board of directors of the ICC's Commercial Crime Services. He has authored, coauthored or edited many books, including *Terrorism and Personal Protection* (1984), *Nation and Identity in Contemporary Europe* (1996), *Countering al Qaeda: An Appreciation of the Situation and Suggestions for Strategy* (2002) and *Aviation Terrorism and Security* (2006).

Mark Juergensmeyer is Professor of Sociology and Director of Global and International Studies at the University of California, Santa Barbara. He is author or editor of 15 books, including *Fighting Fair: A Nonviolent Strategy for Resolving Everyday Conflicts* (1986), *Religious Nationalism* (2000), *Global Religions: An Introduction* (2003), *Terror in the Mind of God: The Global Rise of Religious Violence* (2003) and *Religion in Global Civil Society* (2005).

Peter Katona, MD, FACP is Associate Professor, David Geffen School of Medicine at UCLA, and a practicing infectious diseases physician. He served as EIS Officer, Viral Diseases, at the Centers for Disease Control and Prevention (CDC), and is Corporate Medical Director of Apria Healthcare, and a certified instructor, Louisiana State University, National Center for Biomedical Research and Training (NCBRT), Academy of Counter-Terrorist Education. He is a member of the Los Angeles County Department of Health Services CDC Bioterrorism Advisory Committee for Public Health Preparedness and Response; the Bioterrorism Work Group Committee of the Infectious Diseases Society of America; the Hospital Association of Southern California (HASC); the Emergency Medical Services Taskforce on Hospital Bioterrorism Preparedness; the Los Angeles County Sheriff's Department's Terrorism Early Warning (TEW) Group; the Los Angeles County Task Force on Preparedness for Bioterrorism; the LA County Department of Health Services Advisory Panel on Biological Terrorism; the UCLA Taskforce on Bioterrorism Preparedness; and the Los Angeles County Homeland Security Advisory Council (HSAC). He was the co-founder of Biological Threat Mitigation (BTM), a bioterror consulting firm, and the founder and president of the Center of Medical Multimedia Education Technology (COMMET). He has authored several review articles on bioterrorism preparedness.

Daniel R. Morris, M.Sc.Econ., is a Strategic Intelligence Analyst with the Criminal Intelligence Service of Canada (CISC). He holds a Master's degree in Intelligence and Strategic Studies from the University of Wales, Aberystwyth (UK) and a Bachelor's degree in Criminology. He is currently completing his Ph.D. in War Studies at King's College, where his research focuses on the issues of strategic surprise and warning for counter-terrorism.

Lars Nicander is Director of the Center for Asymetric Threat Studies at the Swedish National Defense College. He was appointed Secretary of the Cabinet Working Group on Defensive Information Operations and Critical Infrastructure Protection. He is a political scientist and has served in various positions within the Swedish national security environment during the last 20 years. He is the project manager for the senior intelligence courses, which were launched under the auspices of the Intelligence Coordinator in August 2003. He is an elected member of the Institute of Strategic Studies in London (IISS), a Fellow of the Royal Swedish Academy of War Sciences and a member of the Board of Advisors to the Terrorism Research Center, McLean, Virginia.

Gregory B. O'Hayon is a Ph.D. in Foreign and Security Policy and is a Strategic Intelligence Analyst for the Criminal Intelligence Service of Canada (CISC). He has written *Big Men, Godfathers and Zealots: Challenges to the*

State in the New Middle Ages (2003). He has argued that traditional security and international relations models are no longer appropriate for understanding the post-Cold War world and is currently responsible for the development of a national strategic early warning system for crime in Canada.

Neal Pollard is Vice President at Hicks & Associates, Adjunct Professor at Georgetown University Schools of Foreign Service, Medicine, and Public Policy; and co-founder and board director of the Terrorism Research Center. He is also a 2005–2006 International Affairs Fellow of the Council on Foreign Relations.

Stephen Sloan is Professor Emeritus at the University of Oklahoma and University Professor at the University of Central Florida. He has been involved in the study of terrorism for 30 years. He pioneered the development of simulations to train police, military and security forces in responding to terrorist threats and incidents. He has also been heavily involved in the development of counter-terrorist doctrine for the military and has consulted widely in the governmental and corporate worlds. He headed a counter-terrorism practice in Washington and was a Senior Research Fellow at the Center for Aerospace Doctrine, Research and Education at the Air University. He is particularly involved in developing intelligence capabilities on the state and local level. He is the author of nine books and numerous articles; his latest books are *Terrorism: Assassins to Zealots*, co-authored with Sean Kendall Anderson (2003) and *Terrorism: The Present Threat in Context* (2006).

Annette Sobel is an Air Force Major General and Director of Intelligence for the National Guard Bureau. Additionally, she is a standing member of the Defense Intelligence Agency's Advisory Board and a member the Senior Officer Advisory Panel for the Joint Military Intelligence College. In Major General Sobel's civilian capacity, she serves as Director of the New Mexico Governor's Office of Homeland Security and Deputy Secretary for Emergency Services. She is currently on loan from Sandia National Laboratories where she serves as a distinguished member of the technical staff. Her work has emphasized weapons of mass destruction information analysis and enabling scientific and technical intelligence. Prior to her current assignments, she served as the National Guard Assistant for Weapons of Mass Destruction and Civil Support to the Chief, National Guard Bureau, after entering the National Guard as state air surgeon, Headquarters New Mexico Air National Guard. General Sobel entered the United States Army in July 1986 as a second lieutenant and was assigned as Director of Undergraduate Medical Education in the Department of Family Medicine, Womack Army Community Hospital, Fort Bragg, North Carolina. While at Fort Bragg, she also served as a visiting instructor at the JFK Special Warfare Center and School and later at the 1724th Special Tactics Squadron, Pope Air Force Base.

John P. Sullivan is a practitioner and researcher specializing in conflict studies, terrorism, intelligence, and urban operations. He is a member of the Los Angeles Sheriff's Department, where he currently serves as a lieutenant. He is co-founder of the Los Angeles Terrorism Early Warning (TEW) Group and coordinates many of its activities. He holds a Bachelor of Arts in Government from the College of William and Mary and a Master of Arts in Urban Affairs and Policy Analysis from the New School for Social Research. He has served as a consultant to Sandia and Lawrence Livermore National Laboratories. He also is a member of the InterAgency Board on Equipment Standardization and Interoperability for terrorism response and a member of the Board of Advisors for the Terrorism Research Center.

Alvin Toffler, author and futurist, is Principal of Toffler & Associates, a consulting company, and author of more than a dozen books, including the bestsellers *Future Shock* (1970), *The Third Wave* (1980), *Power Shift* (1990), *War and Anti-War* (1993) and *Revolutionary Wealth* (2006). He has written about the proliferation of computers, the advent of cable television, niche markets, the shift to work at home, corporate restructuring, the break-up of the Soviet Union, the shift to a global economy and dozens of other scenarios long before they appeared in our nomenclature.

Abdullah Toukan received his Ph.D. degree from the Massachusetts Institute of Technology in Theoretical Nuclear Physics. He spent 20 years as Science Advisor to the late King Hussein of Jordan, at a time when Jordan was at the crossroads of war and peace. He has been the head of the Jordanian Middle East Peace Negotiations Delegation to the Multilateral Arms Control and Regional Security Working Group and a member of the Jordanian Bilateral Middle East Peace Negotiations Delegation to the Jordanian–Israeli Peace Negotiations dealing with borders, territorial matters and security. He was Minister of Telecommunications and co-author, with Shai Feldman, of *Bridging the Gap: A Future Security Architecture for the Middle East* (1997). He has written a number of articles on issues related to arms control and regional security in the Middle East region.

Gregory F. Treverton is a senior analyst at the RAND Corporation. Earlier, he directed RAND's Intelligence Policy Center and its International Security and Defense Policy Center, and he is Associate Dean of the Pardee RAND Graduate School. His most recent post in government was Vice Chair of the National Intelligence Council, overseeing the writing of America's National Intelligence Estimates (NIEs). He holds an AB summa cum laude from Princeton University and an MPP (Master's in

Public Policy) and Ph.D. in Economics and Politics from Harvard. His latest books are *Reshaping National Intelligence for an Age of Information* (2001) and *New Challenges, New Tools for Defense Decisionmaking* (edited, 2003).

Abraham R. Wagner, JD, Ph.D., is engaged in the private practice of law, is Adjunct Professor in the School of International and Public Affairs at Columbia University and was Visiting Professor of International Relations at the University of Southern California. He is also engaged as a consultant on national security and intelligence matters to the Departments of Defense and Homeland Security, serving on the Defense Science Board and other advisory panels. Following 9/11 he was the chairman of a special task force in the Department of Defense looking at technology responses to evolving terrorist threats. Previously he has held positions at the National Security Council Staff; the Intelligence Community Staff; the Department of Defense; and the Defense Advanced Research Projects Agency (DARPA). He is the author of several books, including the four-volume *Lessons of Modern War* series (1996), and numerous articles, professional papers and book chapters.

Jeremy M. Wilson is an Associate Behavioral Scientist at the RAND Corporation and specializes in internal security, police administration, and violence. Dr. Wilson has worked closely with many police organizations throughout the US to help them address internal policy issues and improve effectiveness. He has recently conducted research about the reconstruction of police and justice systems in nation-building operations, the implementation and measurement of community policing, and urban gun violence. He received his Ph.D. in Public Administration from the Ohio State University and his MA in Criminal Justice from Indiana University.

FOREWORD

The Confusion Bomb

Alvin Toffler

In recent years terrorists have done more than crash airplanes into sky-scrapers or bomb bistros and buses. They have exploded a "confusion bomb" in our midst.

Terror today comes in such varied sizes, colors, styles and fashions, accompanied by so many pseudo-justifications, that even nations directly confronted with it can scarcely agree on a definition. Nor is there widespread agreement on the scale of today's terror threat.

Does the world face fringe fanatics or, as some warn, a "long war" between an ascendant Islam and the rest of the world? Alternatively, is today's wave of Islamist terror just a deadly side effect—essentially collateral damage arising from a far war within Islam itself?

Are the Islamist terrorists' main motivations pragmatic and political, or are they religious?

Should countries victimized by terrorists mobilize for war, drastically changing their way of life in response, or is that a form of surrender—in which case, going on with one's routine existence may be the best answer to the attackers?

What risks should be addressed first and, given limited resources, which are too unlikely to bother with?

To what degree does terrorism demand a military, as distinct from a police/judicial, response?

The only confusion that equals that now reigning among the targets of terrorism is that which divides Islamist terrorists themselves. Should Islamist militants target the so-called "far enemy"—the United States—or the "near enemy"—the Muslim regimes closer to home who are insufficiently fanatic? Are beheadings counterproductive because they generate bad publicity? Are innocent Muslims killed in terrorist attacks assured of "martyr status" and entry to heaven? Or are they not?

No single book can answer all these questions definitively and thus defuse the confusion bomb, but this volume, by some of the world's most experi-enced and thoughtful experts from different parts of the world, can clarify many of these issues and move the entire study of terrorism forward. The oook's contributors are drawn from academia, the military, police, medicine and other relevant fields, and therefore helpfully "surround" the central issues from different perspectives.

Two themes, however, run through these pages.

One is clear: the plague of twenty-first-century globalized terrorism cannot be reversed by any single state or nation. It cannot be cured or reversed without far better international cooperation and coordination and the development of a dense global network linking both formal and informal counter-terrorism efforts. The book provides valuable insights into how that goal might be achieved.

The second, and ultimately more important, theme in the book is shared by almost all the authors and has to do with history.

One result of the detonation of the confusion bomb has been a rush to the history books. That is valuable because failure to set a problem in historical context restricts our thinking and limits our imagination. But history, even when accurate, is slippery and dangerous.

The most common approach to yesterday stresses historical continuity— the similarities between past and present. But "continuism" is often accompanied by a tacit assumption that yesterday's remedies are applicable today. However, as a single example demonstrates, nothing could be more intellectually insidious—or more dangerous—especially when confronted with terror.

Thus, the close study of al Qaeda's killers—the bus-bombers in London, the mass murderers in Madrid, and Palestinian suicide bombers, as well— has given rise to a widely shared image of a "typical" terrorist. As a consequence, we often hear that these Islamist terrorists:

- have insisted on "just one interpretation of God's commands";
- have "presented a nostalgic view";
- were "never very numerous";
- recruit "from the privileged classes";
- impose "a systematic emotional and intellectual discipline" and build "a select, cohesive, extroverted band of zealots";
- establish "hundreds of schools to teach boys how to define and defend authoritative dogmas";
- are "successful at persuading others to fight because they fought so courageously themselves."

We hear opinions like these expressed at almost every conference on twenty-first-century terrorism.

It so happens, however, that these quotes do not refer to al Qaeda or Islam at all. They are, in fact, drawn from *The Age of Religious Warfare 1559–1715* by Richard S. Dunn, and they describe the fanatic fighters in the wars between Catholics and Calvinists that ravaged Europe in the mid-sixteenth century.

Past and present seem amazingly similar—but only when context is torn away.

The terrorists Dunn describes could not, no matter how they tried, gain control of nuclear weapons. With the threat of contemporary weapons of mass destruction, continuity in the history of terrorism comes to an end.

In the decades ahead, we face new, cheaper, smaller, even more destructive weapons of mass destruction based on convergent biological, chemical, nuclear and nano-technologies. This is the radically different, high-speed, global context in which twenty-first-century terrorists operate, armed—as their sixteenth-century predecessors were not—with global reach, internet access, mobile communications and, oh, yes, jet airliners.

The continuities of history are shrinking in number and kind. We are living through one of the biggest, most disorienting discontinuities in history. We therefore live every day with the previously "unimaginable."

Starting with that as a premise, the pages that follow help us clear away much of the dangerous fog of terror released by the confusion bomb.

© 2006 by Alvin Toffler

PREFACE

While terrorism has a long history, it has become the focus of worldwide attention as a result of the September 11, 2001 strikes on the US that resulted in an unprecedented number of later strikes by al Qaeda and other terrorist organizations on a worldwide basis ranging from Indonesia to Tunisia to Spain to the Philippines and Saudi Arabia. Subsequently, there have been various attempts to counter this latest wave of terrorism, including the US strikes against Afghanistan and Iraq, President George W. Bush's declaration of a "war against terrorism," the creation of the US Department of Homeland Security, and a 9/11 Commission. Despite these events and reactions, we believe that there is a need to consider how to consolidate and expand efforts against terrorism through the creation of an informal global counter-terrorism network. The purpose of this book is to explain the need for such a network and to elaborate how it could be formed. As we see it, the world is changing today far more rapidly than ever before, and our current public and private institutions cannot keep up with this new paradigm.

While US foreign and domestic policy has evolved over the last decades, more changes are required for it to be effective against non-nation terrorist groups. Since the end of the Cold War we have gone from security via deterrence, using nuclear weapon threats, to a period of blissful belief that threats we had lived with for decades were gone forever, to the wake-up call of 9/11. In the last few years we have adopted a new strategy of pre-emption and regime change, with the intent to democratize the Middle East, along with the urgent need to prevent terrorists and rogue states from gaining access to weapons of mass destruction. Yet many, including us, believe that we are no safer now than before 9/11, with the nation wide open to terrorist strikes and with little done to break the backs of the various terrorist organizations. Hence we see the need for a wider and global counter-terrorism network that would enhance the security of ourselves, our friends and our allies.

Part of our perspective on countering these new terrorism threats stems from our work at the local level in Los Angeles, the most populated county in the US. Here, we have worked on establishing the first distributed Terrorism Early Warning (TEW) intelligence fusion network, coordinating police, fire, hospital and public health preparedness, and seeing the potential of the academic community for new research insights. We believe that we can learn from many of these local efforts, and they can be implemented on a wider level, whether in the region, in the state, in the nation, or ultimately globally. These local efforts provide a type of laboratory for much wider efforts, up to and including the global level.

We firmly believe that to accomplish our counter-terrorist goal, partnerships are now needed that were previously uncalled for or unnecessary and, in many cases, never even considered. Contemporary terrorists act and organize globally. Rather than seeking limited political objectives in single regions or states, they now act as "networks of networks." They link together into constellations of terrorist groups, increasingly converging with other transnational criminal organizations. The precursor events for an attack in New York, Los Angeles, or Paris may be planned, organized and equipped in the Middle East, South Asia or Europe. Groups recruit, plan and train globally. Local intelligence and enforcement efforts working alone are not enough. Success in countering global insurgency requires not only top-down national security intelligence, but bottom-up local and peer-to-peer intelligence and coordination.

Contemporary terrorists exploit the global information and commercial infrastructure. They exploit the gaps and seams between our national frontiers and bureaucratic constructs. Using modern information technology to communicate their message globally, they quickly and agilely adapt to our traditional responses. The US and its allies are trying to counter a networked opponent like al Qaeda with a hierarchical national bureaucracy like the recently created US Department of Homeland Security or Interpol, but much more is required. We must be even more nimble than our terrorist adversaries and build multilateral counter-networks. The distinction between "domestic" and "international" terrorism is increasingly an anachronism. Likewise, the national security paradigm may be dated, as nation-states increasingly transform their internal and external dynamics. Global civil society (and its counterparts global terrorism and crime) demands global security mechanisms to ensure the stability of post-modern "market-states."

Local officials, for example, cannot continue to talk only to their local counterparts and others at the local level. Higher government officials need to communicate better with local authorities as well as their peers worldwide. Law enforcement especially needs to talk to the intelligence community. Just as jihadis, white supremacists, organized gangs and international organized crime syndicates are collaborating more and more and on a wider and wider basis, law enforcement, first responders, intelligence, public health, the medical community, the military and the media all have to collaborate proactively, not only with each other but with their equivalent organizations elsewhere, on a global networked basis.

This book compiles the opinions of experts from many disciplines, including clinical medicine and public health, economics, political science and public policy, and law enforcement, which are our own fields, with those of the military, politics and intelligence to try to give some coherence and direction to this process of forming a global counter-terrorism network. The book identifies the nature of a global counter-terrorism network, shows how such a global network could be created and provides some guidelines for gauging its future effectiveness.

The book is divided into four parts: I. The historical perspective on terrorism; II. Protecting critical infrastructure; III. The changing dynamics of post-modern terrorism; and IV. Fusing terrorism preparedness and response into a global network. We will focus on the paradigm shift required to create the security stemming from a future counter-terrorism network. This development, which is not being currently fully considered, will require both vertical and horizontal integration of these often disparate entities. We will discuss the historical perspective on terrorism in general and weapons of mass destruction (WMD) in particular; the effectiveness—or lack of effectiveness—of these weapons; fitting the local preparedness and response into the regional, national and international levels; the interaction of public and private sectors; the role of bio- and information technology; supporting local as well as global intelligence fusion cells; and the compatibility—or incompatibility—of diplomacy, security and defense. The book should be of interest to professionals in the fields involved in countering and analyzing terrorism, as well as to the wider public.

Terrorism cannot be completely vanquished; it can only be lessened in intensity. The creation of a functioning global counter-terrorism network will not be easy. Funding needs to be allocated, but in a responsible manner. Even more important, it will require a change in attitudes, shifting from a go-it-alone to a more global cooperative approach. Political bickering on all levels needs to be overcome. Organizational paradigms will need to be shifted. The potential erosion of civil liberties will need to be addressed and contained. Despite all this, we feel that this can be accomplished, and this book, we hope, will be a contribution in that direction.

INTRODUCTION

Part I: The historical perspective on terrorism

1 The historical impact of terrorism, epidemics and weapons of mass destruction (Peter Katona)

Terrorism has been evolving as a political, religious and criminal tactic for thousands of years. Initially, acts were local and non-networked, and involved few people. Over time, capabilities increased and acts became more violent, destructive and global. Today, we are seeing the proliferation of biological, chemical, radiological and nuclear weapons of mass destruction (WMD). This chapter will review the history of terrorism, the devastation of the great plagues, and how the use of WMD by both nations and non-state actors in the past offers insight into how they might be used in the future. The history and consequences of terrorism, as well as local, regional and national responses to acts of terrorism, including the resultant societal disruption, are the focus of this chapter.

2 Developing a counter-terrorism network: back to the future? (Lindsay Clutterbuck)

This chapter aims to trace the development of networks and other fundamental concepts in counter-terrorism. The two strands, police-led counter-terrorism as applied on the UK mainland and military-led counter-insurgency as applied in other countries, show both overlaps and elements of cooperation. The chapter draws these together into a European and EU context, detailing new developments and outlining current mechanisms for cooperation across borders. Many of the most effective of these work precisely because they are not formal but are the result of a pragmatic approach by the practitioners of counter-terrorism. The chapter shows how these complement the more formal structures that exist, such as EUROPOL, Schengen Information System, and G8. Finally, some suggestions are put forward on how practical and effective global cooperation can be achieved at

1

the practitioner level without requiring the creation of new layers of bureau-cracy or relying on prior international political agreement.

3 Warning in the age of WMD terrorism (Gregory B. O'Hayon and Daniel R. Morris)

The stated goal of a strategic early warning system is to provide clients with contextualized and over-the-horizon forecasts. The need for such a system is borne out by two realizations: 1) that the contemporary threat environment is characterized by great fluidity and change; and 2) that, in order to be effective in this new context, law enforcement needs to develop a strategic focus in order to act as a force multiplier. In other words, in an environment rich in real and possible threats, we must be diligent and focused in the allocation of scarce law enforcement, intelligence, military and development resources. Also, awareness is key to combating these post-modern threats. In order to accomplish this, we must be able to gauge the nature and level of existent threats, as well as anticipate (to the best of our abilities) the emer-gence of new ones. As risk management tools evolve, modern states and their institutions must be able to cope with changing risk environments; the citi-zenry expects nothing less. A major impetus for developing strategic early warning was the realization that risk environments can change very quickly, in often surprising ways. Thus, our need for anticipation and planning grows. That is why the concept of intelligence-led policing has been gaining converts and credence with a number of law enforcement communities. In a world of limited resources and numerous possible threats, it is no longer enough to "put out fires." Rather, we have to strategically focus resources in order to avoid and/or contain potential conflagrations. Thus, the point of strategic early warning is not to help guide preemptive attacks (although that could be a side benefit), but rather to help harden potential targets. It should also aid those using it to become more nimble and quick in their responses to new challenges. The chapter will introduce the concept of strategic early warning, describe its application in the field of law enforcement, and compare and contrast its uses in terms of organized crime and terrorism. What are the lessons to be learned from both? The chapter presents the challenge that getting buy-in from the wider community, as well as its implementation, represents to managers and practitioners, and finally discusses the solutions found to some of the most important challenges.

2

4 Terrorism and weapons of mass destruction (Michael D. Intriligator and Abdullah Toukan)

It is entirely predictable that WMD will become available to a terrorist group in the near future, and we must be prepared to deal with this likely eventuality. Given the loose constraints on WMD and the inadequate checks on the trade of existing nuclear arms caches or technologies stemming from the former Soviet Union, for example, terrorists will likely acquire and use such weapons. A successful global counter-terrorist network must deal with the terrorists' motivation for such potential acquisition of WMD, within a network of international data sharing and coordinated prevention. Following a presentation of some scenarios for terrorist use of WMD and a discussion of the distinctions between these weapons, the chapter treats recent trends in both terrorist incidents and WMD. While terrorists are more likely to use conventional weapons, the likelihood of terrorists acquiring WMD, while low in the short run, is probably high in the long run. Using the concept of expected loss, the very low probability of a remote possibility, such as a terrorist group gaining access to and even using nuclear weapons, is offset in terms of expected loss by the extraordinarily high loss it entails. It is also a serious mistake to underestimate the ability of terrorists to innovate new techniques of terror, such as their combining airplane hijackings with suicide bombings as occurred on 9/11. The chapter notes that we are in a stage of denial with regard to such a possibility, which must be overcome, and active steps must be taken to deal with potential terrorist threats to national and global security using WMD. It emphasizes that the problem is a global one that must be dealt with on a global scale, including the use of international organizations and international cooperation. The chapter treats terrorist access to WMD as a supply and demand problem. Supplies of WMD must be limited, especially Russian fissile material that could be used to make nuclear weapons, and also its tactical nuclear weapons and chemical weapons and biological stockpiles. At the same time, to reduce demand a new form of deterrence must be developed, with a multi-prong approach based on international cooperation with a credible enforcement mechanism. The chapter includes relevant portions of bin Laden's statements from 1996 up to 2005, showing their consistency since 1996. Included are his statements on how he regards it as a religious duty to acquire nuclear weapons and his stated goal of killing 4 million Americans in retribution for this number of Muslims that the US and its allies, including Israel, have killed. In addition, the chapter provides information on various forms of WMD and counter-proliferation efforts and uses this information to determine the likelihood of WMD acquisition and use by terrorists. It concludes that we should understand al Qaeda's political and strategic aims and objectives, in addition to the events taking place in the Middle East, to reach the conclusion that a WMD attack by a terrorist group is bound to happen. As a result, we must

COUNTERING TERRORISM AND WMD

establish clear priorities for counter-terrorism policies and make serious improvements in them so as to deal with potential terrorist attacks using WMD.

Part II: Protecting critical infrastructure

5 *The Trojan horse of the Information Age (Lars Nicander)*

How has the way we handle conflict on the national and international level changed with the advent of the Information Age? Is the relevance of military conflicts decreasing? Will bayonets, bombs and bullets be replaced by bits, bytes and "morphed" images? Will the need for military deployments due to international missions even decrease in favor of more IT-based solutions, including use of such electronic weapons as computer network attack? Will we see cyber-terrorists who use highly effective IT-based—and less bloody—means and methods to cripple the fabric of Western societies? This chapter focuses—with a Swedish perspective—on the sometimes blurred concept of information operations (IO), which can be considered, to a large extent, as a mirror image of the increased dependencies on electronic information, networks and IT systems in modern civilian societies. The chapter discusses the weaknesses and strengths of a greater and greater expansion of these networks for the sharing of information.

6 *Suicide bombers, soft targets and appropriate countermeasures (Robert J. Bunker)*

This chapter discusses post-modern terrorism dynamics, which include the crime and war operational environment, fifth-dimensional battlespace, bond-relationship targeting, and network organizational forms. Forms of combat power (disruptive and destructive) are then covered along with rates of combat power change. Suicide bomber tactics and techniques are highlighted along with the tactical and operational advantages over traditional bombings. Terrorist group suicide bombing matrixes are included which profile different group targeting preferences and delivery modes. The soft target environment is discussed, as is the concept of what are acceptable casualties. The appropriate countermeasures section covers the tactical and operational at the pre-incident, incident and post-incident timeframes. Strategic-level countermeasures are also discussed.

7 *Terrorist use of new technologies (Abraham R. Wagner)*

While the religious and philosophic underpinnings of the current terrorist movements may be rooted in the Middle Ages, the various technologies that they employ certainly are not. Indeed, terrorist organizations operating in the Middle East, Asia and elsewhere have embraced a range of modern technologies to support their operations in areas including communications, targeting and recruitment, as well as in the design and use of weapons against military and civilian targets. At the same time, the intelligence services, military, and law enforcement agencies engaged in counter-terrorism are operating in an era where a host of new technologies exist, and continue to evolve, that are of potential use in combating terrorism. The focus of this chapter is on how technology has served as an "enabler" on both sides of the terrorism problem, as well the problems created for counter-terrorist operations in terms both of making actual operations more difficult and of the legal impediments presented by new technologies.

Part III: The changing dynamics of post-modern terrorism

8 Responding to religious terrorism on a global scale
(Mark Juergensmeyer)

Perhaps no form of political violence is more conducive to the need for a global response than religious terrorism. The movements associated with such acts in the last decade of the twentieth century and the first decade of the twenty-first are themselves often global in character: their networks of followers are transnational, and they often see the world in terms of a global confrontation. The war against terrorism has been from the point of view of military and diplomatic leaders a kind of global anti-guerilla war. It has been difficult to fight with weapons designed for warfare that is waged in a more conventional and technological way. Many supporters of al Qaeda and similar movements see these conflicts not in military but in theological terms. For them, religious terrorism has been an aspect of a larger spiritual struggle, a cosmic war, one that need not be won in ordinary history and one in which they are convinced eventually they will triumph. This means that the usual way of viewing terrorism—as a political and military strategy—is insufficient for understanding and combating religious violence. The author's study of the subject, based on interviews with religious activists associated with incidents such as the bombing of the World Trade Center, suicide attacks in Jerusalem and Tel Aviv, and the Tokyo nerve gas assaults, has led him to the conclusion that many of these acts were not so much strategic as symbolic. Even those that had specific goals were also to some extent forms of "performance violence"—dramatic acts meant to call attention to a vast, albeit hidden, war. Any response to religious terrorism, therefore, will have to take cognizance of this spiritual war and the human aspirations that often accompany it, including the desire for a renewed role for religion in public life. Faced with such motivations, how can authorities appropriately respond? Put simply, how can religious terrorism come to an end? The possibilities are: 1) destroying violence with violence; 2) choking off terrorism through legal means; 3) infighting within violent groups; 4) self-destruction; 5) terrifying terrorists; 6) terrorists terrifying themselves; 7) violence transformed into political leverage; 8) separating religion from politics; 9) removing the underlying sources of tension; and 10) taking the moral high road and co-opting the terrorists' claim to moral politics.

9 Terrorism in Algeria: the role of the community in combating terrorism (Anneli Botha)

The war on terrorism cannot be won by a single nation alone or even by a limited number of nations. Fighting terrorism has to be a global objective with the participation of the international community as a whole. The assistance of local communities is even more important than the participation and cooperation of nation-states. Countries that have experienced the impact of terrorism have established and nurtured formal and informal relationships with their citizens to prevent and combat terrorism. The establishment of state structures and legislation is inadequate in the absence of local community support. Members of security forces are not omnipresent: they need the eyes and ears of each and every citizen to prevent and combat terrorism on all levels.

This chapter uses Algeria as an example of the positive impact community members can have in counter-terrorism operations. Essential in this commitment is the community members' realization that they have a responsibility to ensure and protect their safety. Under these conditions, "partnerships" are formed between state structures and community members. This chapter provides insight into the development of a national counter-terrorism strategy that includes the local population. Since both parties to a conflict depend on both the local and the international community for recognition and support, acquiring the assistance of the local population is more difficult that it appears. History has proven that the survival of domestic campaigns requires domestic support in the form of a willingness to provide new recruits, financial assistance, weapons and a safe haven. In the case of Algeria, the initial support counter-government groups received from the local community declined dramatically as the campaign developed. Strategy and tactics from both government forces and extremists often determine the success and survival of both sides. In Algeria, strategies and tactics led to the realization that citizens need to participate in countering terrorism. Lessons learned from Algeria could also be applied on the macro- or international level in the "war against terrorism" as a strategy with positive results. Governments, policy makers and security forces could benefit from Algeria's experience in combating terrorism.

This chapter begins with a brief overview of the role-players and impact of the campaign of terror in Algeria. Counter-terrorism efforts in Algeria provide important lessons, including the role of community members in taking and accepting responsibility for their own safety, and communication between community members and security forces. The example set through the role of the "patriots" or local self-defense units in Algeria personifies this ideal. Algerian influence and contribution have inroads towards a transnational terror network in Europe.

10 Cooperation issues in the global war on terror
(Barry Desker and Arabinda Acharya)

More than two and half years into the US-led "war on terror," the international coalition against terrorism remains fragmented despite many initial successes on the military front. Arrests and killings of many of the key leaders of al Qaeda, as well as worldwide disruption of its bases, financial infrastructures and networks, have not been commensurate with the regenerative and adoptive capabilities of the group. Al Qaeda remains resilient enough—mutating into new forms and adapting itself to the changing operational environment—to continue with its campaign of terror, targeting the interests of not only the United States, but its allies and supporters worldwide. On the other hand, however, many obstacles have emerged in counter-terrorism efforts, and the international coalition against terrorism seems to be weakening. At the strategic level, the spirit of cooperation has been undermined by some of the policies of the United States. At a more tactical level, the failure can be attributed to two major factors. One is the failure to understand the nature of the threat, especially the "al Qaeda phenomenon" in its entirety, including the vision, capabilities and acumen, and organizational skills of Osama bin Laden. Second and most important is the failure to address the core issues and the ideology that underlie the militant Islamist threat. With Iraq emerging as the new epicenter of global terrorism, it is incumbent upon the international community to continue to work together to roll back the threat of radical Islamic terrorism by helping the "collective Muslim mind" recover from the "moral and ideological crisis" that has brought transnational Islamist groups into the center-stage of conflict against the West in the first place, and helped sustain their campaign.

Part IV: Fusing terrorism preparedness and response into a global network

11 From combating terrorism to the global war on terror
(Brian M. Jenkins)

This chapter focuses on a review of the evolution of global terrorism since the 1970s. Local, as well as national and international, terror and counter-terror networks have emerged and mutated. This has presented several challenges and opportunities to convert from the current hierarchical structure to a future global networked approach.

12 The new terrorist threat environment: continuity and change in counter-terrorism intelligence (Stephen Sloan)

This chapter addresses changes in organizational doctrine that are needed to prevent, preempt and respond to emerging terrorist threats by emphasizing the need to develop counter-terrorism cellular structures that can mirror-image those employed by terrorists. The development of these counter-terrorism cellular structures is discussed in the context of the need for changing intelligence sources and methods. In particular, the focus is on the importance of refining analytical capabilities by recognizing that an asymmetric approach to counter-terrorism should be emphasized rather than continuing reliance on large-scale hierarchies. Such an approach should draw on the expertise and flexibility of a wide variety of governmental and non-governmental groups, ensuring that they enhance flexibility in intelligence collection, analysis, and interagency cooperation. This should extend to regional and international cooperation. Within the context of the threat, while there is very understandable focus on WMD, it is important to note that conventional threats will continue to remain a major challenge, especially as new techniques of explosive making, design, etc become increasingly available and sophisticated.

13 Actionable intelligence in support of homeland security operations (Annette Sobel)

Terrorism is an amorphous beast, and the methodology to defeat the beast is also amorphous and highly adaptive. This chapter describes the significance, process and objectives of deriving actionable intelligence: intelligence products that support and enable the full spectrum of tactical homeland security operations. The spectrum of operational challenges includes such diverse activities as: pre- and trans-threat senior decision-making, law enforcement, incident command, preventive health measures spanning vaccination and epidemiologic early warning, and countering transnational threats prior to their impact on the homeland. The intelligence process described in this chapter includes an overview of the types of relevant intelligence, an understanding of the power of intelligence fusion products to support and counter threats of terrorism and, finally, a series of recommended approaches to open-source intelligence collation, analysis, fusion, and support to operations. Emphasis is placed on the generation of a set of information requirements. This process is most effective if driven by the end-user of the information, whether policy maker or first responder.

14 The terrorist war on the market-state: a plague in a time of feast (Philip Bobbitt)

Terrorism is not unchanging, but mimics the constitutional order that it attacks. Thus the nation-state terrorism of the twentieth century was organized around national struggles. The terrorism of the twenty-first century will mirror the change in the constitutional order brought about by the emergence of market-states. Like those states it seeks to destroy, it will be networked, global and decentralized, and it will outsource operations.

15 Counter-terrorism in cyberspace: opportunities and hurdles (Neal Pollard)

This chapter looks at the role of information technology in countering terrorism in an era of globalization. Modern terrorism has been characterized as a negative response to globalization, but, at the same time, terrorism has become effective by exploiting the engines of globalization. This chapter outlines the role of information technology in fighting terrorism, especially in intelligence analysis. There are at least three opportunities where the United States must leverage information technology in its war against terrorism: intelligence analysis, critical infrastructure protection, and cyber-conflict. Intelligence analysis in particular presents a watershed opportunity for the US to prevail over terrorism by successfully interdicting it, rather than responding to the consequences of its attacks. However, the use of information technology in all three of these areas is fraught with legal issues that must be addressed before the US can effectively use information technology against terrorism.

16 The global challenge of operational intelligence for counter-terrorism (Gregory F. Treverton and Jeremy M. Wilson)

Military measures have dominated the United States' response to the terrorist threat. Given the stunning success in Afghanistan followed by the travails in occupying Iraq, it could hardly have been otherwise. Yet, over the long run, if terrorism is to be contained, military measures, while relevant, will not be the most important. If the war on terrorism is to be won, it will be won by patient police and intelligence work, across nations. Yet the United States, like its fellow democracies, had drawn a sharp line between intelligence and law enforcement, and for good reason, fearing that concentration of power would harm the liberties of its citizens. The two realms were very different, with different missions, operating procedures and standards of evidence. Moreover, policing is organized by geography, while the terrorist threat recognizes no such boundaries.

Bringing national intelligence and law enforcement together in ways that minimize the risk to liberty is the first task. Unlike the United States, most of its fellow democracies have chosen to create a domestic intelligence service apart from law enforcement, so the comparison of approaches is instructive. The second challenge is cooperating across international boundaries, and insulating that cooperation from the ups and downs of diplomatic interchange. On the whole, that quiet cooperation in the war on terror has been good: witness the number of al Qaeda suspects arrested in Europe. Yet international institutions lag behind the facts of cooperation, and there may be opportunity to reinforce both international norms and institutions. The final challenge in assembling the network is cooperating with "ourselves," that is, sharing information with the 18,000 law enforcement agencies in the United States, not to mention with private citizens. There, the challenge is to play on the comparative advantage of the different levels of government: local law enforcement, for instance, has lots of eyes and ears but little capacity to do special collection operations independent of law enforcement or sophisticated analysis of collected information.

17 Fusing terrorism security and response (John P. Sullivan)

Addressing the security and response needs to effectively negotiate the contemporary global terrorist threat is going to require new approaches and structures. Contemporary terrorism relies upon networks of interrelated terrorist groups, social movements and criminal organizations to conduct operations, secure funds and influence their audiences. Global society, including both the society of nations and emerging organs of civil society, are challenged by these new networked social forms. Traditionally, terrorists and criminal organizations were subnational or, occasionally, transnational threats. Rarely did they truly challenge the nature of the nation-state or the modern state's monopoly on violence and war. Today's global threats (such as al Qaeda's jihadi international) blur the distinctions between crime and war and challenge the structures of the nation-state. These post-modern networks may be emerging "market-state" actors engaged in netwar against existing states and entities. This chapter discusses options for fusing security and response to address this transition. Options for conducting counternetwar through the adoption of early warning approaches, embracing global civil and security networks, and linking multilateral law enforcement and intelligence structures with response (medical, humanitarian and security) organizations will be explored. The chapter also examines the need for networked approaches to combat terrorism, and contains a discussion of early approaches such as the Terrorism Early Warning group model. Finally, the chapter advocates a shift from national security approaches to broader, multilateral, global security approaches for current and emerging conflicts.

Part I

THE HISTORICAL PERSPECTIVE ON TERRORISM

1

THE HISTORICAL IMPACT OF TERRORISM, EPIDEMICS AND WEAPONS OF MASS DESTRUCTION

Peter Katona

Introduction

Throughout history, academics, politicians, security experts and journalists have employed a variety of definitions of terrorism. Some focus on terrorist organizations' modes of operation while others emphasize the motivations and characteristics of terrorism or the modus operandi of individual terrorists. Before we look at the history of terrorism, we need to understand what the term "terrorism" actually means. There is no universally accepted definition of terrorism; even when people do agree on a definition, they may disagree about whether or not the definition fits a particular incident. The National Memorial Institute for the Prevention of Terrorism uses the definition "premeditated, politically motivated violence perpetrated against non-combatant targets by sub-national groups or clandestine agents, usually intended to influence an audience."[1] The United Nations definition also excludes states that sponsor terrorism, while the US definition includes them and names specifically Cuba, Iran, Libya, North Korea, Sudan and Syria, as noted in Table 4.2 (p. 74).

As discussed by Boaz Ganor, Schmidt and Youngman, in their book *Political Terrorism*, noted more than 100 definitions of terrorism from a survey of leading academics, and isolated certain recurring elements and unresolved issues.[2]

The recurring elements in the definition of terrorism were:

1   violence or force (appeared in 84 percent of the definitions);
2   political (65 percent);
3   fear or emphasis on terror (51 percent);
4   threats (47 percent);
5   psychological effects and anticipated reactions (42 percent);
6   discrepancy between the targets and victims (38 percent);

7   intentional, planned, systematic, organized action (32 percent);
8   methods of combat, strategy, tactics (31 percent).

The unresolved issues in the definition of terrorism were:

1   The boundary between terrorism and other forms of political violence.
2   Whether government terrorism and resistance terrorism are part of the same phenomenon.
3   Separating "terrorism" from simple criminal acts, from open war between "consenting" groups, and from acts that clearly arise out of mental illness.
4   Is terrorism a sub-category of coercion, violence, power or influence?
5   Can terrorism be legitimate? What gains justify its use?
6   The relationship between guerrilla warfare and terrorism.
7   The relationship between crime and terrorism.

Terrorism has three important elements: First, there is violence. An activity that does not involve violence or a threat of violence will not be defined as terrorism (including non-violent protest—strikes, peaceful demonstrations, tax revolts, etc.). A violent activity against civilians that has no political aim is, at most, an act of criminal delinquency, a felony or simply an act of insanity unrelated to terrorism. Second, the goal is to attain political objectives: changing the regime, changing the people in power, changing social or economic policies, etc. There may also be ideological or religious aims. The concept of "political aim" is sufficiently broad to include these goals as well. The *motivation* behind the political objective—whether ideological, religious or something else—is irrelevant. Finally, do the targets of terrorism have to be civilians? Terrorism is thus distinguished from other types of political violence such as guerrilla warfare or civil insurrection. Terrorism exploits the relative vulnerability of the civilian "underbelly"—the tremendous fear and anxiety, and the intense media reaction evoked by attacks against civilian society's seams and soft targets. Fear creates uncertainties that government will not protect its citizens and that quality of life will be adversely affected. With all this in mind, the phrase "global war on terror" is inconsistent and misleading since it implies that terror is an enemy rather than a tactic.

Historically, terrorism has taken many forms. Initially, there were individual acts, or small-scale acts committed during wars. The motivation of religion entered early and has persisted, recently adding a more virulent fanaticism. The terrorist and non-terrorist use and proliferation of all forms of weapons of mass destruction (WMD) emerged in the twentieth century:

• Chemical weapons were used by both sides in WWI, by Iraq against Iran and its indigenous Kurd population, and in Tokyo's Aum Shinrikyo sarin gas attack.

- Nuclear weapons were used by the US against Japan to end WWII.
- Biological weapons were used by Iraq against the Kurds (aflatoxin),[3] by unknown perpetrators against US senators and media executives (the anthrax letters), by the Japanese against the Chinese (anthrax, cholera, plague, salmonella and other agents) and by the Bhagwan in the salad bar contamination in Oregon (salmonella).

Civilization has also been coping with the turbulence and terror of natural epidemics such as smallpox, plague, typhus and influenza for thousands of years. The first influenza epidemic was recorded by Hippocrates in 412 BC,[4] although the domestication of chickens may have brought outbreaks as far back as 2,500 BC in the Orient.[5] Since then, influenza epidemics have periodically caused widespread morbidity and mortality. The Spanish influenza outbreak of 1918–1919, for example, caused tens of millions of deaths and may have killed more people than any other epidemic in history. Now, a century later, we are facing a similar issue: the possibility of the avian (H5N1) influenza virus mutating to spread from human to human, causing a global pandemic of enormous proportions.

The great plagues of the Middle Ages caused massive destruction and disruption medically, sociologically and economically. Jared Diamond in *Guns, Germs and Steel* noted that, during modern and ancient wars, infectious diseases killed or wounded more soldiers and civilians than bombs or bullets.[6] Smallpox, for example, killed about 300 million people during the twentieth century—three times the number directly killed in all wars during that century.

Biological agents intentionally used as weapons may become the agents of choice in the future. They are cheap and easy to use; the recipes are easy to acquire; and they can cause great fear and panic. Despite this, there are also drawbacks that may have prevented their use: their effects are highly unpredictable; they depend on the cooperation of weather conditions such as temperature, humidity, wind and inversions; they are socially unacceptable to most of us; and the production of true "weaponized" or genetically altered biological agents requires experienced microbiologists with knowledge of advanced "weaponization" technology.[7] An efficient dispersion device requires knowledge of physics, weather and aerobiology to be effective. As technology advances, the pros of their use will overcome the cons.

Early history: the Sicarii, Zealots, Assassins and Thugees

The Greek historian Xenophon (*c.* 431–*c.* 350 BCE) first wrote of the effectiveness of psychological warfare against enemy populations.[8] Roman emperors such as Tiberius and Caligula, who reigned 14–41 CE, used banishment, expropriation of property, and execution as a means of terror to discourage opposition.

Ancient acts of bioterrorism and biowarfare were local events and consisted of contaminating water supplies, hurling snakes on to the decks of enemy ships[9] or deliberately introducing infectious diseases among enemy forces. Roman armies used the bodies of animals and humans to contaminate water supplies.[10] It is unclear just how effective these and similar efforts were. Given the problems of sanitation in those times and the natural prevalence of waterborne and other infections, a further degradation of water quality probably contributed only marginally to general morbidity, and it is doubtful that such efforts significantly altered the course of history.

More organized terrorist history goes back to the times of the Sicarii, Zealots, Assassins and Thugees. The Sicarii and the Zealots were Jewish terrorist groups active during the Roman occupation of the Middle East during the first century. Their favored weapon was the *sica*, or hort dagger, which gave them their name, literally meaning "dagger men." They murdered those (mainly Jews) they deemed apostate. The Zealots, who targeted Romans and Greeks, give us the modern term "zealot," meaning "a fanatical partisan." Just as terrorists today seek media attention, so the Zealots usually killed in broad daylight in front of witnesses, sending a clear message to the Roman authorities and the Jews who collaborated with them. Their most famous act—committing mass suicide to avoid capture—occurred at Masada near Jerusalem in 70 CE.[11]

Adherents of other religions also resorted to methods that today might be termed terrorism. The Assassins, founded by Hasan ibn al-Sabbah as an offshoot of the Isma'ili Shia Muslim sect, were active between 1090 and 1272. In 1090, Hasan and a band of followers seized a mountain fortress in current Iran known as Alamut. From there, Hasan, who became known as the "Old Man on the Mountain," sent out secret agents to kill his enemies. The tactic was to find young, impressionable youths to be hired killers, analogous to some of today's Muslim suicide terrorists. They were drugged, transported to "paradise" by the intoxication, and upon awakening were promised the afterlife. These young men were then assigned to go to a destination and wait for instructions, much like today's "sleeper agents." Like the Zealots and Sicarii, the Assassins were also given to stabbing their victims, who were generally politicians or clerics who refused to adopt the purified version of Islam the Assassins were forcibly spreading. Today's Taliban and al Qaeda have similar aims and tactics.

The Assassins gave us the modern term "assassin," which originally meant "hashish-eater," a reference to the ritualistic drug-taking they were rumored to indulge in prior to undertaking missions. The Assassins' deeds were often carried out openly at religious sites on holy days, a tactic intended to publicize their cause and incite others. Like many of the religiously inspired suicide attackers of today, they viewed their deaths as sacrificial and a guarantee that they would enter paradise in the afterlife.

Sacrifice was also a central element of the killings carried out by the

Thugees, who gave us the word "thug." They were an Indian religious cult who ritually strangled their victims as an offering to Kali, the Hindu goddess of terror and destruction.[12] Their victims were often travelers chosen at random. The intent was to terrify the victim, a vital consideration in their ritual, rather than to influence any external audience. Active between the seventh and mid-nineteenth centuries, the Thugees are reputed to be responsible for as many as one million murders. They were perhaps the best example of religiously inspired terrorism until the late twentieth century. Although Thugs never attacked British travelers, the British government of India decided to eliminate them, and over 3,000 Thugs were captured during the 1830s. This resulted in 483 of them giving evidence against the rest, and consequently 412 were hanged, with the rest imprisoned or rehabilitated.[13] Despite this, the Thugees' actions impacted trade routes and eventually helped to crush the British rule in India. As an example of state-sponsored terrorism, the Spanish Inquisition used terrorist tactics such as arbitrary arrest, torture and execution to punish what it viewed as religious heresy. With examples from Judaism, Islam, Christianity and Hinduism, religion before the nineteenth century provided the primary justification for terror.[14]

Figure 1.1 The goddess Kali

Biological terrorism gains ground during the Middle Ages

Other than Emperor Barbarossa's brief poisoning of water wells with human bodies in Tortona, Italy in 1155, biological weapon use was not well recorded until 1346 at Kaffa, a seaport on the Crimean coast in current Ukraine. During this siege, an outbreak of plague occurred within the surrounding Tartar army. Thinking they had little to lose, the Tartars hurled the corpses of dead plague victims over the walls of besieged Kaffa. Some historians feel that this may have caused the subsequent Black Death pandemic in Europe between 1347 and 1351, although prevalent unhygienic conditions in the city may have done far more harm. Plague eventually spread via Mediterranean ports to Genoa and beyond, killing an estimated 25 million people *or 50 percent of the European population.*[15] Since the fleas that carry plague are species-specific and don't stay long on dead bodies, this may have been a very inefficient way to spread plague. Like the earlier Justinian Plague of the sixth century, the Black Death of the fourteenth century killed tens of millions. Trade was disrupted due to a shortage of labor, serfdom diminished in importance, and there was widespread fear and panic regarding any social interactions. Jews were accused of causing the pandemic, and their "suspicious" better hygiene practices, which may have helped protect them from getting infected, made them even more suspect. Anti-Semitism eventually led to a migration of western European Jews eastward to Poland and Russia.[16]

Catapulting dead corpses over walls was not unique to the Tartars. In 1340, dead horses were catapulted at Thun L'Eveque (in current France). It was reported that "the stink and the air were so abominable, they could not long endure" and a truce was negotiated.[17] During the Bohemian battle of Carolstein (Karlstein) in 1422, Lithuanian soldiers catapulted the bodies of their slain comrades plus 2,000 cartloads of excrement over the castle walls on to the ranks of the defenders, where deadly fevers quickly broke out. Despite this, defenses held, and after five months the siege was abandoned. Russian troops practiced similar methods in 1710 and 1718, hurling plague-infected corpses over the city walls of Reval during their wars with Sweden.[18] It is still unknown if these tactics were effective.

Conquistadors used biological weapons to terrorize during the Middle Ages. Smallpox was inadvertently brought to America by Spaniards in the early sixteenth century. During Pizarro's conquest of South America he intentionally presented to the natives, as gifts, clothing laden with smallpox. Cortez introduced smallpox to Mexico in 1520, and in two years it reportedly killed 3.5 million Aztecs.[19] During the sixteenth century the population of Mexico went from 22 million to 2 million, presumably from three large outbreaks (smallpox and cocoliztli or hemorrhagic fever).[20] With no prior immunity, the smallpox Cortez and Pizarro brought to the New World quickly spread to the local Indian population, while their soldiers, having already been infected, were immune. Smallpox served the conquistadors initially as an unwitting and later as a deliberate act of biological warfare and terror. Both the intentional and unintentional spread of these diseases among Native Americans killed about 90 percent of the pre-Columbian population.[21]

These acts were not the only European bioterrorist exploitations. In 1495 in Naples, the Spanish infected French wine with blood from leprosy patients (an ineffective ploy).[22] In 1650, a Polish military general reportedly put saliva from rabid dogs into hollow artillery spheres for use against his enemies (another ineffective ploy).[23] In 1675 these events eventually led to the first agreement between German and French forces not to use "poison bullets."

The eighteenth century: terrorism gains formal respect

The actual word "terrorism" comes from the *régime de la terreur* that prevailed during the French Revolution between 1793 and 1794, with terrorism originally an instrument of the state. The new regime now consolidated power within the newly installed revolutionary government, protecting it from elements considered "subversive." To the new regime, terrorism was now actually considered a *positive* term. This overt use of terror was openly advocated by Maximilien Robespierre as a means of encouraging revolutionary virtue, leading to the period of his political dominance. Robespierre viewed terrorism as vital if the new French Republic was to survive its infancy, proclaiming in 1794 that: "Terror is nothing other than justice, prompt, severe, inflexible; it is therefore an emanation of virtue; it is not so much a special principle as it is a consequence of the general principle of democracy applied to our country's most urgent needs."[24] Robespierre may have influenced Napoleon, who flooded the plains around Mantua, Italy in 1797 to enhance the spread of malaria.[25]

During the French and Indian War between France and Britain (1754–1767), both sides relied heavily on the support of Native American Indians. A British general, Sir Jeffery Amherst, surreptitiously provided the Indians loyal to the French with blankets infected with smallpox. The resulting epidemic decimated the Indians. Here, the intentional terrorist use of smallpox probably played a significant role in the British victory.[26] Smallpox, at least in the Americas, was seen to be a particularly useful terrorist weapon. Native Americans, lacking any immunity, experienced case-fatality rates as high as 70 percent.[27]

During the American Civil War (1860–1865), General W.T. Sherman's memoirs contain an account of Confederate soldiers poisoning ponds by dumping the carcasses of dead animals into them. In 1863 Dr. Luke Blackburn, a Confederate and the future governor of Kentucky, attempted to infect clothing with both smallpox and yellow fever sold to Union troops. One Union officer's obituary stated that he died of smallpox contracted from his infected clothing.[28] Yellow fever would not have done much harm since it is spread not by humans but by infected mosquitoes.

The late nineteenth to the mid-twentieth century: ineffective treaties and a shift in tactics

The Italian revolutionary Carlo Pisacane's theory of the "propaganda of the deed" recognized the utility of terrorism to deliver a message to an audience *other than the target*, drawing attention and support to a cause. This typified a new form of terrorism since the French Revolution.[29] Pisacane's thesis would probably have been recognizable to the Zealots, Sicarii and Assassins, but it was put into practice by the Narodnaya Volya (NV), a Russian anarchist populist group whose name translates as "The People's Will."[30] NV was formed in 1878 to oppose the Russian Tsarist regime. From 1865 to 1905 they had killed a number of kings, presidents, prime ministers and other government officials. The group's most famous deed, the assassination of Alexander II in 1881, effectively sealed their fate by incurring the full wrath of the Tsarist regime. Unlike many contemporary terrorist groups, the NV went to great lengths to avoid "innocent" deaths, carefully choosing their targets—usually state officials who symbolized the regime—and often compromising operations rather than incurring collateral damage or the killing of innocent civilians. Author David Rapoport called this the first wave of modern terrorism.[31]

Biological weapons proliferated during the twentieth century. During WWI, the Germans were alleged to have attempted to spread cholera in Italy, plague in St. Petersburg and biological bombs over Britain. Germany also had an ambitious biological weapons project aimed at its enemies' livestock. Anthrax and glanders (both excellent biological weapons) were used to try to infect sheep shipped to Russia. In 1915 a German-American, Dr. Anton Dilger, grew cultures of anthrax and glanders with seed stock supplied by the German government. This seed stock, along with an inoculation device, was given to sympathetic dockworkers in Baltimore with the aim of infecting 3,000 head of horses, mules and cattle headed for the Allied troops in France. It was even alleged that several hundred troops were secondarily affected. Although horsepower was a major component of logistics during World War I, the German use of biological agents was not successful in affecting the war.[32] The best-documented use of biowarfare by Germany came several decades later in 1945, when the Germans used sewage to poison a large reservoir in Bohemia.[33]

In 1918, the Japanese army formed an offensive biological weapons section called Unit 731, which studied the effects of bioweapons on both prisoners of war and civilian communities.[34] By 1931, Japan had expanded its territory into Manchuria and made available "an endless supply of human experiment materials" (prisoners of war) for Unit 731. Between 1932 and 1945, Japan employed more than 3,000 scientists and support staff in its biological weapons project in occupied China.

During an infamous attack in 1941, the Japanese released an estimated

150 million plague-infected fleas from airplanes over villages in China and Manchuria, resulting in several plague outbreaks, tens of thousands of civilian deaths, and many terrorized communities. As many as 11 Chinese cities were eventually attacked with anthrax, cholera, plague, salmonella and other agents. A post-World War II autopsy investigation of 1,000 victims revealed that most were exposed to aerosolized anthrax. Biological weapons experiments on prisoners in Harbin, Manchuria, directed by General Shiro Ishii, continued until 1945. It is estimated that thousands of prisoners and Chinese nationals died from these experiments.[35]

A 1941 bioattack on Changteh, China that had gone wrong killed at least 1,700 Japanese troops, demonstrating that biological weapons are tricky to use and can backfire—a lesson to future terrorists. Nonetheless, the terror these bioweapons caused was felt almost entirely by the Chinese.[36] By 1945, the ambitious Japanese military program had stockpiled 400 kilograms of anthrax to be used in a specially designed fragmentation bomb.

The Japanese did not stop here. During an investigation of Japan's seizure of Manchuria in 1931, Japanese military officials were found to have unsuccessfully poisoned members of the League of Nations' Lytton Commission by lacing fruit with cholera.[37] In 1939, the Japanese military also poisoned Soviet water sources with typhoid at the former Mongolian border.

The first confirmed evidence of biological weapons production by the Soviet Union was provided by Ken Alibek.[38] He believed that tularemia was used at Stalingrad in 1942 against German panzers and Q fever was used in 1943 among German troops on leave in the Crimea. The Soviet Union's Biopreparat facilities continued bioweapons production after WWII on a massive scale, with production measured in *metric tons* and close to 60,000 employees in at least six production facilities.

The Geneva Protocol for the Prohibition of the Use of Asphyxiating, Poisonous or Other Gases, and of Bacteriological Methods of Warfare was signed in 1925 in response to biological and chemical attacks during World War I. This unenforceable treaty prohibited the use of such weapons, but not their development or storage, and provided no inspection provisions.[39] In other words, the treaty was totally ineffective.

The 1930s saw a fresh wave of political assassinations deserving of the word "terrorism." This led to proposals at the League of Nations for conventions to prevent and punish terrorism, and to establish an international criminal court. Neither of these proposals was successful, as they were overshadowed by the events which eventually led to World War II. The Geneva Protocol was signed by the Soviet Union, France, Great Britain and even Iraq. As it did not serve their own strategic goals, the Soviet Union, South Africa and Iraq blatantly violated the provisions of the treaty, and the United States did not ratify the Geneva Protocol until 1975.

In 1972, the United States and 143 other countries signed the Biological Weapons Convention on the Prohibition of the Development, Production and Stockpiling of Bacteriological (Biological) and Toxin Weapons and on Their Destruction, commonly called the Biological Weapons Convention. This treaty prohibited the stockpiling of biological agents for offensive military purposes, allowed production of bioagents only in small quantities for defensive purposes, and forbade research into the offensive use of biological agents. While the US has been faithful to the convention, the George W. Bush administration has been opposed to any enforcement plan, backed by the European Union, requiring "no knock" inspections of a country's biosafety facilities, which they feared would allow others to openly spy on American sites. As of 2005, the US has expanded Biocontainment Level-4 laboratories for the study of the world's deadliest germs from two to five, keeping within the provisions of the convention.

The breaking of these international treaties should not surprise anyone. North Korea terrorizes its citizens, exports weapons of terror and is building its own nuclear weapons, yet the North Korean Constitution states that "the state shall effectively guarantee democratic rights and liberties" as well as "the material and cultural well-being of its citizens." Saddam Hussein, who used chemical and biological weapons on his own as well as other populations, had an Iraqi constitution that stated that "public office is a sacred confidence." Under Stalin, the Soviet Union's Constitution of 1936 recognized freedom of expression as well as pledging "printing presses, stocks of paper, public buildings, the streets, communications facilities and other material requisites for the exercise of these rights." Under these "constitutional rights," Stalin orchestrated the police-state killings of millions of internal political enemies and sent millions more to concentration camps under one of the best examples of state-sponsored internal terrorism.[40]

Narodnaya Volya (NV) was symbolic of the transformation of terrorism during the nineteenth century, coming to be associated, as it still is today, with non-governmental groups. Like the French, they used the word "terrorist" proudly. They developed ideas that would become the hallmark of subsequent terrorism in many countries. Like the Israelis, they believed in the targeted killing of the "leaders of oppression" and were convinced that the developing technologies of the age—symbolized by bombs and bullets—enabled them to strike directly and discriminately. Analogous to the Islamists' view of the Western world of today, NV believed that the Tsarist system against which they were fighting was fundamentally evil. They propagated what has remained the common terrorist goal (or delusion) that those violent acts would spark revolution, or at least some social change.

The NV's actions also inspired radicals elsewhere. Anarchist terrorist groups were particularly enamored by the example set by these Russian populists, although not by their keenness to avoid casualties among bystanders. Nationalist groups such as those in Ireland and the Balkans have adopted terrorism as a means to their desired ends. As the nineteenth century gave way to the twentieth, terrorist attacks were carried out in places as far apart as India, Japan and the Ottoman empire, with two US presidents and a succession of other world leaders victims of assassination by anarchists and malcontents—often loosely affiliated with specific terrorist groups, but operating without their explicit knowledge or support, much as al Qaeda operates today.

Over the next century, the Jacobin spirit infected Russia, Europe and the United States.[41] Radical anarchists such as Leon Czolgosz, who killed William McKinley in 1901, and Alexander Berkman, who shot steel magnate Henry Frick in 1892, and the Russians who assassinated Tsar Alexander II in 1881 all targeted powerful leaders to foment terror and popular revolution.[42] Alongside bombings such as the one at Chicago's Haymarket in 1886, and without benefit of today's widespread media influence, these killings created publicity and popular panic for a revolution that never came.

The United States was not immune to internal terrorism. After the American Civil War (1861–1865) defiant Southerners formed the Ku Klux Klan to intimidate blacks and supporters of Reconstruction. Ku Klux Klan members terrorized blacks, Jews and other groups they deemed inferior. Sometimes Klan members burned crosses on victims' lawns; at other times they brutalized or killed innocent people. At its height in the 1920s, the Klan had more than 2 million members, while today the group has dwindled to only a few thousand.

Then there are the bioterrorism incidents of the past few decades. The first was in 1984, when the Bhagwan religious cult put a bacterium, *Salmonella typhimurium*, into the salad bars of ten restaurants in The Dallas, Oregon. Their plan was to win a local election by incapacitating voters with a gastro-intestinal illness. There were no deaths, but 715 people developed diarrhea. The US anthrax letter attack was the second major bioincident and will be discussed later in this chapter.

In the early 1970s, the leftist terrorist group Weather Underground report-edly attempted to blackmail an official at the US Army Medical Research Institute for Infectious Disease (USAMRIID) to supply organisms which would be used to contaminate municipal water supplies in the US.[43] The plot was discovered when the officer requested several items that were unrelated to his work. In Chicago, in 1972, members of the right-wing terrorist group Order of the Rising Sun were dedicated to creating a new master race. They were found in possession of 30 to 40 kilograms of salmonella typhi, the agent of typhoid. Their plan was to contaminate the water supplies of sev-eral mid-western cities. Dilution would have made their plan untenable. In 1975, the Symbionese Liberation Army (of Patti Hearst fame) was found in possession of technical manuals on how to produce bioweapons. There were also several ricin incidents: the FBI arrested two brothers in 1981 for posses-sion of ricin, and in 1995 two other men were convicted under the Biological Anti-Terrorism Act of 1989 for production of ricin. They were members of the Minnesota Patriots Council, and had planned to poison federal agents by placing ricin on doorknobs. Several Islamic militants were also arrested in Britain in 2004 for attempted ricin use. One of these terrorists fatally stabbed an unarmed arresting police officer.

During the early decades of the twentieth century, terrorism continued to be associated primarily with the assassination of political leaders and heads of state. This was symbolized in 1914 by the killing of the Austrian Archduke Ferdinand by a 19-year-old Bosnian Serb student, Gavril Princip, in Sarajevo, Yugoslavia.[44]

Long before the beginning of World War I in Europe in 1914, what would later be termed "state-sponsored terrorism" had already started. During the interwar years, terrorism increasingly referred to the oppressive measures imposed by various totalitarian regimes, most notably those in Nazi Germany, fascist Italy and Stalinist Russia.

The mid- to late twentieth century: post-colonial terrorism

There was a preponderance of non-state groups involved in the terrorism that emerged in the wake of World War II. The immediate focus for such activity mainly shifted from Europe to its colonies. Across the Middle East, Asia and Africa, nascent nationalist movements resisted European attempts to resume colonial business as usual after the defeat of the Axis powers. That the colonialists had been so recently expelled from or subjugated in their overseas empires by the Japanese provided psychological impetus to indigenous uprisings by dispelling the myth of European invincibility.

In the half-century after World War II, terrorism broadened well beyond the assassination of political leaders and heads of state. In certain European colonies, terrorist movements had two distinct purposes. The first was to put pressure on the colonial powers (such as Britain, France and the Netherlands) to hasten their withdrawal. The second was to intimidate the indigenous population into supporting a particular group's claims to leadership of the emerging post-colonial state. At times these strategies had some success, but not always.

Nationalist and anti-colonial groups often conducted guerrilla warfare, which differed from terrorism mainly in that it tended towards larger bodies of "irregulars" operating along more military lines than their terrorist cousins, and often in the open in a defined geographical area which they partially or totally controlled. Examples are the insurgencies in Iraq, China and Indochina. Elsewhere, such as the fight against French rule in Algeria, these campaigns were fought in both rural and urban areas and by terrorist and guerrilla means.[45]

Still other struggles, like those in Kenya, Malaysia, Cyprus and Palestine, all involved the British, who, along with the French, bore the brunt of this new wave of terrorism. These terrorist groups quickly learned to exploit the burgeoning globalization of the world's media. Following the writings of Pisacane, Hoffman put it as: "They were the first to recognize the publicity value inherent in terrorism and to choreograph their violence for an audience far beyond the immediate geographical loci of their respective struggles."[46] The political slant inherent in modern journalism is no longer unexpected and is even tolerable when social and political issues are the topic of debate. In some cases (such as in Algeria, Cyprus, Kenya and Israel) terrorism arguably helped in the successful realization of terrorist goals, arousing public debate in democratic societies.[47] The exploitation of these "seams" of society is an essential component of today's fourth-generation or asymmetric warfare.

Terrorism continued after the wind-up of the main European overseas empires in the 1950s and 1960s in response to many circumstances. In Southeast Asia, the Middle East and Latin America, there were killings of policemen and local officials, hostage-takings, hijackings of aircraft and bombings of buildings. Civilians often became targets. In some cases, governments became involved in supporting terrorism, mostly at arm's length so as to be deniable. The causes espoused by terrorists encompassed not just revolutionary socialism and nationalism, but also religious doctrines. Bioweapons production also increased. Laws, even the modest body of rules setting some limits in armed conflict between states, would now be ignored in the name of a higher cause.

Saddam Hussein is the first world leader to have brutally used chemical and bioweapons to terrorize his own people. His goals were to systematically terrorize and exterminate the Kurdish population in northern Iraq, neutralize the Shi'ites in the south, win the war with Iran, silence his critics, and test the effectiveness of his chemical and biological weapons. He launched chemical attacks against thousands of innocent civilians in 40 Kurdish villages in 1987–1988, using them as testing grounds. The worst of these attacks devastated the population of Halabja on March 16, 1988.[48]

Despite the events in Iraq and Japan, the bomb has continued to be the terrorist's weapon of choice. Bombs were familiar and reliable devices. WMDs, on the other hand, had too many unknowns. But at the same time, technology and availability were quickly making these weapons more appealing to the modern-day terrorist.

The modern terrorism leading into the twenty-first century

By the end of the twentieth century, there was a new form of "grand" terrorism typified by non-state terrorists such as Islamist suicide attacks in the Middle East and the Tamil Tigers of Sri Lanka. Rapoport called this the fourth wave of modern terrorism.

With possible Western bias, we classify terrorists today in many different ways: domestic or international, state-sponsored or non-state-affiliated, formal or informal, loosely affiliated or formal, religious or non-sectarian, etc. Over the past few decades most of these terrorist movements have increasingly been associated with the indiscriminate killings of civilians. A more detailed classification is as follows:

- special interest (anti-abortion groups or ELF);
- non-ideological (freelancers like the Jackal);
- insurgent (any guerrilla group as in Iraq today);
- political (right or left wing);
- establishment (Saddam Hussein or Fidel Castro);
- religious (Aum Shinrikyo, Hezbollah or al Qaeda).

In 1970, when Palestinian terrorists hijacked several large aircraft and blew them up on the ground in Jordan but let the passengers go free, terrorist acts were viewed by many with as much fascination as horror. This changed in 1972 when 11 Israelis were murdered in a Palestinian Black September attack on Israeli athletes at the Olympic Games at Munich. This event showed a much greater determination to kill in a dramatic way, and now the revulsion felt in many countries was growing stronger. The school killings at Beslan in Russia in 2004 are another example. Here, 25–30 armed terrorists seized a school and took over 1,000 hostages, eventually resulting in over 300 deaths. Attitudes have changed, and now the international media were paying greater attention. Even the Chechen guerrilla leader Shamil Basayev, the now deceased mastermind behind the Beslan school massacre and the 2002 Moscow theater hostage-taking, was unremorseful of what he had done.[49]

A justification offered by the perpetrators of these and many subsequent terrorist actions in the Middle East was that the Israeli occupation of the West Bank and Gaza, beginning in 1967, was an exercise of violence against which counter-violence was legitimate. The same was said in connection with the suicide bombings by which Palestinians attacked Israel during the second Intifada starting in 2000. In some of these suicide bombings there was a new element of viciousness which had not been evident in the Palestinian Islamic religious terrorism of previous decades.

Suicide terror poses a very potent psychological weapon, but the direct impact has been relatively minor, despite the fact that each suicide terror act can cause a much greater number of deaths than a non-suicide terrorist incident. No preventive measures can uniformly succeed against determined suicide murderers, although the direct impact can be tempered. They can be driven by money, humiliation, disenfranchisement, power or cold-blooded murderous incitement, as well as true fanatic religious beliefs. Some may also be unwittingly duped into committing the act. Suicide terror has been steadily rising because terrorists have learned that it pays. Examples are the Israeli withdrawal from Lebanon in 1985; Sri Lanka agreeing to an independent Tamil state in 1990; Turkey granting limited autonomy to its

Kurdish minority in 1990; and the US pulling its military from Lebanon in 1983. These concessions over the last 20 years have encouraged terrorist groups to pursue even more ambitious suicide campaigns.

By the 1960s and 1970s, the numbers of terrorist groups swelled to include not only nationalists, but those motivated by ethnic and ideological considerations. The former included groups such as the Palestinian Liberation Organization (and its many affiliates), the Basque ETA, and the Provisional Irish Republican Army (which has now agreed to lay down its arms), while the latter included organizations such as the German Red Army Faction and the Italian Red Brigades.

In 1978, the Bulgarian exile Georgi Markov was stabbed in the leg with a steel ball or pellet attached on the end of an umbrella packed with ricin while waiting for a bus in London. He died several days later. This assassination was carried out by the communist Bulgarian government, and the technology to commit the crime was supplied by the Soviet Union. This incident represented the first case of multistate-supported (bio)terrorism.

As with the emergence of modern terrorism almost a century earlier, the United States was not immune from this latest wave, although there the identity-crisis-driven motivations of the white middle-class Weathermen starkly contrasted with the ghetto-bred discontent of the Black Panther movement. The Oklahoma City bombing of 1995, killing 168 people, was committed by Timothy McVey, an American, and was the largest terrorist act committed against the US till 9/11.

Like their anti-colonialist predecessors of the immediate post-war era, many of the terrorist groups of this period readily appreciated and adopted methods that would allow them to publicize their goals and accomplishments internationally. Forerunners were the Palestinians who pioneered the hijacking of a chief symbol and means of the new age of globalization—the jet airliner—as a mode of operation and publicity. They provided the inspiration, mentorship and training for many of the new generation of terrorist organizations.

Despite their unproven nature, some European terrorist groups were also anxious to get their hands on biological weapons. The left-wing Red Army Faction/Baader Meinhof Gang, active between the 1970s and 1998, experimented with botulinum toxin in Paris.[50] Religious groups such as Aum Shinrikyo are better known for their use of the chemical agent sarin in the famous Tokyo subway attack of 1995, but they were also very interested in the use of biological weapons such as anthrax and Ebola. They invested $80 million in a biological weapons program with two biological research centers. They purchased a 48,000-acre farm in Australia to test biological agents on livestock and sent members to Africa to look for Ebola samples, not appreciating that the natural reservoir for Ebola was unknown and dormant between epidemics. They even attempted to use aerosolized anthrax, not realizing that they were using a harmless vaccine strain. In 1995, while seeking to establish a theocratic state in Japan, they released deadly sarin gas in several Tokyo subway stations. Their attacks killed 12 people, injured 5,510, and greatly stressed the existing Japanese health care system.

Previously, Aum Shinrikyo had used sarin on another Japanese community, developed and attempted to use other chemical agents (VX gas and hydrogen cyanide) and used biological agents (anthrax, Q fever, Ebola virus, and botulinum toxin) on at least ten other occasions against the mass population and authority figures in Japan. In 1990, they outfitted a car to disperse botulinum toxin through an exhaust system. In 1993 they attempted to disrupt the wedding of Prince Naruhito by spreading botulinum in downtown Tokyo. That same year, they spread anthrax via a sprayer system from the roof of a building. Aum Shinrikyo is the only known example of a non-state terrorist group using WMD for mass murder.

By the mid-1980s, state-sponsored terrorism had re-emerged as the catalyst for the series of attacks against American and other Western targets in the Middle East, with countries like Iran, Iraq, Libya and Syria becoming the principal state sponsors. Falling into a related category were those countries, such as North Korea, that not only bought and sold terrorist weapon technology, but directly participated in covert terrorist acts. Many of these terrorist organizations have today declined or ceased to exist altogether, while others, such as the Palestinian and Spanish Basque groups, motivated by more enduring causes, still remain active. Some, like Hamas, Hezbollah and the IRA, have made moves towards political goals.

Such state-sponsored or -supported terrorism remains a concern of the international community today although it has been somewhat overshadowed in recent times by the re-emergence of the religiously inspired terrorist, an enemy that poses much greater retribution problems. The latest manifestation of this trend began in 1979, when the revolution that transformed Iran into an Islamic republic led it to the use and support of acts of terrorism as a means of propagating its ideals beyond its own borders. Before long, the trend had spread beyond Islam to every major world religion, as well as to many minor cults. From the sarin attack on the Tokyo subway to the Oklahoma City bombing the same year, religion and fanaticism were again added to the complex mix of motivations that led to acts of terrorism. The al Qaeda suicide attacks of September 11, 2001 brought home to the world, and particularly to the United States, just how dangerous to global stability this latest mutation of terrorism was.

The anthrax letters sent to US political and media figures following the terrorist attacks of September 11, 2001 resulted in 23 cases and five deaths over a one-month period. Cases were linked to mail passing through New Jersey, New York and Washington, DC. An estimated 10,000 people were placed on antibiotics. Attorney General John Ashcroft released the text from the anthrax letters sent to Senate Majority leader Tom Daschle, NBC anchor Tom Brokaw, and the *New York Post*. The letters, dated September 11, contained the phrases "Take penacilin [sic] now," "Death to America," "Death to Israel" and "Allah is Great." Homeland Security Director (later to be Secretary) Tom Ridge reported that the anthrax strain sent to Daschle's office was highly weaponized. This terrorist attack caused great panic and anxiety in the community, especially in postal workers. It resulted in significant capital outlays, and left many unanswered questions about the nature of the disease.

Today, terrorism influences events on the international stage to a degree hitherto unachieved, largely due to the media attention of the attacks of September 11, 2001 and the subsequent reaction. Since then, in the United States at least, terrorism has largely been equated with the threat posed by al Qaeda—a threat inflamed not only by the spectacular and deadly nature of the September 11 attacks themselves, but by the fear that future strikes might be even more deadly and employ WMD.

Whatever global threat may be posed by al Qaeda and its "franchises" or affiliates, the US view of terrorism nonetheless remains, to a degree, largely egocentric—despite the George W. Bush administration's rhetoric concerning the "global war on terrorism." This is far from unique. Despite the implications that al Qaeda actually intends to wage a global insurgency, the citizens of countries such as Colombia or Northern Ireland or Israel (to name a few of those long faced with terrorism) are likely more preoccupied with when and where the next FARC or Irish Republican Army or Palestinian attack will occur rather than where the next al Qaeda strike will fall.

As such considerations indicate, terrorism goes beyond al Qaeda, which it not only pre-dates but will also outlive. Given this, if terrorism is to be countered most effectively, any understanding of it must go beyond the threat currently posed by that particular organization. Without such a broad-based globally networked approach, not only will terrorism be unsolvable but it will be unmanageable.

Islamist jihadists see the United States as a symbol and stronghold of satanic Western values. Some, such as Osama bin Laden, made the US their actual target. Unlike the violence of the 1960s and 1970s, the attacks on the US are not secular or Marxist. Unlike the nationalist terror of the IRA or the FLN, they aren't aimed to achieve a negotiated political settlement. They are neither part of a war of rebellion, nor a form of left-wing anarchism, nor a barbarous exercise of power. This new terrorism springs from an unswerving conviction that to destroy evil America is to do God's work. Since it doesn't play to world opinion, world opinion cannot act as a brake upon it. And since it failed to achieve its ultimate objective in America, we should expect it will strike again and again.

The globalized and increasingly more complex nature of today's terrorism needs an infrastructure. It needs money for funding travel, acquiring and planning the use of explosives, dispatching, safe houses and the search for soft vulnerable targets. It needs supporters—planners, preachers and commanders. It needs supporting organizations—religious, educational and welfare systems that brainwash a new generation, often with hatred, lies and ignorance.

Since 1968, threats or actual use of chemical or biological weapons account for only 52 cases out of more than 8,000 in the RAND Chronology of International Terrorism.[51] But over time, terrorism and its subordinate, WMD terrorism, have taken a similar path. They have evolved from state sponsorship to a much more unpredictable terrorist non-state movement. The fundamentalists, fanatics and extremists have been around for thousands of years. What is new is the potential to use WMDs on an unprecedented scale with today's biotechnology, misinformation, religious fervor and lack of restraints.

Today's terrorists have a changing face with a new variety of networked actors and organizations. Their strategic success is not the same as their mission or tactical success. Many may have a religious nexus, or involve a rogue state or transnational organized crime. Their short-term goals include obtaining recognition, vengeance, the freeing of prisoners, turning the tide in guerrilla war, tying up government resources, destroying or seizing facilities, discouraging foreign investment or assistance, causing governments to over-react, and exposing a government's inability to provide security to its citizens. They desperately seek media attention and want to exploit the ethical, political and cultural seams of a democratic society. Their long-term goals include destroying the existing social systems, replacing existing governments, using power to impose their will on their enemies, and setting up religious supremacy.

Success will depend on the effectiveness of counter-intelligence, effective education and indoctrination at a young age, and the ability to convince the local constituency that, in the end, terrorism will not accomplish its stated goals and is not in their best interest in the long term. While terrorists continue to innovate, victory in defeating them will depend on the ability not only to collect, analyze and share intelligence wisely, but to reorganize not in a hier-archical way but rather in a global networked manner. With today's rapidly changing world, can Karl von Clausewitz's statement that "War is the con-tinuation of policy [politics] by other means"[52] also be said about terrorism?

Notes

1 National Memorial Institute for the Prevention of Terrorism, http://www.mipt.org/terrorismdefined.asp, and 22 U.S.C., section 2656f(d).
2 "The Definition of Terror in the Eyes of Radical Islamic Leaders," http://www.ict.org.il/articles/articledet.cfm?articleid=523
3 Personal communication, Richard Spertzel, former Iraqi weapons inspector.
4 B. Murphy and M. Webster, "Orthomyxoviruses," in B.N. Fields, D.M. Knipe, P.M. Howley et al., Fields Virology, 3rd edn. (Philadelphia, PA: Lippincot Williams and Wilkins, 1996), pp. 1397–1445.

5 Kenneth F. Kiple (ed.), *The Cambridge World History of Human Disease* (Cambridge: Cambridge University Press, 1993), p. 393.
6 J. Diamond, *Guns, Germs and Steel* (New York: W.W. Norton, 1999).
7 This term "weaponization" only began to have significance during the second half of the twentieth century when technology was able to produce a much more efficient bioweapon.
8 Xenophon, *The Cavalry General*, http://www.sonshi.com/xenophon.html
9 "Biological Warfare," http://www.absoluteastronomy.com/encyclopedia/b/bi/biological_warfare.htm
10 "History of Biological Terrorism," Fort Worth Public Health Department, http://www.fortworthgov.org/health/threats/bio_history1.asp
11 "Zealotry," http://en.wikipedia.org/wiki/Zealot
12 "Origins of the word 'Thug,' " http://www.bbc.co.uk/religion/religions/hinduism/features/thugs/
13 Ibid.
14 D. Rapoport, "Modern Terror: The Four Waves," *Current History*, Dec. 2001, pp. 419–25.
15 "The Black Death: Bubonic Plague," http://www.themiddleages.net/plague.html
16 "The Virtual Jewish History Tour," http://www.jewishvirtuallibrary.org/jsource/vjw/Poland.html
17 S.K. Lewis, "History of Biowarfare. Medieval Siege," Nova Online, http://www.pbs.org/wgbh/nova/bioterror/history.html#
18 "History of Bioterrorism: A Chronological History of Bioterrorism and Biowarfare throughout the Ages," http://bioterry.com/HistoryBioTerr.html
19 "The New World," http://www.strategypage.com/articles/smallpox/3.asp
20 R.A. Acuna-Soto, D.W. Stahle, †M.K. Cleaveland and M.D. Therrell, "Megadrought and Megadeath in 16th Century Mexico," *Emerging Infectious Disease*, vol. 8, no. 4, April 2002.
21 "The New World," http://www.strategypage.com/articles/smallpox/3.asp
22 "Biological and Chemical Terror History," http://my.webmd.com/content/article/61/67268.htm
23 "History of Biowarfare and Bioterrorism," http://www.azdhs.gov/phs/edc/edrp/es/bthistor2.htm
24 "A Brief History of Terrorism," Center for Defense Information, http://www.cdi.org/friendlyversion/printversion.cfm?documentID=1502
25 "History of Biological Terrorism," http://www.fortworthgov.org/health/threats/bio_history1.asp
26 Ibid.
27 "Vaccines and the Power of Immunity," http://www.postgradmed.com/issues/1997/06_97/vetter.htm
28 J.K. Smart, "History of Chemical and Biological Warfare: An American Perspective," http://www.usuhs.mil/cbw/history.htm
29 "A Brief History of Terrorism," Center for Defense Information, http://www.cdi.org/friendlyversion/printversion.cfm?documentID=1502
30 "Narodnaya Volya," http://www.absoluteastronomy.com/encyclopedia/n/na/narodnaya_volya.htm
31 D. Rapoport, "Modern Terror."
32 E.M. Eitzen and E.T. Takafuji, "Historical Overview of Biological Warfare," in *Medical Aspects of Chemical and Biological Warfare*, ed. F.R. Siddel, E.T. Takafuji and D.R. Franz (Washington, DC: Office of the Surgeon General, 1997), pp. 415–23.
33 http://www.freedominion.ca/phpBB2/viewtopic.php?p=352705&

34 "A Preliminary Review of Studies of Japanese Biological Warfare and Unit 731 in the United States," http://www.centurychina.com/wiihist/germwar/731rev.htm

35 "History of Chemical Warfare and Current Threat," http://www.nbc-med.org/SiteContent/MedRef/OnlineRef/FieldManuals/medman/History.htm

36 P. Williams and D. Wallace, *Unit 731* (New York: Free Press, 1989).

37 "History of Biowarfare and Bioterrorism," http://www.azdhs.gov/phs/edc/edrp/es/bthistor2.htm

38 K. Alibeck, *Biohazard* (New York: Random House, 1999).

39 Protocol for the Prohibition of the Use in War of Asphyxiating, Poisonous or other Gases, and of Bacterial Methods of Warfare (Geneva Protocol), http://www.nti.org/e_research/official_docs/inventory/pdfs/geneva.pdf

40 J. Balzar, "Words on Paper," *Los Angeles Times Magazine*, July 24, 2005, pp. 23, 32–5.

41 "Terrorism History," http://www.findarticles.com/p/articles/mi_qa3695/is_200201/ai_n9049189

42 "Emma Goldman: A Guide to Her Life and Documentary Sources. Chronology (1901–1919)," http://sunsite.berkeley.edu/Goldman/Guide/chronology0119.html

43 USAMRIID was the major facility involved in US offensive weapons production until 1969.

44 A. Roberts, "The Changing Faces of Terrorism," http://www.bbc.co.uk/history/war/sept_11/changing_faces_02.shtml

45 Chapter 9 and Conflict Map, http://www.nobel.se/peace/educational/conflictmap/

46 "A Brief History of Terrorism," Center for Defense Information, http://www.cdi.org/friendlyversion/printversion.cfm?documentID=1502

47 See Chapter 9.

48 US Department of State, "Saddam's Chemical Weapons Campaign: Halabja, March 16, 1988," http://www.state.gov/r/pa/ei/rls/18714.htm, Aug. 9, 2005.

49 Nightline interview, ABC News, July 28, 2005.

50 "Red Army Faction," http://en.wikipedia.org/wiki/Baader-Meinhof_Gang, Aug. 9, 2005.

51 RAND Chronology of International Terrorism, http://www.rand.org/ise/projects/terrorismdatabase/

52 K. von Clausewitz, *On War*, vol. 1 (Berlin: Dümmlers Verlag, 1832), Ch. 1, section 24.

2

DEVELOPING A COUNTER-TERRORISM NETWORK

Back to the future?

Lindsay Clutterbuck

Introduction

In the aftermath of the tragic events in the USA on September 11, 2001 and the subsequent bomb attacks in Bali, Madrid and London, and with the daily litany of attacks on police, military personnel and civilians as the insurgency in Iraq continues, a particular perception of the phenomenon known today as terrorism has become prevalent. At first sight it seems that terrorism now manifests itself in ways that Western people have not encountered before and hence it must pose a new threat to society. In the face of this, counter-terrorism concepts and strategies appear to be inadequate or unsuitable for dealing with it, and hence a process of modernization is needed. Current counter-terrorism organizational systems and structures are hierarchical and bureaucratic, whilst the terrorists now operate by using highly adaptable, flexible and resilient network forms of organization. Counter-terrorism practitioners must therefore organize themselves in a similar way if the threat is to be mitigated successfully.[1]

However, a question must be asked. Is this an accurate description of the current situation? Any assessment of what is new and different must inevitably begin by examining the paradigm that existed before it allegedly changed to become "new" and "different." There are key areas that seem to characterize terrorism in its new form and that appear to be innovative, unique and a product of the twenty-first century, e.g. the desire of terrorists to cause mass casualties and their use of chemical and, potentially, biological weapons. A close examination of the history of terrorism in the nineteenth century shows that both of these, along with many other concepts, are not "new" at all but have their origins in the past. For example, the intention of terrorists to kill large numbers of people has not recently intensified. That desire was articulated long before they possessed the means to even begin to achieve it. Today, it is their capacity to kill on an unprecedented scale that has taken a quantum leap forward.

In 1849, the German radical democrat Karl Heinzen contributed an essay to the journal *Die Evolution*.[2] It was entitled "Der Mord" (Murder) and in it he argued that murder, even that of hundreds of thousands of people, was a historical necessity if revolutionary progress was to be achieved. The ability of revolutionaries to cause death on such a scale required new inventions: explosives, poisons, missiles and mines which "could destroy whole cities with 100 000 inhabitants."[3] Fortunately, the world that Heinzen inhabited lacked the means to achieve this level of death and destruction except during full-scale war or popular revolution. Sixty years after the explosion of the first nuclear bomb and at a time when the artificial creation of new and deadly biological organisms is a reality, the gap between the terrorists' intent and their potential capabilities has narrowed dramatically.

A close examination of history is equally revealing if it is applied to the development of counter-terrorism. Many of the systems, structures and methods that are now taken for granted as mainstays of counter-terrorism also had their origins in the nineteenth century. Perhaps more extraordinarily, so do several of the "new" ideas that are currently being put forward as significant innovations: close cooperation and coordination between police and intelligence agencies, the ability to communicate effectively between police forces and their overseas counterparts and the effective identification and monitoring of suspect individuals at the nations' borders. Not only were all of these identified from the 1880s onward as critical if the growing international problem of terrorism was to be dealt with but systems and structures were then put in place to achieve them.

The main focus of this chapter will be on the historical antecedents of one particular aspect of terrorism and counter-terrorism, that is their utilization of network forms of organization. It falls into two main parts. Firstly, it shows that networks were already a feature of terrorism in the nineteenth century and how police counter-terrorism in the UK also utilized the concept as a means to combat it. Secondly, in the context of a police perspective, it reviews the current state of progress towards the proposed concept of a "global counter-terrorism network."

Network forms of organization in terrorism and counter-terrorism

Networks have been fundamental to the way we interact with each other since time immemorial. At the most basic level, they consist of the people we know and are close to; our immediate family, friends and neighbors. Gradually, they have expanded and developed beyond this until they are now fundamental to the way we conduct business (and conflict) in an increasingly complex world.

34

Some of the earliest research work to recognize that network forms of organization underpinned much terrorist activity, and that therefore the military needed to base their "operations other than war" on a similar concept, was carried out by John Arquilla and David Ronfeldt.[4] They recognized that there was an increasing probability that at least one of the protagonists in an Information Age conflict would be a non-state, paramilitary or other form of irregular force and hence likely to be organized as a network rather than as a traditional, military-style hierarchy. They termed it "netwar" and characterized it as follows:

> The term "netwar" denotes an emerging mode of conflict (and crime) at societal levels, involving measures short of war, in which the protagonists use—indeed, depend on using—network forms of organization, doctrine, strategy and communication ... It differs from traditional modes of conflict and crime in which the protagonists prefer to use hierarchical organizations, doctrines and strategies.[4]

They described the likely theoretical structure of a group that has adopted netwar as its preferred means of organization:

> An archetypal netwar actor consists of a web (or network) of dispersed, interconnected "nodes" (or activity centers)—this is its key defining characteristic. The organizational structure is quite flat. There is no single central leader or commander; the network as a whole (but not necessarily each node) has little or no hierarchy. Decision making and operations are decentralized ... allowing for local initiative and autonomy.[5]

However, networks can be categorized in ways other than those based on their structure. When delineated in terms of their functionality, three types of counter-terrorism network can be discerned. Each network can then further be described in terms of how they were established (their status), the type of contacts that occur between them (their nature) and the geographic span of their operations (their spread). In the first instance, the status of each organizational node can be formal, semi-formal or informal. A formal network, including each node and its main linkages, is one that is recognized by governments, e.g. Europol and Interpol. Semi-formal networks occur when an organization with a mandate to carry out its counter-terrorism role sets out to create and maintain contacts with other organizations wherever it serves its purpose to do so, e.g. police to police, intelligence agency to intelligence agency or police to intelligence agency. Informal networks tend to exist between smaller, often specialized, groups and individuals who know and trust each other through regular contact built up over the course of their counter-terrorism duties.

In addition, all three types of network can also be categorized according to the ways in which they communicate with each other. It can be officially, where requests and communications are made through agreed channels or in a legally specified format, semi-officially, where contacts are routine but occur in a mutually agreed manner, and unofficially, where they are conducted on an informal person-to-person basis. In the latter case, they are once more based on mutual trust and understanding between individuals. Finally, each type of network and each form of contact can operate on a series of five geographic levels: locally, nationally, regionally, internationally and globally.

A complex series of interrelationships can therefore exist in a counter-terrorism network, based on the origin and status of each node, the ways that they interact and the scope of their contacts. When viewed in this way, they are more prevalent than at first may be imagined. They also appeared very early on in counter-terrorism. This was in reaction to a campaign of political violence that first saw the phenomenon we now refer to as terrorism emerge in a form that grew increasingly familiar in the closing decades of the twentieth century and that continues to this day.

The origins of terrorism

Any attempt to determine the origins of terrorism shows that its use to further political ends has its roots deep in the past. They lead back to the Nizari Isma'ili sect of eleventh-century Persia (more generally known as the "Assassins"), the Sicarii of first-century Palestine and on into antiquity in the writings of various Chinese military strategists.[6] However, terrorism as it is generally recognized today first appeared during the latter half of the nineteenth century. It had three early exponents, the Russian social revolutionary movement, the "physical force" Irish Republican groups and the Anarchists. All three, to a greater or lesser degree, existed as networks. The Russians (of Narodnaya Volya) were a closely knit conspiratorial group with little outside contact, whilst the Anarchists who were prepared to kill rather than talk existed as even smaller knots of like-minded individuals.[7] By far the most influential of the early groups that developed the use of terrorism were the extreme Irish republican groups of the 1880s that became established in the USA: Clan na Gael, the Irish Republican Brotherhood and the Skirmishers. Their public face was as a well-organized, overt political and social movement but this concealed a series of terrorist cells and all the support mechanisms that were essential to maintain them with explosives, finance and personnel.

I have argued in detail elsewhere that the true progenitors of modern terrorism are these Irish republican groups and hence the concentration here is on how they organized themselves to carry out their self-imposed mission.[8] Their objective from the 1870s onward was to force the British government to withdraw from Ireland by the use of physical force. They intended to achieve this by conducting a series of attacks in major cities on the British mainland using improvised bombs initiated by time-delay mechanisms. The strategy, tactics and techniques that they employed, even their very concept of mounting a sustained campaign of terrorist attacks conducted over a long period of time in order to force political concessions, became the standard model for future terrorist groups to adopt, irrespective of their ideology or ultimate objectives.

The campaign began on January 14, 1881 with the detonation of a bomb placed by the Skirmishers in a ventilation shaft in the wall of Salford Infantry Barracks, Manchester. An adjacent butcher's shop and rope factory were destroyed, a 7-year-old boy was killed and three people were injured.[9] Three elements of this attack were innovative and unprecedented. Firstly, the selection of the target was not for any practical purpose but because of its symbolic significance. In 1867, Salford was the scene of the public hanging of three Irishmen, Allan, Larkin and O'Brien, all of whom had been convicted of taking part in the rescue of two leaders of an earlier extreme Irish group known as the Fenian Brotherhood. In the course of the rescue attempt from the police van that was conveying them to prison, a police officer was shot dead and several bystanders injured. The first bomb attack therefore took place in Salford as a consequence of what the town itself had come to represent in the eyes of the extreme Irish republican movement. The "Manchester Martyrs," as they quickly became known, would not be forgotten. Secondly, the way that the bomb was detonated meant that from the outset the plan was for this event to be the first of many and not just a one-off attack.[10] The bomb consisted of gunpowder and a time-delay fuse, thus allowing the perpetrator the maximum opportunity to plant the device, escape and strike again. The effect of the attacks was therefore meant to be cumulative. Finally, the perpetrators were at best reckless or careless as to whether death or injury might be caused to innocent people or, at worst, they considered it of no consequence.

By the time that the campaign was effectively over in 1885, the Skirmishers had long been neutralized. They had been forced to withdraw after a series of arrests of their cell members in the UK and after a bitter feud in the USA with the other groups. The attacks had then been continued by the better funded, more technologically advanced combined forces of Clan na Gael and the Irish Republican Brotherhood. Bombs had exploded or been found in London, Liverpool and Glasgow.[11] They had been detonated in the Chamber of the House of Commons, the Tower of London and outside Scotland Yard itself. The London Underground system on October 30, 1883 saw two bombs dropped from passing trains, one on the District Line as it left Charing Cross station and one on the Metropolitan Line as it approached Edgware Road station. They exploded almost simultaneously, causing severe damage to the carriage and tunnel but no injuries. On January 2, 1885, a further bomb exploded inside a passenger carriage on the Metropolitan Line near Gower Street, almost destroying it but, once more, causing no injuries.[12] Suitcase bombs consisting of over 20 pounds of dynamite and a mechanical timing mechanism had been placed in the left-luggage offices in London mainline railway stations. Only the smallest of errors in their construction prevented all of them from detonating with an almost inevitable loss of life. Other attacks had been mounted against an industrial gas holder, an aqueduct, government buildings and the office of the *Times* newspaper.

By great good fortune, no one had been killed in any of the attacks except for three Clan na Gael members who were trying to attach their bomb underneath an arch of London Bridge when it detonated prematurely. How could such things be achieved at the very heart of the British Empire and under the nose of Scotland Yard itself?

The terrorist network

The campaign of bombing waged by the extreme Irish groups from 1881 onwards was not only unique and unprecedented in terms of both its concept and how they carried out their attacks, but it was also unique and unprecedented in terms of how it was organized and supported.[13] Their objective was to win independence for the island of Ireland and, to achieve this, Americans of Irish descent travelled from the USA to carry out attacks against targets on the British mainland. To make this possible there was an established infrastructure in the USA that supported them by raising the necessary finance and recruiting the bombers. A well-established logistics operation developed the bomb-making technology, procured the explosive and smuggled it into the UK. Over the years since 1848 when they first established themselves, Irish republican groups had shipped arms and American Civil War veterans to Ireland to foment and take part in an uprising. They had bought a ship and sailed to Australia to free their leaders from a British penal colony and on more than one occasion they had crossed from the USA into Canada (then a British colony) to try to seize and hold territory by force of arms to use as a bargaining chip to force the withdrawal of the British from Ireland.[14] Already able to carry out their activities on a global scale, these Irish-Americans were to become the first international terrorists.

Clan na Gael was by far the biggest of the three main Irish-American "physical force" groups. It had many members at all levels of society and presented a very public face. Internally, it was organized along Masonic lines, with an oath of allegiance, ritual and the use of specific handshakes and "hailing signs."[15] Members were divided into numerous "Centers" of up to a dozen individuals led by a "Head Center" who was the only one who knew the identities of all the members. Each had autonomy of action but they were linked through their Head Center to the higher levels of the organization. Fundraising and canvassing for political support were carried out openly and often through the pages of the so-called "Dynamite Press." This was the collective name given to a number of rival Irish-American newspapers whose main objective was to advance the cause of Irish independence by advocating the use of any and all means possible, including that of physical force.[16]

Two other, smaller, groups were also involved in "physical force" operations. The Irish Republican Brotherhood worked in close cooperation with Clan na Gael through a "Revolutionary Directory" consisting of seven men selected by both organizations. Initially, they had been established in 1857 as the Irish-based arm of Clan na Gael, and their focus was on an internal Irish "rising."[17] The Skirmishers were a breakaway faction of Clan na Gael who independently initiated their own campaign in Britain. They were responsible for the Salford bomb and several others until their teams were arrested and broken up in 1883.[18] It was only then, once their rivals had been removed from the scene, that the greater resources and better organization of the "Revolutionary Directory" turned to unleashing its own, potentially far more lethal, campaign.

As a consequence of the need for the greater than normal secrecy needed to plan and mount successful terrorist attacks, the planning and operational elements of the bombing campaign were undertaken by a small group of trusted individuals. The members of the operational "cells" traveled independently from the USA to the UK by sea and met up again on their arrival. "Safe houses" and facilities where the bombs could be made and stored were then obtained, sometimes with the help of a local contact but more usually through their own efforts. By using merchant seamen sympathetic to their cause, they smuggled explosives, detonators and timers to the UK aboard transatlantic passenger liners or dispatched them separately as freight to await collection from the dockside.[19] Money was entrusted only to the cell leader, a security measure that in one case would cause great difficulties to the other members of the cell when he was arrested.[20]

On the face of it, all the advantages lay with the terrorists. Initially, they benefited from both strategic surprise and the impact of their tactical innovations. They could dictate the pace and tempo of the campaign by their ability to attack at a time and place of their choosing and hence they could develop a cumulative effect through the use of a series of bomb attacks. They utilized modern technology in the form of the new explosive known as dynamite and developed a series of chemical and mechanical time-delay mechanisms to detonate it when they were safely clear of the area. The source of their funds and technological expertise was a sympathetic community where they could operate with impunity and which lay safely in another country on the other side of the Atlantic Ocean. In London, the formidable task of finding and apprehending the bombers lay with the Metropolitan Police.

The origins of counter-terrorism

The first organized police force to be established on the British mainland was the Metropolitan Police Force in London. They came into being on September 29, 1829 as a disciplined, unarmed force that operated solely as a uniformed, preventive body.[21] Its Detective Department was not formed until 1842 and for many years it remained small in number (usually about 20 officers). Eventually, it was expanded and reorganized in 1878 and became the Criminal Investigation Department (CID). The Metropolitan Police were not unaccustomed to dealing with Irish extremism or to the task of monitoring the numerous foreign refugees fleeing revolution or political persecution in their own countries who had sought refuge in London, but the events of the 1880s posed them a completely new set of challenges.[22]

In 1880, the chief superintendent in charge of the CID had given Detective Inspector John Littlechild the task of specializing in matters relating to extreme Irish nationalism, and subsequently all relevant matters were referred to him.[23] Consequently, whilst the Salford bomb attack was not predicted, the potential for a new period of extreme Irish republican attacks was at least foreseen. By March 1883, several bombs had exploded in Liverpool and Glasgow and an unexploded device had been found outside the Mansion House, the official residence of the Mayor of London. The final straw came on March 15 when the offices of the Local Government Board in Whitehall and the office of the *Times* newspaper suffered damage from explosive devices. Two days later, on March 17, 12 extra officers with Irish backgrounds were brought in to reinforce the existing detectives by forming a new "Irish Bureau."[24] Eventually numbering some 30 detective officers who were based at Scotland Yard, they quickly became known within the Force as the "Irish Brigade." They formed the core of the first coordinated, systematic counter-terrorism effort in Britain.

To effectively perform this function, a variety of new tasks had to be undertaken. Physical security was needed inside and outside buildings at risk and, to carry this out, armed uniformed officers from the Royal Irish Constabulary were drafted in to assist the hard-pressed Metropolitan and City of London Police Forces. For the first time, close personal protection by armed detectives was given to prominent individuals at risk. These included both Queen Victoria and the Prime Minister. The planned recruitment of an extra one thousand officers was brought forward in order to provide increased patrols in central London where the heart of government was located.

The detectives of the Irish Bureau set about applying to the new problem the tried and tested methods that they regularly used to detect and arrest criminals. They were extremely busy as suspects were placed under surveillance, and investigative and intelligence leads were followed up, whilst informers were actively sought out and recruited. What was different was that, for the first time, this approach needed to be both systematic and on an unprecedented scale. In order to secure the safety of London and to deal with a threat that was national in its scope, the detectives also needed to operate nationally and internationally. There were many local forces to deal with, and most of them apart from those in large cities such as Glasgow and Manchester had no detectives of any sort. In addition, success could only be achieved by enlisting the help of police forces in the USA and in continental Europe.[25]

Constructing an international counter-terrorism network

It soon became apparent that the source of the attacks lay in the USA. As there was no federal police entity in the USA to interact with, the UK detectives had to devise their own ways of ensuring that vital enquiries were carried out. Where possible, they corresponded with local police forces and, when the matter was important enough, one or two detectives at a time traveled to the USA to carry out enquiries themselves. However, most of the work was not conducted directly with state or local police forces but with Pinkerton's Detective Agency, which was engaged to undertake the enquiries on their behalf. A regular stream of correspondence eventually developed on the background details of suspects known to be or suspected of being in the UK, intelligence reports on activity and individuals who might be relevant and investigative leads that had been followed up. The requests for information were part of a two-way process and, after the murder in Chicago during 1889 of Dr. Patrick Cronin, a prominent member of Clan na Gael, Robert Pinkerton requested assistance from Scotland Yard for one of his detectives when his investigation took him to London.[26]

Investigative work was not confined to the USA, and records from the late 1880s show enquiries being initiated in Africa with the Griqualand West Police and the Kimberley Police and even as far afield as Australia with a request for assistance to the Melbourne Police.[27] Information was not provided by police forces alone but also came into the Irish Bureau from UK government representatives based in Europe and the USA.

By November 1886, the threat of bombers and explosives arriving at British ports was judged so severe that nearly 60 Metropolitan Police officers were deployed around the country to seaports where transatlantic or continental passengers disembarked. This also included a contingent posted to Charing Cross, the railway station in London that acted as the terminus for trains arriving from the main cross-Channel port at Dover. Their role was described by John Sweeney, an Irish Brigade detective who later rose to be head of the organization that developed directly from it in 1887: the Metropolitan Police Special Branch. He stated that "All foreigners, especially Irish Americans, were carefully shadowed. Men were sent from headquarters to various ports, to board all incoming vessels, and scrutinize everyone who disembarked, watching all suspected persons, and making sure that all baggage was properly examined by the Customs Officials."[28]

Beyond this national perimeter, a further line of defense was opened up at the seaports in Europe where the transatlantic liners docked. Officers were permanently posted to Le Havre in France, The Hague in Holland, Christiansand in Norway, Copenhagen in Denmark and Cuxhaven in Germany. France was seen as a center for Irish extremist activity, both at the Channel ports and in Paris. As well as being frequent visitors, two Irish Bureau detectives were stationed in Paris for long periods at a time, where they worked with at least the knowledge of, if not always in full cooperation with, the local police. Maurice Moser, an inspector in the Metropolitan Police Irish Bureau, reveals how the French police assisted him to drug the drink of an Irish suspect, relieve him of his recently collected mail, steam it open, copy it and return it before he regained consciousness![29]

The physical surveillance of suspects was an essential tool for the detectives. From 1883 onwards, suspects were "watched—not intermittently—but continuously, by the system now called shadowing [where] the policeman engaged in watching hands his suspect over to another policeman who takes up the duty before the first man can be relieved."[30] By 1890, the system had been refined and formalized. In Ireland itself, named suspects were categorized centrally as either "A," "B" or "C." Each category related to the specific instructions as to the actions that officers should undertake if they came across the relevant individual: " 'A'—place under immediate continuous surveillance, 'B'—Telegraph movement in cipher to the police at their destination where they were to be watched and 'C'—Police to note the activity of their local suspects."[31] At least two members of the Irish Parliamentary Party were categorized as "A" and, as they would have traveled frequently to the Houses of Parliament in London, the system must have operated in Ireland, Britain and, by implication, where the detectives were stationed in Europe as well.

The intelligence-gathering network was by no means confined to the sources of potential information already described. Great efforts were put in by the detectives to gather intelligence at the public and private meetings of organizations of interest. These included the Irish National League, the Anti-Coercion League and the Fenian Brotherhood, as well as those held by Clan na Gael and the Irish Republican Brotherhood. Secondly, a considerable amount of effort was put into monitoring items of relevant interest in the local and national press in Britain, Ireland and the USA. Not surprisingly, the "Dynamite Press" in the USA came in for close scrutiny, especially the *United Irishmen*. A major influence on this paper and a regular contributor to it was Jeremiah O'Donovan Rossa, whose other role in physical force Irish republicanism was as the strategist and organizer of the Skirmishers. Newspapers published locally in Ireland were also scrutinized regularly.

Another element in the overall surveillance network involved keeping a close watch on the visitors of prisoners convicted of carrying out bomb attacks or other activity such as arms procurement. This surveillance also extended to monitoring the correspondence that they both received and sent, with copies of the letters being sent directly from the prisons to the detectives at Scotland Yard.

From March 1884, a new counter-terrorism organization began to operate, at first in London and then in the north of England. Whilst it had no official name, it was sometimes known as the "Secret Service" and was an ad hoc mixture of Royal Irish Constabulary officers already working covertly in Britain and newly recruited civilian "agents." Its head was Edward Jenkinson, the Assistant Under Secretary for Police and Crime at Dublin Castle in Ireland, who was seconded to the Home Office in London to fulfill this role.[32] In overall terms, it was not a great success. This mainly stemmed from its predilection for regularly expanding its remit beyond that of gathering intelligence and passing it on to police forces to decide what action needed to be taken. A fierce battle over its activities ensued between Jenkinson and the Assistant Commissioner of the Metropolitan Police CID, James Monro. Jenkinson accused the detectives of incompetence and interference in his work whilst Monro constantly complained that Jenkinson did not pass on the intelligence he gained, preferring to act on it unilaterally. It culminated in January 1887 when the Home Secretary relieved Jenkinson of his responsibilities as "Secret Agent." This role was then personally taken over by Monro.[33] From then on, countering Irish republican terrorism on the British mainland was unequivocally recognized by the Home Office as a police function.

Over a period of about six years, from 1881 onwards, a comprehensive system designed to counter the world's first sustained terrorist bombing campaign was devised and implemented in the UK. There were many facets to it and this brief review has concentrated only on the aspects relevant to the development of its systems and structures. Much was also achieved in terms of operational tactics, investigative methodology and intelligence-gathering and -processing techniques. A strong argument can be made that, during the 1880s, not only were many of the diagnostic features of terrorism today devised and implemented but so too were many of the essentials of counter-terrorism that we now rely on as key factors in defeating it. Foremost amongst these is the ability to build, sustain and integrate local, national and international police networks for the gathering and exchange of information and intelligence on terrorism and, where required, to act in a coordinated way to mitigate it. With the case study of the 1880s in mind, it is instructive to review contemporary progress towards that end.

Towards a global counter-terrorism network: existing systems and structures

Regionally

Across the world today, many more police counter-terrorism networks exist at the local level than ever before. Some are rudimentary and some are sophisticated, but the main achievement lies in the very fact of their exist-ence. The arguments over whether they are necessary or not generally have been won. In addition, great efforts are being made in many countries not only to link together their relevant national intelligence assets but also to enable them to interact in an effective way with these local networks. This work is vital but, as terrorism is now more than ever a global phenomenon, it must also be carried out simultaneously with the creation of international networks that also can function at the regional and global levels.

A good example of a regionally active counter-terrorism network can be seen in the 26 states of the European Union (EU). Particular progress has been made at the formal level as the EU has developed new structures whose role encompasses police counter-terrorism. The largest of these is Europol, a predominantly police-based organization that was originally formed in 1991 to help member states to improve their effectiveness against drug trafficking and organized crime.[34] Its remit was extended to terrorism as from January 1, 1999 but it was not until after the attack on the World Trade Center in New York that significant resources were allocated to the task.

A second, even newer, formal body is the Situation Centre (SITCEN). This is "an integrated group of analysts from our external intelligence services and the internal security services to jointly assess the terrorist threat as it develops, both inside Europe and outside."[35] Cooperation at the European level is only during the analytical process and any resulting operational cooperation is conducted at the national level on a bilateral or trilateral basis.

The concept of international cooperation between police forces to combat terrorism is not new. After the success of the Metropolitan Police Special Branch in dealing with the threat from extreme Irish republicanism in the 1880s, many approaches were made to them by their continental counter-parts to ask for assistance to deal with the growing problem of Anarchist assassination attacks. Requests came from France, Italy, Germany and even Russia.[36] These contacts were both unofficial and semi-official in nature. Some of the unofficial, police-to-police ones were carried out but for domestic political reasons most official requests from foreign governments for assistance were turned down by the UK government.

With the re-emergence of terrorism in Europe during the mid-1970s, the police agencies responsible for counter-terrorism in the countries where it was most prevalent formed a loose, informal body known as the Police Working Group on Terrorism (PWGT). Its main objective was to facilitate the exchange of information at the operational level. By 1975, a semi-formal structure known as the Trevi Group had been created to provide a wider forum for practitioners working in the areas of terrorism, public order, drugs and organized crime.[37] The PWGT became part of the Trevi framework. Today, the functions of the Trevi Group have been absorbed into the EU "Third Pillar" intergovernmental criminal justice arrangements. The PWGT still continues and forms the basis for a network of police counter-terrorism liaison officers (CTLO) that operates across the EU. Its survival over so many years during the numerous organizational upheavals of the EU itself is testimony to its value.

Despite the progress that has been made, it has recently been acknowledged that the arrangements in the EU for counter-terrorism work are not ideal.[38] There are too many bodies involved and there is still too much reluctance for states to routinely share intelligence and information within EU structures. It is to be hoped that the appointment of the first European Council Counter-terrorism Coordinator, Gijs de Vries, may help to rectify the situation. However, despite its shortcomings the EU is setting the pace for regional cooperation between states in countering terrorism.

Internationally and globally

As the wave of Anarchist assassinations of heads of state and prime ministers swept over Europe and the USA from the 1890s onward, there were moves towards securing international legal cooperation against the Anarchists. The UK government felt compelled to reject all of them on the grounds that public opinion would not tolerate new domestic laws generated by international agreement as they did not feel threatened by this particular terrorist threat.[39] Little further progress was made internationally on this issue, despite a rallying call a few years later by the President of the USA, Theodore Roosevelt. He stated that "Anarchism is a crime against the whole human race, and all mankind should band together against the Anarchist. His crimes should be made a crime against the laws of all nations . . . declared by treaties among all civilized powers."[40] However, fearing any increased involvement in European politics, the USA decided not to attend the conference that had been arranged to agree a treaty and, after opposition from other nations as well, no further progress was made. Gradually, as the threat began to wane, so did the interest of governments in international measures to counter it. It was not to become a pressing issue once more until the resurgent threat from terrorism became widespread again from the end of the 1960s onwards.

Today, much of the cooperation at the international level is done bilaterally or trilaterally. The G8 nations show how this can be developed even further. As well as the use of summit meetings of the G8 to highlight strategic and policy-level declarations on terrorism and counter-terrorism, an increasing amount of cooperation and joint initiatives are taking place on a more practical level. They involve both "officials," i.e. government level, and "experts," i.e. practitioners from the police and intelligence services. As a consequence, a new international network is slowly being created.

There is only one network that both deals with counter-terrorism and can realistically claim to be global in its span, and that is the International Criminal Police Organization (ICPO), more popularly known as Interpol. The

182 countries that are members of Interpol are theoretically constrained by Article 3 of its constitution which prohibits it from taking action in any case that has a political, military, religious or racial character.[41] Despite this, it has made some significant practical advances in its capacity to deal with terrorism before, during and after a terrorist attack.

The foremost vehicle for this is the Fusion Task Force (FTF), a body created in September 2002 with the objectives of identifying active terrorist groups and their members, encouraging and supporting the collection, analysis and sharing of information and intelligence and enhancing the capacity of countries to deal with both terrorism and organized crime.[42] The FTF works through four regional task forces, covering Southeast Asia, Central Asia, Central America and Africa, and its primary role is to support two global initiatives. The first of these is Project Passage, whose aim is to disrupt the movement of terrorists across national borders, and consequently it focuses on the organized crime and logistical networks that make this possible by the provision of false or illegally obtained documents. The second is Project Trent. Its objective is to identify individuals who have attended terrorist training camps and then to notify the fact to the country where they are currently residing.

A further recent Interpol initiative in March 2005 seeks to draw attention to the potential interest shown by an increasing number of terrorists in acquiring and using biological weapons. A dedicated bioterrorism unit has been established at Interpol headquarters and work is under way to devise an international program to build both national and international capacities to detect and counter the threat.[43]

The recent initiatives put into place by Interpol show what can be done when there is sufficient international will to achieve it. If it is accepted that national reservations concerning the general sharing of sensitive operational intelligence will remain a fact of life for many years to come, then the provision of global and regional thematic support by Interpol in the areas outlined above may be an effective route for the organization to follow.

Conclusion

When examined in the light of terrorism and counter-terrorism today, the campaign of bomb attacks launched against targets on the British mainland from 1881 onwards by extreme Irish republican groups based in the USA and the measures that were taken to counter them show that many aspects of what are now considered to be new and innovative concepts are in fact firmly rooted in the past. However, there are key differences, most notably in terms of the scale of death and destruction that terrorist groups seek to achieve and the impact that modern technology can make in preventing and detecting them.

In organizational terms, there are many parallels between the way that

Clan na Gael, the Irish Republican Brotherhood and the Skirmishers structured themselves internally and then operated as networks. This use of network forms of organization is seen even more strikingly when the counter-terrorist response that they engendered is examined. The key point of learning here is that an international terrorist threat requires international counter-terrorism action and cooperation. This was recognized and acted upon very early on, as was the realization that a reliance on trying to stop the bombers when they were already in the country was too little and too late. The need for a layered defense was identified and acted upon, starting at the ports where suspects were likely to arrive and then extending out into other countries whose seaports were also arrival points for transatlantic passengers. In this way, a network to enable internal police cooperation was first established and then extended to become a network with international reach.

Today, the world is more complex and interconnected than ever before, and the coordination of both internal and external information and intelligence-gathering mechanisms is even more vital. The challenge for today and the future is twofold. The main task must be to connect up in the most effective and efficient way the existing relevant nodes. The urge to create new or additional organizational nodes should be resisted unless they demonstrably serve a useful purpose. In this way, duplication and unnecessary complexity are avoided.

The second challenge is to build on existing unofficial and informal counter-terrorism networks in order to transform them into robust systems and structures that will survive the departure of individuals and other inevitable privations. Only by doing so will we be able to deal with a real and deadly threat that is increasingly global in its reach and, as a consequence of the Information Age that we live in, already global in its impact.

Notes

1 John Arquilla and David Ronfeldt, *The Advent of Netwar* (Santa Monica, CA: RAND, 1996).
2 Carl Wittke, *Against the Current: The Life of Karl Heinzen* (Chicago, IL: University of Chicago Press, 1945), pp. 73–75.
3 Karl Heinzen in *Die Evolution*, quoted by Walter Laqueur in *Terrorism* (London: Weidenfeld & Nicolson, 1977), p. 40.
4 Arquilla and Ronfeldt, *Advent of Netwar*, p. 5.
5 Arquilla and Ronfeldt, "The Advent of Netwar," in *In Athena's Camp: Preparing for Conflict in the Information Age* (Santa Monica, CA: RAND, 1997), p. 280.
6 For an account of the Assassins, see Edward Burman, *The Assassins: Holy Killers of Islam* (London: Crucible, 1987) and, for the Sicarii, see Stewart D'Alessio and Lisa Stolzenberg, "Sicarii and the Rise of Terrorism," *Terrorism*, vol. 13, no. 4–5, July 1990.
7 For a first-hand account of their activities, see Vera Figner, *Memoirs of a Revolutionist* (New York: Greenwood Press, authorized translation, 1968).

8 See Lindsay Clutterbuck, "The Progenitors of Modern Terrorism: Russian Revolutionaries or Extreme Irish Republicans?," *Terrorism and Political Violence*, vol. 16, no. 1, Spring 2004, pp. 154–81.

9 K.R.M. Short, *The Dynamite War: Irish-American Bombers in Victorian Britain* (Dublin: Gill & Macmillan, 1979), p. 50.

10 The Russian approach differed. With two notable exceptions, every time that Narodnaya Volya used explosives in order to try to assassinate Tsar Alexander II it involved electrical detonation by command wire. They all failed. When they differed and used a fuse, as in the Winter Palace, they came close but they eventually succeeded with contact-initiated home-made grenades. The use of a mechanical timer was never attempted. See Clutterbuck, "The Progenitors of Modern Terrorism."

11 A variety of time-delay fuses were employed, ranging from chemical to mechanical methods. The most common were detonators consisting of a clockwork mechanism that caused a small pistol to fire into a block of dynamite, thus triggering the explosion. They were known as "infernal machines." See Short, *Dynamite War*, p. 176.

12 Just over one hundred years later, in July 2005, the London Underground system was targeted during two waves of terrorist attacks. In the first wave, on July 7, three "suicide bombers" detonated their explosives on separate Underground trains, killing themselves and 52 passengers. In a further ironic coincidence, one bomb was detonated on a Circle Line train just as it left Edgware Road station. Seven passengers were killed. A second, unsuccessful series of attacks took place on July 21 when the devices of all of the alleged "suicide bombers" failed to detonate. On both these dates, an additional, linked attack also took place above ground on different London buses. Thirteen people died on the Number 30 bus on July 7. No one was injured in any of the second-wave attacks and, by July 30, all of the would-be suicide bombers had been arrested. On May 18, 1885, James Gilbert Cunningham was found guilty of planting the bomb on the Metropolitan Line train earlier that year and with an accomplice, Harry Burton, of carrying out several other attacks. He was sentenced to life imprisonment. (For events in 1883 and 1885, see Short, *Dynamite War*, pp. 205–10. For the events of July 7, 2005 see "Report of the Official Account of the Bombings in London on 7th July 2005" published in London by The Stationery Office, May 11, 2006.)

13 Prior to the campaign on the British mainland that had begun in 1881, the emphasis of the Fenian Brotherhood (predecessors of Clan na Gael) was on fomenting a "rising" in Ireland that would drive the British out by force. All their preparations were geared to this end and, in February 1867, a concerted attempt was made. Despite the presence of Irish-American Civil War veterans in Ireland, it made little impact and was soon defeated. It was not until ten years after this that the ideas of attacking British interests abroad and then in Britain itself were taken up and developed into a concept of operations. See *The Green Flag* by Robert Kee (London: Weidenfeld & Nicolson, 1972) for an account of Irish republicanism during this period.

14 See, respectively, T.D., A.M. and D.B. Sullivan, *Speeches from the Dock* (New York: P.J. Kenedy and Sons, believed 1900), pp. 291–312; Short, *Dynamite War*, pp. 32–4; and Ian Hernon, "The Fenian Invasion of Canada, 1866," in *Blood in the Sand: More Forgotten Wars of the 19th Century*, ed. Ian Hernon (Stroud, Gloucestershire: Sutton Publishing, 2001), pp. 175–92.

15 The details as they appertain to Clan na Gael are described in the memoir of Henri Le Caron, *Twenty Five Years in the Secret Service* (London: William Heinemann, 1894), pp. 107–17 and, for the Irish Republican Brotherhood, see

Leon O'Broin, *Revolutionary Underground* (Dublin: Gill and Macmillan, 1976), pp. 7–10.

16 For example, the paper the *Irish World* under the editorship of Patrick Ford was instrumental in developing the concept of a "physical force" campaign against the British government to be carried out by a small band of dedicated "skirmishers" and in raising $23,350 by March 1877 to implement it. The majority of this was spent on developing and constructing a submarine, the *Fenian Ram*, with the intention of using it to sink British shipping in US harbors. See Short, *Dynamite War*, pp. 33–44.

17 See the memoirs of Joseph Denieffe, a founder member, in *A Personal Narrative of the Irish Revolutionary Brotherhood* (New York, 1906; reprinted in Ireland by Irish University Press, 1969).

18 This was achieved as a result of a combined operation between the detectives of the Metropolitan Police and the Birmingham City Force. It was the first time a joint police operation had ever been attempted. See Short, *Dynamite War*, pp. 125–41.

19 Detective Inspector Maurice Moser describes the discovery at Liverpool docks of six "infernal machines" in casks purporting to contain cement. Maurice Moser and Charles Rideal, *Stories from Scotland Yard* (London: George Routledge and Sons, 1890).

20 Le Caron, *Twenty Five Years in the Secret Service*, pp. 294–8.

21 See Gary Mason, *The Official History of the Metropolitan Police* (London: Carlton Publishing, 2004).

22 Bernard Porter, *Plots and Paranoia* (London: Unwin Hyman, 1989).

23 John Littlechild, *Reminiscences of Chief Inspector Littlechild* (London: Leadenhall Press, 1894), p. 10.

24 Bernard Porter, *The Origins of the Vigilant State: The London Metropolitan Police Special Branch before the First World War* (London: Weidenfeld & Nicolson, 1987).

25 Chief Constables Register 1888, Metropolitan Police archive.

26 Ibid.

27 Ibid.

28 John Sweeney, *At Scotland Yard* (London: Grant Richards, 1904), p. 51.

29 Moser and Rideal, *Stories from Scotland Yard*, pp. 149–53.

30 H. Thynne to Irish Under Secretary, undated (1890?), Balfour Papers, Public Records Office, PRO 30/60/7.

31 O'Broin, *Revolutionary Underground*, pp. 42–3.

32 For a fuller account of its role, activities and impact, see Lindsay Clutterbuck "Countering Republican Terrorism in Britain: Its Origins as a Police Function," *Terrorism and Political Violence*, vol. 18, 2006, pp. 95–118.

33 The British mainland-based functions of the Home Office "Secret Service" passed to the newly formed Metropolitan Police Special Branch in February 1887. However, it continued to exist and to act as the central collecting point for intelligence on extreme Irish republicanism gathered from Ireland itself and from overseas.

34 Michael Santiago, *Europol and Police Cooperation in Europe* (London: Edwin Mellen Press, 2000).

35 Interview with Gijs de Vries, European Council Counter-Terrorism Coordinator, Mar. 4, 2005. Reported online at http:/www.euractiv.com/Article?tcmuri=tcm; 29–136245–16&type=Interview

36 David Rapoport, "Attacking Terrorism: Elements of a Grand Strategy," in *The Four Waves of Modern Terrorism*, ed. Audrey Kurth Cronin and James E. Ludes (Washington, DC: Georgetown University Press, 2002), pp. 46–73.

37 See John Benyon, Lynne Turnbull, Andrew Willis, Rachel Woodward and Adrian Beck, *Police Cooperation in Europe: An Investigation* (Leicester, England: Centre for the Study of Public Order, University of Leicester, 1993).

38 House of Lords European Union Committee, "After Madrid: The EU's Response to Terrorism; Report with Evidence," 5th Report of Session 2004–05, Mar. 8 (London: Stationery Office, 2005), pp. 27–8.

39 Porter, *Origins of the Vigilant State*, pp. 111–13.

40 Rapoport, "Attacking Terrorism," p. 52, quoting Jensen.

41 For an examination of the implications of this, see Paul Swallow, "Of Limited Operational Relevance: A European View of Interpol's Crime Fighting Role in the 21st Century," *Transnational Organised Crime*, vol. 2, no. 4, 1996, p. 116.

42 http://www.interpol.int/Public/FusionTask Force/default/.asp

43 Interpol Fact Sheet TE/02, "Enhancing the Response to Bio-terrorism Threat," accessed via www.interpol.int

3

WARNING IN THE AGE OF WMD TERRORISM[1]

Gregory B. O'Hayon and Daniel R. Morris

Anyone who looks at America's acts of aggression against the Muslims and their lands over the recent decades will permit this [the use of weapons of mass destruction] based only on the section of Islamic law called "Repayment in Kind," without any need to indicate the other evidence.

If a bomb was dropped on them [the Americans] that would annihilate 10 million and burn their lands to the same extent that they burned the Muslim lands—this is permissible, with no need to mention any other proof.

Sheikh al-Fahd, prominent Saudi cleric, *A Treatise on the Legal Status of Using Weapons of Mass Destruction against the Infidels*, May 2003

Warning analysis is plagued by a particularly ominous paradox: success depends on the realization of the null hypothesis—the nonoccurrence of an anticipated event. Indeed, the veracity of warning may only truly be revealed through disaster—when warning is not heeded or given. The true significance of al-Fahd's fatwa, for instance, cannot be known until after the catastrophe; in the absence of divine foreknowledge, its significance can only be estimated through analysis. Clausewitz outlined the problem succinctly: "Whatever is hidden from full view in this feeble light has to be guessed at by talent, or simply left to chance. So once again for lack of objective knowledge one has to trust to talent or to luck."[2] Intelligence is rooted in the origins of human civilizations because even the earliest societies were not prepared to entrust all matters to good fortune.

The self-negating prophecy, as Michael Handel called it, is a paradox that reflects the methodological and epistemological challenges inherent in intelligence work. Betts captures the essence of intelligence, noting, "It is the role of intelligence to extract certainty from uncertainty and to facilitate coherent decision in an incoherent environment."[3] Through the ages, the principal function of intelligence has been to provide leaders with sufficient forewarning of impending plot or military surprise attack. This chapter will underline that surprise is one of the terrorist's principal and distinguishing weapons. The central place accorded the mechanism of surprise in terrorism renders counter-terrorism, first and foremost, an intelligence mission.

The prospect of terrorists acquiring and deploying WMD elevates the urgency of the warning mission for counter-terrorism. With domestic and foreign groups determined to inflict mass casualties and immense economic disruption, we no longer have the luxury to deride terrorism as a tolerable nuisance. It is in this context that we examine terrorism, surprise and warning. This chapter seeks to contribute towards a conceptual understanding of the terrorist threat, the environment it inhabits, and the role of warning in countering it.

Managing uncertainty in an age of WMD

Much academic and popular discourse since 1989 has centered on the "unbundling"[4] of strategy, the state and the international system, but it was not until September 11, 2001 that many saw the Janus-face[5] of what had once been hailed as the beginning of a peaceful and prosperous new world order. This dualistic nature, whereby two seemingly conflicting forces coexist in the same time and place, explains how the world is being altered by the broad forces of globalization/integration and fragmentation—what Rosenau has termed fragmegration.[6] Al Qaeda and the global jihadist insurgency it spearheads are emblematic of this duality as they resist globalization, perceived as American-led Westernization, in favor of a resurrected and global Caliphate.

Al Qaeda is symptomatic of our new threat environment. Now, states and their hierarchical institutions must face networked adversaries who are as comfortable in the world's major metropolises as in its war zones and underdeveloped hinterlands. They do, however, inhabit what we could term *terrae incognitae*,[7] zones they often share with criminals and warlords. These are geographies that have escaped the formal (i.e. state) means of governance. Although referring primarily to areas where the state has failed or collapsed, and where the rules of war are increasingly ignored, *terrae incognitae* are also found in the slum and marginalized areas of major cities. These zones have progressively detached themselves from the rest of the world except in one crucial way: they plug into the international political economy of crime and terrorism.

The events of September 11 reinforced one, if not the principal, role of the state: protection of the homeland and citizenry. Over the years, the state has taken on increasingly more responsibilities as the notions of protection and security spilled over into economic and social realms. The modern state has endowed itself with myriad rules, regulations and institutions to fulfill these responsibilities. Osama bin Laden claims that we are as in love with life as he and his ilk are in love with death, highlighting the reality that modern states have erected the most sophisticated means of risk mitigation ever developed. Life is no longer short, brutish and violent for most of the citizenry of these states and, as a result, its notions of risk and time have also changed. Life is now expected to be long, interesting and peaceful.

Intelligence is one of the principal means through which states can fulfill their protective function. Intelligence is not a crisis management tool, but rather seeks to reduce risk and uncertainty in order to avoid or, at the very least, mitigate the potential impacts of threats. Intelligence must focus not only on clear and present threats, but perhaps more importantly on those that have yet to emerge, that are still incubating. Intelligence offers decision-makers (political, military and law enforcement) the peripheral vision needed to identify over-the-horizon threats.

Traditionally, states have responded to organized crime and terrorist and military threats by utilizing counter-leadership targeting: essentially, attempting to decapitate a threatening organization. Other strategies include containment and isolation, or tacit understandings that establish parameters for coexistence beyond which conflict will ensue. Such understandings usually emerge when it becomes evident that the costs of threat elimination far outweigh the benefits. However, the threat discussed in this volume—the use of WMD by terrorists against the United States and its allies—precludes any such tacit understanding. It is the WMD aspect of the terrorism threat that has transformed it from one we could live with[8] into an existential one. It is therefore imperative to understand the nature of the threat and the environment it inhabits.

As with organized crime, the organizational structure of terrorist groups varies greatly, and often reflects the environments in which they were born. Their structures also evolve as a result of their operational environments. Therefore, many groups have adopted a networked structure because: "the [plausible] deniability built into a network affords the possibility that it may simply absorb a number of attacks on distributed nodes, leading an attacker to believe the network has been harmed and rendered inoperable when, in fact, it remains viable and is seeking new opportunities for tactical surprise."[9] Therefore, decapitation strategies are unlikely to be effective against a networked organization.

Short of infiltrating all groups with the potential of using WMD, we must set up warning systems at critical junctures where these terrorist groups are likely to be most visible and vulnerable. This necessitates the development and implementation of several warning systems to augment the capability of traditional counter-terrorism. There are two locales where such tripwires could be set: 1) in locations where WMD, especially nuclear materials, could potentially be obtained; and 2) in the criminal underworld, where terrorists are most likely to be active for the purposes of fundraising and service procurement. The latter refers to a whole gamut of services, ranging from the securing of identity documents to the use of established trafficking networks and infrastructures to ferry men and materiel.

Hierarchical organizations, such as law enforcement and security agencies, have a difficult time fighting networks because they tend to project themselves on to their adversaries (i.e. they assume the adversary acts and thinks in a similar fashion). To change this, hierarchies must adopt certain network design principles to reform their organization and doctrine, because "whoever masters the network form first and best will gain major advantages."[10]

Although the threat of Islamist terrorism currently dominates the headlines and our imaginations, it is imperative to underline that non-Islamist domestic terrorist and religious groups have demonstrated their intent and ability to use WMD in countries such as Japan and the United States. We will return to these threats in subsequent sections, but suffice it to say that we ignore them at our peril. Our discussion now shifts from the general nature of the threat environment to the nature of the terrorist threat itself, and the inherent difficulties of rendering intelligence warning against this formidable, asymmetric adversary.

Surprise and the warning problem[11]

[T]he surprise attack will come from the other country, one of those attacks you will never forget. It is something terrifying that goes from south to north, east to west. The person who devised this plan is a madman, but a genius. He will leave them frozen [in shock].[12]

Al Qaeda operative, intercepted communication, August 2001

The intelligence environment has been likened to the reality of war as Clausewitz envisaged it: "rife with political friction and contradictions, an environment in which uncertainty is the only certain thing."[13] Terrorism depends on and thrives in these conditions of inherent uncertainty, for it is the enigmatic reality of this environment alone that makes terrorism a conceivable method of political violence. The logic of terrorism assumes its coherence only through the surprise of the adversary. In the analysis that follows, surprise emerges as the terrorist's tactical mechanism of necessity, and his strategic weapon of choice. Understanding the critical role of surprise in terrorism is important for developing a better understanding of the warning problem for counter-terrorism.

Terrorism and the role of surprise

The art of war in the narrower sense must now in its turn be broken down into tactics and strategy. The first is concerned with the form of the individual engagement, the second with its use . . . Tactics and strategy are two activities that permeate one another in time and space but are nevertheless essentially different. Their inherent laws and mutual relationship cannot be understood without a total comprehension of both.[14]

Carl von Clausewitz

The relationship between the manner in which force is used and the strategic objectives sought is discussed in the writings of such disparate strategic thinkers as Clausewitz, Lenin and Sun Tzu. This nexus of tactics and strategy is central to our understanding of the logic of terrorism. While terrorism is a method, it cannot be understood solely in operational terms; terrorism is also a strategy to realize political goals. The strategic dimension of terrorism is inextricably linked with the tactics that distinguish it as a mode of operation; in terrorism, surprise conceptually bridges the tactical with the strategic.

Surprise has long been regarded as a potent force multiplier, as "increments of forces provide an arithmetical advantage, but the effects of successful shock are geometrical."[15] Surprise allows an adversary to contemplate bold military feats that would otherwise be well beyond his capabilities.[16] For this reason, surprise has been described as a weapon of necessity that is frequently employed by the weaker adversary to a conflict.[17]

For James Wirtz, the "geometrical" effects of surprise can be understood in terms of how the mechanism transforms the dialectical nature of war. Surprise, in its ideal form, removes an opposing force from the battlefield,[18] allowing the attacker to control the key conditions of the engagement by enabling him to dictate where, when and how hostilities will be initiated.[19] In so doing, the attacker can essentially determine the initial outcome of the engagement.[20] In effect, Wirtz argues, "surprise . . . transforms war from a strategic interaction into a matter of accounting and logistics."[21]

The operational art of terrorism depends heavily on the element of surprise. Surprise is the equalizer that largely negates the asymmetric advantage of the defender. Returning to Wirtz's notion, the operational use of surprise allows terrorists to inflict damaging blows upon an inactive adversary. Without surprise, the terrorist's plans are likely to be thwarted by the defender's security services, or even the intended victims themselves. The operational success of the September 11 plot, for instance, rested significantly on the hijacker's ability to achieve surprise. When passengers on a fourth plane learned of what had just occurred in New York and Washington, they actively resisted their hijackers, making the plane crash before reaching its intended target. "Without the surprise needed to prevent the passengers from realizing that they were engaged in a conflict, the terrorists lacked the forces necessary to maintain control of the aircraft."[22] In other words, when the terrorists failed to achieve surprise, they no longer faced a static opponent; war's dialectic was restored and the terrorists were forced into combat.

In nearly all forms of asymmetric warfare, surprise is a key operational enabler. Wirtz links the element of surprise to the logic of special operations, noting that commandos can be a decisive fighting element only through surprise: "[S]pecial forces are lightly armed, poorly supplied, and generally outnumbered by their adversaries . . . To achieve their objectives, they have become experts in unconventional modes of transportation and operations to enable them to appear and disappear in unexpected ways and at unanticipated times and places."[23] The basic principle behind the commando raid does not fundamentally differ from that of the terrorist operation: the surgical concentration of force to strike a stronger enemy at decisive points. For both, surprise is the enabling mechanism of the operation.

The role of surprise in terrorism, however, goes well beyond the function of operational enabler. One of the defining features of terrorism is that it seeks to affect a wider audience than the immediate victims of the attack. The terrorist's immediate victims are often incidental, akin to extras in a theatrical performance. The dramatic murder of innocents provides terrorists with a vehicle to realize their strategic objectives. This vehicle is a media interface. The "theatrical" nature of terrorism elevates the mechanism of surprise beyond that of a mere enabler. At the operational level, the gun and the bomb are used to inflict physical damage upon the immediate victims. At the strategic level, surprise is used to inflict psychological shock upon the broader adversary. These processes are inextricably linked, though it is the latter instrument that gives the terrorist hope of achieving his political goals.

Bruce Hoffman notes how terrorists calculate the degree and manner of force to produce the desired effects: "Terrorists . . . plan their operations in a manner that will shock, impress and intimidate, ensuring that their acts are sufficiently daring and violent to capture the attention of the media and, in turn, of the public and government as well."[24] This calculated application of violence is reflected in the pattern of escalation in terrorism.[25] As the public becomes inured to a particular level or manner of violence, new, more violent methods must be devised to sustain its attention. The terrorists access this audience through the media coverage of their deeds. The media link the operational and strategic settings of terrorism, enabling the tactical surprise of the attack to be experienced on a grand scale. For instance, on the morning of September 11, 2001, media coverage of the event ensured that the effects of surprise were experienced by millions of people in near-real time.[26] By amplifying the shock effect of the event through the media, tactical surprise can produce strategic effects.

The strategic dimension of surprise in terrorism is revealed in the broader context of asymmetric warfare. In asymmetric conflicts, victory can rarely be measured in military terms.[27] The object of surprise attack may be political victory more than any military impact on the battlefield situation. As Betts explains, "Where potential for political compromise exists, either because of war weariness or constraints of the international system (such as pressure from allies), it may be rational for a surpriser to be willing to lose the battle for he may win the war by invalidating the opponent's confidence in the durability or acceptability of the status quo."[28]

In effect, this is what happened as a consequence of the Vietcong offensive against US forces during the Tet holiday of 1968, which sought to employ surprise to terminate war on the attacker's terms. Despite initial success in the opening days of the offensive, Tet was ultimately a failure on the battlefield.[29] Nevertheless, the shock of the offensive was profoundly felt by Americans at home, where support for the war had been waning amid increased disillusionment with the Johnson administration and its intervention in Southeast Asia.[30] The surprise during Tet was more tactical than strategic—the offensive itself had long been expected, but US forces failed to predict the scope, intensity, targets and timing of the offensive.[31] However, the attacks were, as Betts points out, "almost a complete surprise to the press and public and this contributed to the psychological and political impact, which was ultimately more important than the severe military setbacks the [Vietcong] suffered by the time the offensive had run its course."[32]

Tet proved decisive not because of the military effects felt on the battlefield, but because of the political effects felt on the home front.[33] Tet discredited the Johnson administration, served as a catalyst that led to a re-evaluation of US policy, and left a lasting impression on American perceptions of the dangers of intervening in far-removed conflicts.[34]

The extraordinary success of the Tet offensive in effecting the desired change in Southeast Asia is undercut by the fact that the outcome was largely fortuitous—"political victory despite military failure" was not the conscious strategy behind the offensive.[35] To be sure, however, the political effects of successful surprise are not lost on the conniving actors of international terrorism. Indeed, the "ripple effects" of the Tet offensive are precisely what most terrorists seek to accomplish: undermine the target government, install a favorable status quo and leave a lasting impression on a target audience. This is the essence of the propaganda-by-deed paradigm—a tradition deeply rooted in the history of contemporary terrorism.[36]

The 1983 terrorist bombing of the US Marine barracks in Beirut illustrates the linkage between surprise at the operational level and political victory at the strategic level. On October 23, a truck bomb destroyed the Battalion Landing Team headquarters building at Beirut International Airport, killing hundreds and causing massive destruction. The physical shock of the attack was considerable—the bomb had an explosive force equivalent to between 12,000 and 18,000 pounds of TNT, making it one of the most powerful conventional bombs ever built.[37] The psychological shock, however, was more far-reaching. The death of 241 US military personnel in a single terrorist attack was unprecedented. As Jenkins notes, "The attack provoked an intense debate in the United States, curtailed the deployment of the U.S. Marines in Lebanon, virtually 'killed' the Multinational Peacekeeping Force, destroyed U.S. policy in Lebanon, and undermined U.S. policy in the Middle East."[38]

Ultimately, the terrorist attack in Beirut was strategically successful because it revealed the asymmetry of political commitment to the conflict by dramatically exposing US vulnerability to this type of enemy. As Betts proposes, "Strategic surprise as a device by the weaker party to bring his enemy to terms can be effective in a limited war only when the stronger party's interests are not absolute and his commitment not open-ended, but are revealed to him—in the shock of surprise—as less worth the cost of perseverance than his opponent's interests are."[39] In Beirut, the shock of the surprise attack was a catalyst for change; surprise at the tactical level translated into political victory at the strategic level.

More specifically, Tet and Beirut reveal the impact of surprise beyond the immediate theatre of operations on the public consciousness of the defender's society. Terrorists have learned these and other historical lessons well.[40] Many terrorists are, for instance, keenly aware of the West's aversion to casualties, and incorporate this perceived vulnerability into their strategic planning.[41] Osama bin Laden has expressed his belief that the US public is still haunted by Vietnam, pointing to the Mogadishu crisis in 1993 as supposed evidence. Bin Laden commented in a 1997 interview that the United States left Somalia "after claiming that they were the largest power on earth. They left after some resistance from powerless, poor, unarmed people whose only weapon is the belief in Allah ... The Americans ran away."[42] Concerning the strategic aims of the September 11 plot, Wirtz suggests that bin Laden apparently expected that public sensitivity to troop casualties would mean an ineffectual US retaliatory response in the Arab world, and that such an intervention would ultimately spark an Islamic revolution against the West.[43] To achieve this, the shock of the attack would need to be sufficient to guarantee a US response—a threshold the attacks clearly met.

What unites the cases discussed here is the function of surprise in generating psychological shock that serves as a vehicle for change—change that could not be brought about through force alone.[44] It is this strategic function of surprise that makes it one of the terrorist's principal and distinguishing weapons. From the preceding discussion, we can conclude that there is an important and unique relationship between the mechanism of surprise and the method of terrorism. Surprise links the operational setting of terrorism with the tactics employed and the strategic objectives of the attack. In a general sense, the purpose of the terrorist operation is to employ a tactic—surprise—that will generate strategic effects. There is, then, an integral relationship between the tactical and strategic dimensions of terrorism: the more a terrorist attack surprises at the tactical level, the wider the effects of surprise will be felt at the strategic level.

The terror warning problem

The central place accorded the mechanism of surprise in terrorism has important implications for countering the threat. If terrorist attacks succeed because they surprise, then counter-terrorism largely becomes a matter of warning and response. While the terror warning problem resembles that in the military domain, there are fundamental differences owing to the nature of the adversary that make the former analytically more challenging.[45]

Compared to the threat of military surprise attack, the preparations for a terrorist attack are less extensive, less complicated and thus far less apparent to the analyst. This problem is exacerbated by the tendency for intelligence indicators of imminent terrorist attack to be often highly fragmented and scattered among government agencies and internal compartments.[46] As the September 11 tragedy revealed, coordinating intelligence efforts across many government services can be highly problematic. Tactical warning for terrorism, then, is impeded by the tendency for indicators to be few in number, diffuse in concentration and subtle in appearance.

Ascertaining an enemy's intention to attack is one of the most difficult challenges of strategic analysis, and its apprehension by the defender is a necessary—but not a sufficient—condition for averting strategic surprise.[47] The amorphous nature of many terrorist organizations makes even the known ones exceptionally difficult to fully comprehend, let alone ascertain where, when and how they intend to strike.[48] There is an inherent difficulty in formulating a clear picture of the threat posed by highly fluid terrorist organizations such as al Qaeda: "Al Qaeda . . . is a network of networks of networks, in which terrorists from various organizations and cells pool their resources and share their expertise. This loose affiliation does not have the clear lines of communication of a centralized command structure, such as that provided by an army general or a corporate chief executive officer."[49] Many of the stable intelligence factors used to determine the intentions of a foreign government are either absent or far less apparent in the terrorist threat.[50]

It is important to note that the intention to attack is often openly expressed by the terrorists well in advance—a cheap and effective practice calculated to create or sustain an atmosphere of fear.[51] Osama bin Laden's notorious fatwa of February 23, 1998 is a pertinent example. It called on "every Muslim who believes in God and wishes to be rewarded to comply with God's order to kill Americans and plunder their money wherever and whenever they find it," in order to drive the US and its "armies out of all the lands of Islam, defeated and unable to threaten any Muslim."[52] Following the near-simultaneous terrorist bombings of US embassies in Kenya and Tanzania in August 1998, and the maritime attack against the USS *Cole* two years later, any lingering doubts in the US intelligence community over bin Laden's intention to attack the United States were dispelled.[53] Bin Laden's fatwa represented a declaration of war on the United States, and was interpreted as such by US intelligence officials after the embassy bombings in Africa.[54]

An ironic corollary of the government's favorable asymmetry of capabilities vis-à-vis the terrorist is an unfavorable asymmetry of vulnerability that accords the terrorist an extraordinary advantage over the defender.[55] The terrorists themselves represent a tangible and discrete, though incredibly elusive, target for the government. As Jenkins points out, terrorists rarely hold territory, have no populations to protect and have no regular economy.[56] In other words, terrorists present few vulnerable targets for the government to engage. Nations, on the other hand, are ubiquitous targets.[57] In theory, the universe of possible terrorist targets is virtually unlimited: almost anything can be attacked, particularly when the terrorist is willing to die in the commission of the assault.[58] In practice, however, terrorists are probably more likely to select targets that will reap high propaganda capital, such as major nodes of public activity, symbols of national prestige, and critical infrastructure.

Even within the more limited universe of "valuable" targets, however, motivated terrorists will have little difficulty identifying and attacking targets in an advanced, democratic state.[59] Richard Betts illustrates: "The United States has almost 600,000 bridges, 170,000 water systems, more than 2,800 power plants (104 of them nuclear), 190,000 miles of interstate pipelines for natural gas, 463 skyscrapers . . . nearly 20,000 miles of border, airports, stadiums, train tracks. All these usually represented American strength; after September 11 they also represent vulnerability."[60] What is more, Brian Jenkins has found that the array of targets being attacked by terrorists has been expanding over the years.[61]

The asymmetry of vulnerability is reflected in the fact that, to paraphrase Jenkins, terrorists can attack virtually anything, anywhere, at any time, whereas governments cannot protect everything, everywhere, all of the time.[62] This significantly complicates the problem of warning and response. General warning of a terrorist attack is of limited use when intelligence consumers are faced with the prospect of defending all potential targets.

It is a central feature of asymmetric warfare in general and terrorism in particular that the attacker, when confronted with effective enemy counter-measures, will actively adapt to maintain his offensive capability. As Bruce Hoffman notes, "The terrorist group's fundamental organizational impera-tive to act also drives [their] persistent search for new ways to overcome or circumvent or defeat governmental security and countermeasures."[63] Hoffman speaks of a "Darwinian principle of natural selection" in terror-ism: the strong and clever adapt and continue fighting; the rest do not.[64] Consequently, each successive generation of terrorists tend to be tougher, smarter and more effective than their predecessors.[65]

One reflection of this adaptive character is a phenomenon that may be called operational displacement.[66] To illustrate, the installation of metal detectors in 1973 significantly reduced the incidence of skyjackings,[67] but was met by an increase in the number of concealed bombs on aircraft.[68] Similarly, the fortification of US embassies in 1976 significantly reduced the number of embassy takeovers, but led to a corresponding increase in the assassination of diplomats and the bombing of embassies.[69] When the government responds to warning with effective countermeasures, terrorists simply change their mode of attack or move on to softer targets. Because of this "hydraulic" effect of government action and terrorist reaction, terrorism is not so much reduced by countermeasures, but displaced from one type of target or tactic to another. This displacement phenomenon, together with the broader asymmetry of vulnerability, complicates the task of anticipating and responding to terrorist threats.

Criminal intelligence and strategic warning

> During the first part of this period, when the System's con-
> ventional military strength greatly exceeded the Organization's,
> only the Organization's threat of retaliation with its more than
> 100 nuclear warheads hidden inside the major population
> centers still under System control kept the System, in most
> cases, from moving against the Organization's liberated zones.
> Therefore, the Organization resorted to a combination of chem-
> ical, biological, and radiological means, on an enormous scale,
> to deal with the problem. Over a period of four years some
> 16 million square miles of the earth's surface, from the Ural
> Mountains to the Pacific and from the Arctic Ocean to the
> Indian Ocean, were effectively sterilized. Thus was the Great
> Eastern Waste created.
>
> Andrew MacDonald, *The Turner Diaries*

In recent years, criminal intelligence analysis has become "a discipline in its own right and . . . the key support function for every aspect of law enforcement."[70] Thus, strategic and tactical decisions are now guided by intelligence which is seen as "a precondition to effective policing, rather than as a supplement."[71]

As part of the push to become intelligence-led, criminal intelligence agencies have begun to develop a strategic early warning capability. This resulted from the increasing complexity of the threat environment for law enforcement, as is the case in the security field. It is an environment inhabited by an increasingly wide array of threats, and one that is increasingly influenced by world events and external phenomena. Moreover, criminal and security intelligence are frequently looking at the same types of groups: functionally and geographically networked criminal conspiracies that often have inter-jurisdictional and international connections. These groups also use similar means and methods of communication, concealment and transportation. The only qualitative difference is that criminals rarely exhibit the same espousal of an ideology as do terrorists.

Strategic early warning (SEW) for organized crime[72] is based on the same methods developed for early warning in the fields of public health and national security. First, it is based on all-source intelligence analysis. Second, it tracks key indicators that serve as tripwires to detect changes in trends and patterns of behavior that could have an impact on the organized crime situation in the coming months and years. Third, it utilizes scenario analysis on which it bases its judgments. The purpose of scenario analysis is to provide the analyst with an outlet to imagine several different futures grounded in the information that is currently available.[73] Scenarios synthesize information, and foster a multi-disciplinary approach that is a requisite in this increasingly complex global context.

Most importantly, SEW is also based on a networked reporting mechanism linking the entire law enforcement community, especially police officers and analysts who act as sentinels. It is they who are most likely to pick up on new, unusual and unexpected occurrences that could have larger implications. However, such a network is not limited to the law enforcement community as it lacks the requisite expertise in every domain of interest. SEW is not only all-source, but multi-disciplinary. Such a mechanism would also encourage the voicing of community members' insights, which could then be used to develop scenarios. These insights are among the most important "value-added" the community brings to the table.

Scenarios are not just constructed around the currently available information and the insights of the intelligence community, but also according to the historical record. Although it is important, especially in the field of counter-terrorism, not to be wedded to the past as a guidepost to the future, it nevertheless points to different ways one could acquire, transport and use WMD. The value-added component criminal intelligence brings to this issue is experience—open-source examples of groups resorting to the development and use of WMD clearly show that they have been dealt with by law enforcement—as well as peripheral vision. Terrorists operate in the same underworld as do criminals, and may first become visible to criminal investigators and analysts.

The most cited historical example is that of the Aum Shinrikyo sarin gas attack on the Tokyo subway system in March 1995. Through luck and happenstance, this millenarian religious group eluded the gaze of Japanese authorities despite its numerous criminal activities, including murder, kidnapping, drug manufacturing and trafficking.[74] The group appeared frequently on the radar screens of Japanese law enforcement; needless to say, red flags should have been raised each time. Although Aum had sought to ingratiate itself with Japanese society by fostering its public image as a group of enlightened scientific truth seekers, it better resembled an organized crime or terrorist group. As it pursued its apocalyptic vision, Aum masterfully navigated the Japanese and Russian underworlds in order to procure the chemicals, biological and nuclear agents it needed to fulfill it. However, it is Aum's involvement in the drug trade (manufacturing LSD and meth-amphetamine) which could have rendered it most visible and vulnerable to law enforcement.

The case of William Krar is another instance in which a terrorist endeavor may have avoided the gaze of intelligence agencies, but was clearly visible to law enforcement and other government agencies. On April 10, 2003, the FBI raided Krar's storage unit in Noonday, Texas, where they seized numerous weapons, ammunition, explosives, bomb-making materials and chemicals such as cyanide and hydrochloric acid.[75] Under "ideal" conditions, Krar would have had enough hydrogen cyanide to kill more than 6,000 people.[76] The white supremacist was caught because of a mail delivery mix-up; a package he had sent to a colleague in New Jersey, containing counterfeit IDs, was delivered to the wrong address. Krar does not appear to have been operational, but rather to have acted as a facilitator and service provider to the wider supremacist/anti-government movement in the US. Although he had been arrested several times and stopped paying his taxes for many years, he nevertheless escaped concentrated law enforcement attention.

The Krar example clearly demonstrates how even a group or individual without the significant financial and technical means of Aum can develop a WMD capability. The equipment and chemicals used by both Aum and Krar were readily available, and the knowledge accessible in the public domain. As Korosec points out, "Unlike the more stringent requirements for production of sarin or other nerve agents, fabricating hydrogen cyanide devices demands no greater skills beyond those needed to construct an ammonium nitrate-anhydrous hydrazine truck bomb like that used in Oklahoma City."[77]

It is also important to underline that, in both instances, the parties resorted to criminal activities in order to achieve their "military" goals. In fact, most of the world's terrorist groups (and all its insurgents) avail them-selves of the profits of crime, be it from extortion (revolutionary tax), pay-ment card fraud, tax evasion, welfare fraud, counterfeit currency, product piracy or bank robbery. This has been especially pronounced as the sources of state sponsorship have dried up. Moreover, even when they are "legitim-ately" financed, terrorists require certain goods and services that are not available on the legitimate open market; primary among these is the need for counterfeit and fraudulent identification documents and weapons. They also require front companies and other cover devices in order to move money, men and materiel from one location to another without raising the alarm.

Since September 11, much has been made of the possible cooperation between organized crime and terrorist groups. As has been argued through-out this chapter, both groups inhabit the same general environment, and it would therefore not be unreasonable to assume that they could meet and possibly work together. That is not to say, however, that they would necessarily build tactical and/or strategic alliances with one another; rather, these could be one-offs, ad hoc encounters where one group provides services to the other. It is more likely that organized crime groups would act as goods and services providers to terrorist groups, rather than the other way around.

For organized crime to facilitate WMD terrorism, there are three possibilities: 1) organized crime is specifically hired to steal WMD materials for a terrorist group; 2) organized crime already has WMD materials for sale; 3) a terrorist group uses the trafficking networks established by organized crime in order to smuggle out materials and/or transport them into the targeted theatre. This sort of outsourcing could be explained by the terrorist group's lack of a proper trafficking network to get the material or weapons from point A to point B, while it is more likely that an organized crime group would have the necessary contacts, in strategic locations such as ports, to guarantee delivery. Knowing this, it is now possible to establish tripwires in specific locations and on specific groups most likely to be involved in such an endeavor. It is important to remember that, for terrorists and criminals alike, "necessity rules the underground."

Conclusion

Strategic warning differs from tactical warning in that it seeks to identify over-the-horizon, precursor events and phenomena that could potentially have an impact on the threat environment. Tactical warning entails more threat-specific information and analysis—more the "when" and "where" of a threat, than the "who," "what" and "why" provided by strategic warning analysis. In terms of the potential use of WMD by terrorist organizations, it is crucial to answer these strategic questions so as to place sensors in the appropriate locations or choke points. One of these points is in the criminal underworld. It is in this context that criminal intelligence and strategic early warning for organized crime may have a role in countering the threat of terrorism—a method of political violence that depends almost entirely on the mechanism of surprise.

Our understanding of the relationship between terrorism and surprise suggests important implications for counter-terrorism. Above all, it illuminates the strategic nature of the threat, dispelling the common misconception of terrorism as primarily a tactical warning problem. When we understand terrorism from the broader perspective of asymmetric warfare, it becomes clear that countermeasures may, at best, frustrate specific modes of attack, but are unlikely to reduce terrorism. Faced with specific countermeasures, terrorists will adapt by tactical innovation or by shifting targets.[78] Moreover, the asymmetry of vulnerability ensures that there will be no shortage of targets for terrorists to choose from. Colin Gray describes the problem succinctly when he writes, "It is of the essence of the irregular, asymmetric threat that it will not comprise a replay of yesterday's outrage."[79]

Anticipating tomorrow's outrage depends, above all, on imaginative threat perception. Dismissing terrorism as a tactical problem ignores the important relationship between tactical and strategic warning. The preparations for a terrorist attack are far less pronounced than those of a military surprise attack, but nevertheless generate detectable warning indicators. As Ermarth points out, the terrorist "must mobilize, motivate, organize, prepare, and execute . . . all the while feeding, fueling, funding, and cajoling his operation."[80] Tactical warning, however, is possible only when analysts know what indicators to look for. This is where strategic warning analysis is of central importance. Similarly, strategic estimates need to be continuously evaluated in light of the tactical picture on the ground. Without a symbiosis between the two types of intelligence information, intelligence practitioners may be faced with the near impossible task of warning against the "unknown unknown."[81]

Notes

1 The views expressed in this chapter are those of the authors and do not necessarily reflect the views or policies of Criminal Intelligence Service Canada.
2 Quoted in Michael Handel, "Intelligence and the Problem of Strategic Surprise," in *Paradoxes of Strategic Intelligence*, ed. R.K. Betts and T.G. Mahnken, 1 (London: Frank Cass, 2003).
3 Richard K. Betts, "Analysis, War, and Decision: Why Intelligence Failures Are Inevitable," *World Politics*, vol. 31, no. 1, 1978, p. 69.
4 Unbundling refers to a world in which the hegemony of the state as the ultimate means of organizing a society's economic, political and military life is being altered and, in some cases, challenged by a wide array of non-state and sovereignty-free actors such as non-governmental organizations (NGOs), transnational corporations (TNCs), transnational organized crime (TOC) and international terrorism. Jean-Marie Guéhenno, "The Impact of Globalisation on Strategy," *Survival*, vol. 40, no. 4, Winter 1998–99, pp. 5–19.
5 John Arquilla and David Ronfeldt, *Swarming and the Future of Conflict* (Santa Monica, CA: RAND/National Defense Research Institute, 2000); and John Arquilla and David Ronfeldt (eds.), *Networks and Netwars: The Future of Terror, Crime, and Militancy* (Santa Monica, CA: RAND, 2001), http://www.rand.org/publications/MR/MR1382/
6 James Rosenau, *Turbulence in World Politics* (Princeton, NJ: Princeton University Press, 1990).
7 These are geographical spaces that escape formal means of governance. Often equated with failed states, *terrae incognitae* encompass the world's war zones—where the rules of war are increasingly ignored—as well as its slum areas, be they in the cities of North or South.
8 We refer here to the historical experience of democracies such as France, the United Kingdom and Germany that have been able to thrive in spite of their struggles with various forms of terrorism largely without undermining the very foundations of their societies and polities.
9 Arquilla and Ronfeldt, *Networks and Netwars*, p. 13.
10 Ibid., p. 15.
11 This section is adapted from research carried out by the coauthor Daniel R.

Morris at the Department of International Politics, University of Wales, Aberystwyth. The coauthor wishes to thank Dr. Peter Jackson and Professor Len Scott for their feedback, guidance and support.

12 Malcolm Gladwell, "Connecting the Dots: The Paradoxes of Intelligence Reform," *New Yorker*, Mar. 10, 2003, quoting from John Miller, Michael Stone and Chris Mitchell, *The Cell* (New York: Hyperion, 2003).

13 Handel, "Intelligence and the Problem of Strategic Surprise," p. 8.

14 Carl von Clausewitz, *On War*, ed. and trans. Michael Howard and Peter Paret (Princeton, NJ: Princeton University Press, 1976), p. 132, quoted in Alan Beyerchen, "Clausewitz, Nonlinearity, and the Unpredictability of War," *International Security*, vol. 17, no. 3, 1992–93, p. 74.

15 Richard K. Betts, *Surprise Attack: Lessons for Defense Planning* (Washington, DC: Brookings, 1982), p. 5.

16 James J. Wirtz, "Theory of Surprise," in *Paradoxes of Strategic Intelligence*, ed. R.K. Betts and T.G. Mahnken (London: Frank Cass, 2003), p. 103.

17 Walter Laqueur, *World of Secrets: The Uses and Limits of Intelligence* (London: Weidenfeld & Nicolson, 1985), p. 258; T.V. Paul, *Asymmetric Conflicts: War Initiation by Weaker Powers* (New York: Cambridge University Press, 1994), p. 29; Betts, *Surprise Attack*, p. 5; Wirtz, "Theory of Surprise," p. 106.

18 Wirtz, "Theory of Surprise," p. 104.

19 Michael I. Handel, "Surprise and Change in International Politics," *International Security*, vol. 4, no. 4, 1980, p. 74; Wirtz, "Theory of Surprise," p. 103; Paul, *Asymmetric Conflicts*, p. 29.

20 Wirtz, "Theory of Surprise," p. 103.

21 Ibid., pp. 103–4.

22 Ibid., p. 105.

23 Ibid., p. 104.

24 Bruce Hoffman, *Inside Terrorism* (London: Indigo, 1999), p. 183.

25 Martha Crenshaw, " 'Suicide' Terrorism in Comparative Perspective," in *Countering Suicide Terrorism: An International Conference* (Herzliya, Israel: International Policy Institute for Counter-Terrorism, 2001), p. 24; Bruce Hoffman, "Intelligence and Terrorism: Emerging Threats and New Security Challenges in the Post-Cold War Era," *Intelligence and National Security*, vol. 11, no. 2, 1996, p. 213.

26 Wirtz, "Theory of Surprise," p. 113.

27 Paul, *Asymmetric Conflicts*, p. 27.

28 Richard K. Betts, "Strategic Surprise for War Termination: Inchon, Dienbienphu, and Tet," in *Strategic Military Surprise: Incentives and Opportunities*, ed. Klaus Knorr and Patrick Morgan (London: Transaction, 1983), p. 167.

29 Ibid., p. 161.

30 James Wirtz, *The Tet Offensive: Intelligence Failure in War* (London: Cornell University Press, 1991), p. 2.

31 Ibid., p. 3; Betts, "Strategic Surprise," p. 161.

32 Betts, "Strategic Surprise," p. 161.

33 Wirtz, *Tet Offensive*, p. 1.

34 Ibid., pp. 2–3; Betts, "Strategic Surprise," p. 167.

35 Wirtz, "Theory of Surprise," p. 109.

36 Martha Crenshaw, "The Logic of Terrorism: Terrorist Behavior as a Product of Strategic Choice," in *Terrorism and Counter-terrorism: Understanding the New Security Environment*, ed. Russell D. Howard and Reid L. Sawyer (Guilford, CT: McGraw-Hill, 2003), p. 62.

37 Neil C. Livingstone, "The Impact of Technological Innovation," in *Hydra of Carnage—International Linkages of Terrorism: The Witnesses Speak*, ed. Uri

Ra'anan, Robert L. Pfaltzgraff Jr., Richard H. Schulz, Ernst Halerin and Igor Lukes (Toronto: Lexington, 1986), p. 140.

38 Brian M. Jenkins, "Defense against Terrorism," *Political Science Quarterly*, vol. 101, no. 5, 1986, p. 782.

39 Betts, "Strategic Surprise," p. 165.

40 Jack Davis, "Strategic Warning: If Surprise Is Inevitable, What Role for Analysis?," Sherman Kent Center for Intelligence Analysis, *Occasional Papers* 2, no. 1, 2003.

41 James Wirtz, "Déjà Vu? Comparing Pearl Harbor and September 11," *Harvard International Review*, vol. 24, no. 3, 2002; Rob de Wijk, "The Limits of Military Power," *Washington Quarterly*, vol. 25, no. 1, 2002, p. 79.

42 Quoted in Steven Simon and Daniel Benjamin, "America and the New Terrorism," *Survival*, vol. 42, no. 1, 2000, p. 69.

43 Wirtz, "Déjà Vu?"

44 Betts, "Strategic Surprise," p. 165.

45 Fritz W. Ermarth, "Signs and Portents: The 'I & W' Paradigm post-9/11," *National Interest*, Oct. 2, 2002.

46 Richard K. Betts, "Intelligence Test: The Limits of Prevention," in *How Did This Happen? Terrorism and the New War*, ed. James F. Hoge Jr. and Gideon Rose (Oxford: Public Affairs, 2001), p. 158.

47 Steve Chan, "The Intelligence of Stupidity: Understanding Failures in Strategic Warning," *American Political Science Review*, vol. 73, no. 1, 1979, p. 179; Betts, *Surprise Attack*, p. 87.

48 De Wijk, "The Limits of Military Power," p. 79.

49 Frank J. Cilluffo, Ronald A. Marks and George C. Salmoiraghi, "The Use and Limits of U.S. Intelligence," *Washington Quarterly*, vol. 25, no. 1, 2002, p. 66.

50 Kenneth A. Luikart, "Transforming Homeland Security: Intelligence, Indications, and Warning," *Air and Space Power Journal*, Summer 2003, p. 1.

51 Ermarth, "Signs and Portents."

52 Daniel Benjamin and Steven Simon, "A Failure of Intelligence?," in *Striking Terror: America's New War*, ed. Robert B. Silvers and Barbara Epstein (New York: New York Review, 2002), pp. 289–90.

53 Wirtz, "Déjà Vu?"; Benjamin and Simon, "A Failure of Intelligence?," pp. 289–90.

54 Wirtz, "Déjà Vu?"; Richard K. Betts, "The Soft Underbelly of American Primacy: Tactical Advantages of Terror," *Political Science Quarterly*, vol. 117, no. 1, 2002, p. 22.

55 Betts, "Soft Underbelly," p. 28.

56 Brian M. Jenkins, *The Lessons of Beirut: Testimony before the Long Commission* (Santa Monica, CA: RAND, 1984).

57 Jenkins, "Defense against Terrorism," p. 777.

58 Colin S. Gray, "Thinking Asymmetrically in Times of Terror," *Parameters*, vol. 32, no. 1, Spring 2002, p. 9; Jenkins, "Defense against Terrorism," p. 777; Betts, "Soft Underbelly," p. 28.

59 Patrick Morgan, "The Opportunity for a Strategic Surprise," in *Strategic Military Surprise: Incentives and Opportunities*, ed. Klaus Knorr and Patrick Morgan (London: Transaction, 1983), p. 197.

60 Betts, "Soft Underbelly," p. 30.

61 Jenkins, "Defense against Terrorism," p. 777.

62 Ibid.; Jenkins, *Lessons of Beirut*; Betts, "Soft Underbelly," p. 28.

63 Hoffman, *Inside Terrorism*, p. 180.

64 Ibid., pp. 178–9.

65 Ibid.

66 This concept is an adaptation of the criminological theory of crime displacement, developed by M. Felson and R. V. Clarke, *Opportunity Makes the Thief: Practical Theory for Crime Prevention*, Police Research Series Paper 98 (London: Home Office, 1998).

67 Charles F. Parker and Eric K. Stern, "Blindsided? September 11 and the Origins of Strategic Surprise," *Political Psychology*, vol. 23, no. 3, 2002, p. 605.

68 Crenshaw, " 'Suicide' Terrorism," p. 24.

69 Walter Enders and Todd Sandler, "The Effectiveness of Antiterrorism Policies: A Vector-Autoregression-Intervention Analysis," *American Political Science Review*, vol. 87, no. 4, 1993, p. 830; Jenkins, *Lessons of Beirut*.

70 Marilyn B. Peterson (managing ed.), *Intelligence 2000: Revising the Basic Elements* (Sacramento, CA: Law Enforcement Intelligence Unit and International Association of Law Enforcement Intelligence Analysts, 2000), p. 13.

71 Ibid.

72 Such a system is currently being developed by Criminal Intelligence Service Canada (CISC).

73 Alexander Fink, Andreas Siebe and Jens-Peter Kuhle, "How Scenarios Support Strategic Early Warning Processes," *Foresight*, vol. 6, no. 3, 2004, pp. 173–85.

74 David E. Kaplan and Andrew Marshall, *The Cult at the End of the World* (New York: Crown Publishers, 1996).

75 Jack Douglas Jr., "Sentencing Will Not Solve Mystery of Texan's Weapons Horde," *Fort Worth Star Telegram*, May 4, 2004; Thomas Korosec, "Gun Dealer Sentenced, but Motive still Mystery," *Houston Chronicle*, May 5, 2004; Michael Reynolds, "Homegrown Terror: A Bomb Is a Bomb," *Bulletin of Atomic Scientists*, vol. 60, no. 6, 2004.

76 Korosec, "Gun Dealer Sentenced."

77 Ibid.

78 Hoffman, *Inside Terrorism*, pp. 182–3.

79 Gray, "Thinking Asymmetrically," p. 11.

80 Ermarth, "Signs and Portents."

81 Davis, "Strategic Warning."

4

TERRORISM AND WEAPONS OF MASS DESTRUCTION

Michael D. Intriligator and Abdullah Toukan

Introduction: motivation, definitions and a historical example

The threat of terrorist use of nuclear weapons and other weapons of mass destruction (WMD), whether biological, chemical or radiological, is a real one that represents a most serious threat to the US and other nations that are potential targets of subnational terrorist groups or networks. Transnational terrorism and the potential acquisition by terrorists of weapons of mass destruction are part of the "asymmetric" dynamics of the various unexpected and new threats that have thrust the international community into a new and uncertain conflict. These dynamics have been witnessed in the 9/11 (September 11, 2001) al Qaeda terrorist attacks on New York and Washington, the 3/11 (2004) terrorist attacks against Madrid, and the 7/7 (2005) terrorist attacks against London. Terrorist acquisition and use of nuclear weapons is an extremely serious problem that must not be dismissed as the subject of works of fiction. Indeed, the US casualties and losses on 9/11 would be seen as relatively minor in comparison to a possible terrorist strike using nuclear weapons. One of the only things that both candidates in the US 2004 presidential elections agreed on was that this is the most serious threat the country faces.

Graham Allison[1] discusses this issue in his 2004 book *Nuclear Terrorism*, emphasizing that, as he puts it in the subtitle of this book, nuclear terrorism is the *"ultimate preventable catastrophe."* Unfortunately, this conclusion may be overly optimistic in that his proposals for strict control over fissile material and the prevention of the acquisition of nuclear weapons by additional nations, while excellent policies, may not work perfectly. It is also possible that terrorist groups have already obtained enough of this material to produce a nuclear weapon or even already possess such a weapon.

It is possible to identify various "nightmare scenarios." Most devastating would be a repeat of 9/11 but this time with a nuclear weapon. If a subnational terrorist group gained access to a nuclear weapon, it could use it or at least threaten to do so. If Osama bin Laden had even a crude nuclear weapon he could have used it on 9/11 or in other al Qaeda attacks. Some information exists about terrorists' intentions to obtain nuclear weapons. Osama bin Laden has specifically referred to the acquisition of nuclear weapons by the al Qaeda terrorist network as a "religious duty," and documents were found in the al Qaeda caves in Afghanistan regarding their intent to use WMD that even included a schematic diagram of a nuclear weapon. After the 9/11 attacks, al Qaeda spokesman Abu Gheith wrote:

> We have not reached parity with them. We have the right to kill 4 million Americans—2 million of them children—and to exile twice as many and wound and cripple hundreds of thousands. Furthermore, it is our right to fight them with chemical and bio-logical weapons, so as to afflict them with the fatal maladies that have afflicted the Muslims because of the [Americans'] chemical and biological weapons.[2]

If this stated goal of retribution were true, the only way that al Qaeda could attain this objective would be to use nuclear weapons or a highly destructive and sophisticated biological agent.

Other nightmare scenarios involving terrorists using WMD include a strike with conventional weapons against a nuclear power plant near a major city such as Indian Point near New York City or a terrorist group placing a nuclear weapon in a container on a freighter entering a major port, such as the Los Angeles/Long Beach port complex, the largest in the US. It would not be difficult to place such a bomb in one of the many containers entering US ports, as almost none of them are inspected. Furthermore, the Los Angeles/Long Beach port represents an important potential target for terror-ists, as it accounts for over 40 percent of all US foreign trade, so knocking it out of commission would have an enormous impact on the economies of the US and all its trading partners, potentially disrupting much of world trade.[3]

Garwin[4] most fears what he calls "megaterrorism," involving thousands of casualties, by means of biological warfare agents or nuclear weapons. He postulates that terrorists could use a nuclear weapon stolen from Russia or an improvised nuclear device based on highly enriched uranium built in the US. Some acts of megaterrorism, including 9/11, were foreseen by the US Commission on National Security in the 21st Century (the Hart–Rudman Commission) in its 1999 report, which stated that:

> Terrorism will appeal to many weak states as an attractive option to blunt the influence of major powers . . . [but] there will be a greater incidence of ad hoc cells and individuals, often moved by religious zeal, seemingly irrational cultist beliefs, or seething resentment . . . The growing resentment against Western culture and values . . . is breeding a backlash . . . Therefore, the United States should assume

that it will be a target of terrorist attacks against its homeland using weapons of mass destruction. The United States will be vulnerable to such strikes.[5]

It is customary to classify nuclear, biological, chemical and radiological weapons as WMD, but there are important differences among these weapons. In fact, it is misleading or even mistaken to lump together all of these weapons as one category of "weapons of mass destruction," since nuclear weapons are in a class all to themselves in view of their tremendous destructive potential, as shown in Table 4.1. While nuclear weapons are not now, as far as we know, in the hands of terrorists, they could be sometime in the future, given that this is an old technology that is well understood worldwide and given that there has recently been a proliferation of WMD-related technologies and material. Furthermore, recent trends in terrorist incidents indicate a tendency toward mass-casualty attacks for which WMD are well suited. There is even a type of rivalry between various terrorist groups to have the largest impact and the greatest publicity, topping the actions of other such groups.

The terrorist attacks on the World Trade Center (1993), the Tokyo subway (1995) and the Murrah Federal Building in Oklahoma (1995) clearly signaled the emergence of this new trend in terrorism mass-casualty attacks. Terrorists who seek to maximize both damage and political impact by using larger devices and who try to cause more casualties have characterized this new pattern of terrorism. There have also been the revelations that A.Q. Khan, the "father" of the Pakistan nuclear weapon, provided nuclear weapons technology to several nations, suggesting the emergence of a type of nuclear weapons "bazaar" that will sell components, technology, fissile material, etc. to the highest bidder, including another nation such as Libya, North Korea or Iran or possibly a well-financed terrorist group. If terrorists had access to the needed funding they could probably easily find another such expert or middleman to provide them the detailed plans and even the components for a nuclear weapon. Even without a full-fledged nuclear weapon they could assemble a radiological dispersal device that could cause massive disruption and also have massive psychological effects on the population.

There are many different types of terrorist groups or networks worldwide; they are not all fundamentalist Muslim or based in the Middle East or South and Southeast Asia. Table 4.2 shows the various terrorist groups that might resort to the use of WMD against the US. Their motivation is probably not merely poverty and ignorance, as is often alleged, but rather revenge for past humiliations and retribution as stated in the above quote from Abu Gheith. As Friedman[6] states, "The single most underappreciated force in international relations is humiliation." Of course, different terrorist groups have different motivations and ideologies so there is no such thing as a "stereotypical terrorist." Furthermore, a terrorist network would not be able to

Table 4.1A The comparative effects of biological, chemical and nuclear weapons delivered against the United States

	Area covered (sq. km)	Deaths assuming 3,000–10,000 people per sq. km
Using missile warheads		
Chemical: 300 kg of sarin nerve gas with a density of 70 milligrams per cubic meter	0.22	60–200
Biological: 30 kg of anthrax spores with a density of 0.1 milligrams per cubic meter	10	30,000–100,000
Nuclear: One 12.5 kiloton nuclear device achieving 5 lbs per cubic inch of over-pressure	7.8	23,000–80,000
1.0 megaton hydrogen bomb	190	570,000–1,900,000
Using one aircraft dispensing 1,000 kg of sarin nerve gas or 100 kg of anthrax spores		
Clear sunny day, light breeze:		
sarin nerve gas	0.74	300–700
anthrax spores	46	130,000–460,000
Overcast day or night, moderate wind:		
sarin nerve gas	0.8	400–800
anthrax spores	140	420,000–1,400,000
Clear calm night:		
sarin nerve gas	7.8	3,000–8,000
anthrax spores	300	1–3 million

operate in the capacity that it does in the absence of the other components that engage in fundraising, recruitment and social support. In contrast to previous forms of transnational terrorism, the support base of transnational terrorist groups has spread throughout the globe rather than in any distinct geographical cluster.

One important consequence of the US invasion of Afghanistan was to eliminate the main base of al Qaeda, destroying its central command structure. In the absence of this central command structure, individual networks appear to have gained greater freedom and independence in tactical decisions than the traditional terrorist cells of the past. This particular trend in terrorism represents a different and potentially far more lethal one than that posed by the more familiar, traditional, terrorist adversaries. The 9/11 attacks have demonstrated that transnational terrorism is now more lethal and that it can have a fundamental political and strategic impact. Further,

Table 4.1B Biological weapons' estimated casualties using aerosol delivery mechanism

	Amount released	Estimated damage/ lethality
Anthrax	100 kg spores released over a city the size of Washington, DC	130,000–3 million deaths
Plague	50 kg Y. pestis released over city of 5 million people	150,000 infected 36,000 deaths
Tularemia	50 kg F. tularensis released over city of 5 million people	250,000 incapacitated 19,000 deaths

Sources:
World Health Organization, *Health Aspects of Chemical and Biological Weapons: Report of a WHO Group of Consultants*, 2nd edn. (Geneva: WHO, 1970).
Centers for Disease Control and Prevention (CDC), Fact Sheets on Biological and Chemical Agents, Atlanta, GA.
Anthony H. Cordesman material in the Office of Technology Assessment, *Proliferation of Weapons of Mass Destruction: Assessing the Risks*, US Congress, OTA–ISC–559, Washington, DC, Aug. 1993, pp. 53–4.

the threat of terrorist use of WMD is still possible and perhaps inevitable given the goals of al Qaeda, which is probably now rebuilding its central command structure.

There has been to date only one example of a terrorist group using WMD. This historical example is the Japanese terrorist group Aum Shinrikyo's release of sarin nerve gas on the Tokyo subway on March 20, 1995. This attack represented the crossing of a threshold and demonstrated that certain types of WMD are within the reach of some terrorist groups. The attack came at the peak of the Monday morning rush hour, right under police headquarters, in one of the busiest commuter systems in the world, and resulted in 12 deaths and over 5,000 injuries. While the number of deaths was relatively small, this was the largest number of casualties of any terrorist attacks up to that time. This number of casualties is exceeded only by the 9/11 attacks on the World Trade Center in New York and the Pentagon in Virginia as well as in Pennsylvania that resulted in about 3,000 deaths and almost 9,000 nonfatal casualties.[7] Even before their attack on the Tokyo subway, Aum Shinrikyo had conducted attacks using sarin and anthrax.[8] Following this 1995 attack in Japan, President Clinton issued Presidential Decision Directive 39 stating that the prevention of WMD from becoming available to terrorists is the highest priority of the US government.

What is the threat of terrorists' use of WMD?

Owing to a number of global developments over the past decade, the threat that terrorists might resort to weapons of mass destruction has received

Table 4.2 Possible terrorist groups that might resort to the use of WMD against the US

Group	Description	Possible reason
Non-state-sponsored terrorists	These are groups that operate autonomously, receiving no significant support from any government. These groups may be transnational, they don't see themselves as citizens of any one country, and groups that operate without regard for national boundaries carry out thereby transnational terrorism. Typical: al Qaeda terrorist organization	Terrorist organization backed in a corner, losing ground and support internationally. Terrorist organization trying to recapture public attention by resorting to higher levels of terrorism, resulting in mass casualties.
State-sponsored terrorists	International terrorist group that generally operates independently but is supported and controlled by one or more nation-states as part of waging asymmetric surrogate war against their enemies. The US has labeled Cuba, Iran, Libya, North Korea, Sudan and Syria as states sponsoring terrorism.	To undermine US policy and influence, and for the US to change its policy.

increased attention from political leaders and the news media. These developments include: the proliferation of WMD-related technologies, materials and know-how; trends in transnational terrorist incidents, suggesting a growing tendency toward mass-casualty attacks for which WMD are well suited; and the interest in WMD that has been expressed by Osama bin Laden and al Qaeda.

The events of September 11 and the wave of anthrax-laced envelopes mailed in the US during 2001—a case that still has not been solved— together constituted a watershed in the perception of the non-conventional terror threat in general and of bioterrorism in particular. These events heightened the potential link between transnational terrorism and WMD, with biological weapons in particular looming as a new and dangerous threat.

Overall, while there has been remarkably little historical use of WMD by terrorists and very few fatalities resulting from their use, one cannot rule ‹ terrorist groups gaining such weapons and using them in the future. Soon or later they could be available to terrorists. As former Secretary of Defense William Cohen stated concerning WMD terrorism: "The question is no longer if this will happen, but when." In addition, other groups have sought to gain access and use nuclear weapons and other WMD, which compounds the problem as new nations and subnational groups seek these weapons. Some terrorist groups may feel that, in order to attract worldwide attention, they should escalate from conventional to biological or nuclear weapons. The likely users of these and other WMD are probably fundamentalist terrorist groups, given both their motivation and their access to funding and expertise.[9]

It should be noted that nuclear weapons are "self-protecting"—they are difficult to acquire, to use and to take care of properly. This has the effect of keeping such weapons out of the reach of most national and subnational groups, including terrorists, and Table 4.3 summarizes some of the technical hurdles for nuclear, biological and chemical weapons programs. Nevertheless, a well-financed terrorist group could have the resources needed to hire the experts who could build and take care of such weapons, as was the case with Aum Shinrikyo. The CIA had predicted copycat phenomena in that case, but they did not in fact materialize, probably due to the difficulties of building and maintaining such a weapon. Also, each weapon is different and, while there are some weapons that can be developed easily, such as ricin, others are extremely difficult to build, including nuclear weapons. Nevertheless, with demand rising and marginal cost falling, as is also the case with other WMD technologies, it is only a matter of time before such weapons, including nuclear weapons, become available to terrorist groups.

Table 4.3 shows that, when addressing the supply and demand sides for WMD, the technical hurdles to produce such weapons should be taken into consideration. Owing to the complexity and expense of the processes needed to develop nuclear weapons, the *supply* side has to be addressed for such weapons. While few states are known to have nuclear weapons capability, those that have nuclear reactors should be addressed. With tight security measures at these plants and export controls, as well as all material under IAEA safeguards, no nuclear material would fall into the hands of terrorist organizations.

By contrast, owing to the relative ease with which biological, chemical and radiological weapons can be produced in a vast number of open laboratories and facilities that are designated as purely civilian, the *demand* side should be addressed for such weapons. This implies the need to identify and to destroy terrorist organizations that are pursuing the production or possession of these weapons.

Overall, the likelihood of terrorist groups acquiring WMDs is probably low in the short run but high in the long run. There is no way to demonstrate

Table 4.3 Technical hurdles for nuclear, biological and chemical weapon programs

	Nuclear	Biological	Chemical
Feed materials	Uranium ore, oxide widely available; plutonium and partly enriched uranium dispersed through nuclear programs, mostly under international safeguards.	Potential biological warfare agents are readily available locally or internationally from natural sources or commercial suppliers.	Many basic chemicals available for commercial purposes; only some nerve gas precursors available for purchase, but ability to manufacture them is spreading.
Scientific and technical personnel	Requires wide variety of expertise and skillful systems integration.	Sophisticated research and development unnecessary to produce commonly known agents. Industrial microbiological personnel widely available.	Organic chemists and chemical engineers widely available.
Plant construction and operation	Costly and challenging. Research reactors or electric power reactors might be converted to plutonium production.	With advent of biotechnology, small-scale facilities now capable of large-scale production.	Dedicated plant not difficult. Conversion of existing commercial chemical plants feasible but not trivial.
Comments	Black-market purchase of ready-to-use fissile materials or of complete weapons very possible.	Biological organisms are less expensive and easier to produce than nuclear material or many of the chemical warfare agents.	Legitimate commercial chemical plants and facilities can produce the required warfare agents.

that terrorists will acquire and use such weapons, but, conversely, there is no way to demonstrate that they will not do so. Given the chance that this might happen and given the magnitude of potential losses involved, it is important to prevent this as well as to be prepared for such an eventuality, even at the cost of many billions of dollars. Using the concept of expected loss, the very low probability of a remote possibility, such as a terrorist group gaining access to and using nuclear weapons, is more than offset in terms of expected loss by the extraordinarily high losses such strikes would entail. It is also a serious mistake to underestimate the ability of terrorists to innovate new techniques of terror, such as using hijacked airplanes as suicide attacks, as occurred on 9/11. Both hijackings and suicide attacks were well known before but never combined effectively in this particular way.

There could be comparable innovations using nuclear weapons in the future, possibly combined with an unexpected delivery system, such as a rental van in a parking garage, a barge in a harbor or a cruise ship with thousands of people aboard. In fact, there are many possible targets. In addition to the typical open-air targets such as urban areas, the center of a city during rush hour, and port areas, smaller-scale attacks confined to enclosed areas would also be potential targets for terrorists. Some of these targets could be shopping malls, convention centers, domed sports stadiums, sealed buildings with central air-conditioning, subways, trains, airport terminals and passenger aircraft. In addition, terrorists could attack by the dispersal of chemical or biological weapons through building ventilation systems and by disabling the cooling systems of nuclear reactors, among a myriad of other possibilities.

It is important to study in detail how truly effective WMDs would be in furthering a terrorist group's ultimate agenda in both the short term and the long term. Of course, terrorist groups must choose among alternatives under constraints, but a well-financed group could choose to develop WMDs, following the model of Aum Shinrikyo, but possibly on an even larger scale, as an ultimate demonstration of its capabilities. It would be a mistake to underestimate the potential of terrorists to attack in new ways.

While certain types of WMD have been available for a long time, including nuclear weapons, these weapons have only been used by one country against another: the US against Japan at the end of World War II. That does not mean, however, that subnational terrorist groups would not use them in the future. The best policy choices for governments in the area of combating WMDs as opposed to combating conventional weapons would be to identify those policies that would best undermine resource availability for terror groups and force them into choice patterns that can be countered at least cost to these governments.[10]

What are the legal, political and other approaches that could be used to prevent terrorist use of WMD?

It is difficult even to consider the challenges presented by potential terrorist use of WMD since we are in a stage of denial, where the nightmare scenarios make even thinking about the problem and its remedies extraordinarily difficult. People find it hard to consider this threat seriously and instead consign it to fictional scenarios, such as in novels and films. It is necessary to overcome this denial mechanism and to take active steps to prevent potential terrorist threats to national and global security using WMD.

This problem is a global one that must be dealt with on that scale; international organizations must be involved and close international cooperation is necessary. Terrorism cannot be addressed by unilateral action when there is little or no support from other states. Each nation should realize that its major responsibility lies in protecting its citizens, but that it cannot do so without international cooperation and reliance on international organizations that may require it to give up some of its sovereignty.

The world's nuclear weapons stockpiles and the world's stockpiles of weapons-grade materials (both military and civilian) are overwhelmingly concentrated in the five nuclear weapon states (United States, United Kingdom, France, China and Russia). Additional nuclear weapons or components exist in Israel, India, Pakistan and, possibly, North Korea. In addition, civilian plutonium for many nuclear weapons also exists in Belgium, Germany, Japan, Switzerland and elsewhere, sometimes in quantities large enough to make a weapon.

Access to WMDs can be treated as a supply and demand problem: supplies must be limited, with the current huge supplies of WMDs safeguarded or destroyed. Russian stockpiles of tactical nuclear weapons should be safeguarded, while Russian stockpiles of chemical weapons and biological weapons, the largest in the world, should be destroyed through an expansion of the Nunn–Lugar Cooperative Threat Reduction program. As discussed by Allison, stockpiles of fissile material, both highly enriched uranium and plutonium, must be safeguarded under the same type of protection that the US gives to its stockpile of gold at Fort Knox, a new type of "gold standard."[11] Furthermore, Allison emphasizes the importance of preventing the acquisition of nuclear weapons by additional states, including Iran and North Korea. He notes that terrorists can obtain nuclear weapons only through theft of such a weapon or acquisition of the necessary fissile material, as they do not have the technical and financial capabilities to produce this material. Thus, preventing them from acquiring such capabilities can help make the problem of terrorist use of nuclear weapons a *preventable* one, although we believe he is too sanguine in this regard.

Allison stresses the supply side, noting, correctly, that the problem of nuclear terrorism would disappear if terrorists were to be denied access to nuclear weapons and to the fissile material necessary to produce them. It may be the case, however, that such access cannot be completely denied and it may also be the case that some of this materiel is already in the hands of terrorist groups, so it is important to treat the demand side as well as the supply side. To reduce the terrorist demand for nuclear weapons, a new form of deterrence must be developed, with a global deterrence system that would be used against any terrorist group using WMD. Such a system should be embodied in a formal agreement with a multi-pronged approach based on international cooperation with a credible enforcement mechanism.[12] An important step in this regard was the unanimous passage by the UN General Assembly of the Nuclear Terrorism Convention of April 13, 2005, which makes criminal the possession or use of nuclear weapons or devices by non-state actors. It calls for an appropriate legal framework to criminalize nuclear terrorism-related offenses, allowing for arrest, prosecution and extradition of offenders, and it will enter into force after it is signed and ratified by at least 22 states.

There are several convincing reasons to acknowledge and address the possibility of loose nuclear materiel or even loose nuclear weapons floating in the international system. Intelligence agents are constantly discovering previously unknown networks through which the materials and information necessary to create a nuclear weapon may have passed undetected. On November 28, 2004, authorities thwarted a near-complete plan to smuggle the necessary elements of a uranium enrichment plant from South Africa to Libya. Another concern is that highly enriched uranium or plutonium may be missing from Russia's ill-protected supply, and that it could already be circulating among subnational networks. All of these concerns point to the need for a viable strategy on the demand side to reinforce treatment of the supply side. Put another way, traditional military analysis considers both *capabilities* and *intentions*. While Allison focuses on capabilities in terms of preventing terrorists from acquiring nuclear weapons, it is also necessary to look at intentions in terms of the demand for these weapons and how terrorists could be deterred from acquiring them or using them. The Cold War deterrence model is probably insufficient to protect national security, so the issue is that of developing a new model that can replace it.[13]

There are clearly serious challenges to adapting existing deterrence models, designed for state-to-state interaction, to non-state actors such as terrorist groups and networks. Thus, traditional concepts of deterrence will have to be reshaped to deal with the issue of how terrorist groups could possibly be deterred. It would be a mistake, however, to dismiss deterrence in this regard. For example, it is sometimes argued that suicide bombers cannot be deterred because they are already sacrificing their lives for their cause. This argument is flawed, however, as it is not the suicide bombers that must be deterred but rather their controllers, who make the decisions. There is the problem of knowing where to find the terrorist group in order to strike back.[14] Terrorists do have something to lose and could thus potentially be deterred. In particular, terrorists have a stake in overall group survival. In general, the possibility of deterring terrorism must be systematically analyzed to determine how that might be accomplished. This situation is somewhat analogous to the beginning of the Cold War before the doctrine of mutual assured destruction (MAD) was developed. The challenge to strategic analysts now is to develop a concept of deterrence that would be effective against terrorists.

Konishi[15] argues that preemptive/preventive military action has the potential to be a highly effective form of deterrence policy that has gone unrecognized because the US had "never [before] attempted to deter terrorists through military force." He states, "The United States has adopted a deterrence strategy that involves overwhelming military force aimed both at terrorists and states that harbor terrorists." He notes that "classic deterrence relies on the ability to convince potential adversaries that acts of aggression will result in greater costs than benefits. It is uncertain whether classic deterrence can succeed against asymmetrical threats such as terrorism. However, such a strategy is worth trying if over time it proves to limit a rapid escalation of terrorist activities." He notes that terrorists face an unprecedented threat from US forces, that such action would teach them to respect US military superiority, and that terrorists cannot sustain armed action in the long term. It is, however, unlikely that American military force would be able to deter al Qaeda in this way, given that the Vietcong were not dissuaded by the US preponderance of force during the Vietnam War and the Mujahedin, who were the base of the Taliban in Afghanistan, were not deterred by the Soviets during their invasion of that country.[16]

There are serious dangers in the current doctrine of preemption/preventive war as enunciated in the Bush administration Office of the National Security Advisor's September 2002 document *The National Security Strategy of the United States of America.* The Bush administration introduced the doctrine of preemption, including striking suspect sites before obtaining "absolute proof." This "Bush doctrine" not only sets a precedent for other nations to follow but it also gives strong incentives for further nuclear weapons proliferation as states threatened by the US seek such weapons for their own protection, cases in point being North Korea and Iran. In 1997, then Secretary of Defense William Cohen stated that US military superiority was so great that potential adversaries, unable to compete in conventional arms, "may feel compelled to use apocalyptic weapons in a struggle against the United States." Along similar lines, after the 1991 Gulf War a Pakistani brigadier was asked what lesson he drew from the war and he responded that it was "don't fight the US with conventional weapons." Terrorist groups could reason the same way and seek nuclear weapons or other unconventional weapons. It should be noted that the European Union adopted a security strategy in December 2003 focusing on "preventive measures" as opposed to the Bush preemptive force doctrine.

Another non-solution would be the elimination of all nuclear weapons or making WMDs illegal. This is wishful thinking in the absence of an enforcement mechanism. Even if there were a treaty along these lines, for example the Non-Proliferation Treaty (NPT), Article VI of which calls for the eventual elimination of all nuclear weapons, a nation could opt out of it, as North Korea did.

How should the US respond after a nuclear terror attack?

An important question that has not been addressed in the open literature and has hardly been discussed at all is how the US ought to respond *after* a nuclear terror attack on a US city. We hope that there is some classified contingency planning for such an event.

In April 2005 the US Department of Homeland Security established the Domestic Nuclear Detection Office (DNDO) to provide a single accountable organization with the responsibility to develop and deploy a system to detect and report attempts to import or transport a nuclear device or fissile or radiological material intended for terrorist use. DNDO has been charged to work in close cooperation and coordination with federal, state and local governments and also with the private sector as part of a multi-layered defense strategy to protect the US from a nuclear or radiological terrorist attack. While it is important to establish such an office to detect these weapons, it is equally important to have contingency plans if this office does in fact detect a terrorist nuclear weapon and plans for what the appropriate response should be to the actual detonation of such a weapon of mass destruction on US soil.

One member of the House of Representatives, Congressman Tom Tancredo of Colorado, said that we should "take out" the Muslim holy sites like Mecca and Medina (or Jerusalem). This, however, is precisely the wrong way to respond, as it would only lead to much more terrorism and unify the Muslim world in a holy war against the US. Probably the best answer to what the US should do is to follow precedent. After 9/11, the US invaded Afghanistan, the host of al Qaeda, and cleared out their bases there. This retaliatory strike against those harboring the terrorists responsible for the attack was useful as a warning to other terrorist groups, and it has provided a brief period of respite from such attacks until al Qaeda rebuilds its command and control structure. Of course, the US may not know the location of the bases of the terrorists who were responsible, nor the particular terrorist group behind the attack. In that case we have to do everything we can to avoid a spasm response, such as the one suggested by Representative Tancredo. Rather we should work with our allies and the UN to identify the source and take out those responsible, including their hosts, their funding sources, etc.

It is important that the US show evidence to the world community before launching a military attack. This will not undermine or weaken the US; on the contrary, presenting the evidence to the world community and going through the United Nations will mobilize an international coalition that will stand with and support any US military actions. If the US immediately assumes that the terrorist group is state-sponsored and launches military strikes against Syria or Iran, these actions will inspire new hatreds against the US and its interests.

An immediate retaliation by the US to a WMD terrorist attack using nuclear weapons will be catastrophic for the US and many of its allies. Without any doubt, the US has the strongest military forces in the world, as well as the biggest nuclear arsenal. For this very reason, the US can wait for a while until it exhausts all avenues before it launches any military strikes. Indeed, US policy makers must address some basic questions before taking actions that could be of high risk, including:

- Is it of strategic interest to immediately strike an Islamic country using WMD?
- Does the US have adequate intelligence information to justify such an attack?
- Have all the other options been considered?
- What objectives will these actions achieve?

In short, the US must develop a comprehensive set of national strategies addressing terrorism, weapons of mass destruction, and homeland security. In close collaboration and cooperation with its close allies and the international community at large, the US just might succeed in deterring and eliminating the threat posed by terrorists and weapons of mass destruction.

Conclusions

First, we are probably not any safer overall now than we were before the implementation of the post-9/11 strategies. In some ways we are safer, possibly in terms of airplane hijackings, but in other ways not, or the situation is even worse, given the avowed goal of some terrorist groups to obtain nuclear weapons. Overall, we tend to be reactive rather than acting proactively.

Second, the major question is how can we prevent a terrorist attack using WMDs, such as a nuclear 9/11? We should recognize and avoid the denial syndrome and begin thinking about "worst-case scenarios" and working on ways to prevent them from happening. We certainly should not ignore the possibility that these events could ever happen, as we did before 9/11, where the weapons the terrorists used were merely box cutters, a far cry from WMD.

Third, there is need for more research on these matters. It is important to study how terrorists think and the nature of their motivation. Terrorists will likely be using the path of least resistance, so tightening up airport security, for example, will mean that they will substitute other vulnerable targets, such as ports, nuclear power plants, bridges, high-rise office buildings and other critical infrastructure. Clearly any protection should have a net benefit after taking into account its direct and indirect consequences. It is important to encourage analysts to engage in thinking as terrorists would, as in the "Red Team" exercises of the Cold War, and to act upon the conclusions of these studies. It is also important to recognize the importance of psychology in deterrence.[17]

Fourth, we must establish clear priorities for US counter-terrorism policies. Part of establishing clear priorities should be serious improvements in intelligence systems that have failed us repeatedly, including better organization, upgrading of capacities, better use of the private sector (including universities), holding intelligence services and individuals that lead them responsible for their failures, and so on. Some of the initiatives that have been undertaken at the local level, such as the Terrorism Early Warning (TEW) group that was initially established in Los Angeles County, should be expanded to the regional, state, national and international levels. Another initiative should be providing more resources for diplomatic efforts that have been starved for funds and personnel. Yet another point that somehow is not emphasized is destroying Russian stockpiles of chemical and biological weapons, which are larger than those of the rest of the world combined. Russia wants to destroy them and has committed itself to the Chemical Weapons Convention, but it cannot afford to build the necessary incinerators, which the US or a consortium of nations could provide. This could be the most cost-effective use of any US defense spending. Similarly, the Russian stockpiles of tactical nuclear weapons and biological weapons should be adequately protected or destroyed, as they could possibly be stolen by or sold to a terrorist group.

Fifth, and finally, new initiatives must be undertaken to deal with the issue of terrorist use of nuclear weapons and other WMD at the global level, including through international institutions. The recent proposals on reform of the UN system, in the Report of the High-Level UN Panel on Threats, Challenges and Change, are valuable in this regard.[18] They must, however, be supplemented by initiatives and reforms that deal directly with this threat, including the sharing of information and the creation of a task force that could take direct action against terrorist groups that could be planning to use WMD.

Acknowledgements

We would like to acknowledge the valuable research assistance of three UCLA graduate students who assisted in the preparation of this paper: Tracey DeFrancesco, Sarah Paulson and Josh Rosenfeld. We have also received useful comments on this paper from Gilbert Kim, another UCLA graduate student. All are students in the Master of Public Policy program in the UCLA School of Public Affairs. We have also received valuable suggestions from Jamus Jerome Lim, a graduate student in the Department of Economics at the University of California, Santa Cruz.

Notes

1 Graham Allison, *Nuclear Terrorism: The Ultimate Preventable Catastrophe* (London: Times Books, 2004).

2 Abu Gheith, " 'Why We Fight America': Al-Qa'ida Spokesman Explains September 11 and Declares Intentions to Kill 4 Million Americans with Weapons of Mass Destruction," MEMRI, No. 388, June 12, 2002, http://www.memri.org/bin/articles.cgi?Area=jihad&ID=SP38802. Al-Jazirah TV Broadcasts: Usama Bin Ladin's 1998 Interview with Jamal Ismail in Arabic, 20 Sept. 2001.

3 For other scenarios, see Anthony Lake, *Six Nightmares: Real Threats in a Dangerous World and How America Can Meet Them* (Boston, MA: Little, Brown, 2000), and Richard L. Garwin, "Nuclear and Biological Megaterrorism," 27th Session of the International Seminars on Planetary Emergencies, Aug. 21, 2002, http://www.fas.org/rlg/020821-terrorism.htm, part of which appears as "The Technology of Megaterror," *Technology Review*, Sept. 1, 2002.

4 See Garwin, "Nuclear and Biological Megaterrorism."

5 See US Commission on National Security in the 21st Century (chaired by former US Senators Gary Hart and Warren Rudman), *New World Coming: American Security in the 21st Century*, Sept. 1999.

6 See Thomas L. Friedman, "The Humiliation Factor," *New York Times*, Nov. 9, 2003. See also Jessica Stern, *The Ultimate Terrorists* (Cambridge, MA: Harvard University Press, 2000), and Jessica Stern, *Terror in the Name of God: Why Religious Militants Kill* (New York: Ecco, 2003).

7 It should be noted that most of the injuries in the Tokyo subway attack were not serious; most casualties were released from the hospital within one day. For the numbers of deaths and injuries from this and other terrorist strikes, see Johnston's archive, http://www.johnstonsarchive.net/terrorism/

8 See Bruce Hoffman, *Inside Terrorism* (New York: Columbia University Press, 1998).

9 See various articles in the journal *Terrorism and Political Violence*.

10 See Todd Sandler, "Fighting Terrorism: What Economics Can Tell Us," *Challenge*, 45, 2002, who argues that the way to prevent terrorism is to remove support for the terrorists in terms of funding, recruits, information, etc.

11 See Allison, *Nuclear Terrorism*.

12 See Graham Allison and Andrei Kokoshin (2002) "The New Containment: An Alliance against Nuclear Terrorism," *National Interest*, Fall 2002, and "US–Russian Alliance against Megaterrorism," *Boston Globe*, Nov. 16, 2001.

13 For a discussion of deterrence of terrorism, noting that deterrence still matters, see Jonathan Stevenson, *Counter-terrorism: Containment and Beyond*, Adelphi Papers (London: International Institute for Strategic Studies, 2004).

14 See Charles D. Ferguson, William C. Potter, Amy Sands, Leonard S. Spector and Fred L. Wehling, *The Four Faces of Nuclear Terrorism* (Monterey, CA: Center for Nonproliferation Studies, Monterey Institute, 2004).

15 See Weston Konishi, "The Case for Deterrence in the Anti-Terror Campaign," June 2002, http://www.weltpolitik.net/Sachgebiete/Internationale%20Sicherheitspolitik/Probilembereiche%20und%20L%F6sungsans%E4tze/Terrorismus/Analysen/The%20Case%20for%20Deterrence%20in%20the%20Anti-terror%20Campaign.html

16 See also Michael A. Levi, "Deterring Nuclear Terrorism," *Issues in Science and Technology*, Spring 2004.

17 See Michael D. Intriligator and Dagobert L. Brito, "The Potential Contribution of Psychology to Nuclear War Issues," *American Psychologist*, 43, Apr. 1988, pp. 318–21.

18 See United Nations, "A More Secure World: Our Shared Responsibility, Report of the High-Level UN Panel on Threats, Challenges and Change," Dec. 2, 2004, http://www.un.org/secureworld/

Part II

PROTECTING CRITICAL INFRASTRUCTURE

5

THE TROJAN HORSE OF THE INFORMATION AGE

Lars Nicander

When the husband and wife futurologists Alvin and Heidi Toffler described the conflicts in the Third Wave—the Information Age—in *War and Antiwar* (1993) they argued that, while the aims of a war or military campaign had not changed, the method of waging it had. A new form of warfare—information warfare—with a whole new doctrine had seen the light of day.

This change can be described in a simplified way as the difference between the theory of the nineteenth-century German strategist Clausewitz (that "war is merely a continuation of politics with different means" and that, consequently, war and peace are two clearly distinguishable conditions) and the theory of the Chinese strategist Sun Tzu (500 BC) (that "the highest art of war is not to win a battle but to win without battle").

In the latter case, the boundaries between peace, crisis and war are dissolved. It is a matter of retaining "the monopoly on formulating the problem," of getting one's adversary to behave as one wishes, perhaps primarily by influencing his will, but if this fails one must also be able to threaten and meet his objective capacity in a credible way.

What, then, are information operations and information warfare? In 1998 in the United States, the original concept of information warfare (IW) was given the new designation of information operations (IO), primarily because the private sector and civil authorities did not want to talk about warfare in peacetime. IW was then given the more limited meaning as "information operations during crisis and war" and primarily within a military framework. Like NATO, Sweden has adopted this change; the most semi-official definition can be found in the Swedish government's information technology bill of March 2000 (1999/2000: 86, p. 36):

> Information operations are combined and coordinated measures in peace, crisis and war to support political or military goals by influencing or exploiting the information and information systems of the adversary or other foreign player. This can occur by using one's own

information and information systems while these assets must also be protected. An important element is the attempt to influence decision-making processes and decision-making.

There are both offensive and defensive information operations. These are carried out in political, economic and military contexts. Examples of information operations include information warfare, mass-media manipulation, psychological warfare and intelligence operations.

Defensive information operations are coordinated and integrated measures in times of peace, crisis and war as regards policy, operations, personnel and technology to protect and defend information, information systems and the ability to make rational decisions.

Other closely related concepts are also used to describe partial methods of information operations. "Overarching information security"—also including policy, organization, etc.—is the IT-based defensive component and is collectively termed information assurance (IA). Perception management is the cognitive form of exerting an influence, with psychological operations ("PSYOPS") as its most organized sub-category. In the context of civil preparedness, there is much talk of critical infrastructure protection (CIP). Within the Swedish armed forces, the current reorganization into a network-based defense (NBD) represents a new way of leading military units in which information can be converted into armed response, almost in real time.

The most spectacular sub-category of information operations is offensive computer network attacks (CNA) in which infological weapons such as computer viruses, Trojan horses, logical bombs and denial-of-service attacks attempt to attack specific information systems. Likewise, electromagnetic pulse (EMP) and high-power microwave (HPM) weapons—for instance hidden in a briefcase—can, without smoke, sound, light or smell, knock out or (at close range) even melt the electronics in vital information systems.

The effect which such attacks could have on such targets as financial systems has led official Russian representatives to draw comparisons with nuclear war and demands from nations such as the US that these systems should be included in arms control. The use of military computer network attacks should also require the same high decision-making level as the use of nuclear weapons. The comparison with nuclear war, however, falters in one important respect, since nuclear war was "threshold raising" in that both superpowers had a mutual, destructive second-strike capacity—the so-called balance of terror.

Today, when the "enemy" exists not only in one direction and conflicts are more multifaceted, with military, economic and religious elements, the anonymity involved in computer network attacks becomes "threshold lowering." How can a state retaliate if it cannot clearly identify the sender? On the other hand, what might someone be capable of doing, even against one's best friend, if there was no risk of discovery?

The other "sensitive" extreme in the spectrum of information operations is psychological operations with media manipulation and perception attacks. In Western society with its strong media institutions it is almost anathema to assert that states should in some way use these methods other than at a relatively low military level.

The fact that Sweden's total defense approach provides this country with an agency which in peacetime makes plans for psychological defense—though only in the event of crisis and war—arouses a delight mingled with terror among the defense planners of other nations. In the US, CNN and other institutions would start talking about a "Big Brother" society if the media there felt themselves likely to be influenced by a similar institution in any formal sense.

A proposal in autumn 2001 to create a special authority within the Pentagon to shape the strategic media image also had to be withdrawn after a media backlash. At the same time, the technical possibilities of virtual image manipulation (morphing) are almost unlimited, something which can be seen in the Hollywood films *Forrest Gump* and *Wag the Dog*. Since seeing is believing, this can be a very effective weapon.

The increased role of PR agencies in creating public sentiment, above all in third-party countries, in favor of one side of a conflict, by such methods as planting video sequences in news programs, was manifested at the start of the conflict in Bosnia. Representatives of the Serbs in Bosnia employed the Saatchi & Saatchi agency, while the Muslim side had their own PR firm. If in a corresponding way (for example during a crisis situation in the Middle East) a morphed video sequence were to be shown in which Israeli units were apparently bombing and burning down Mecca, this could have instant and irreversible effects on events before there would be time to make any denials.

Particularly in peacetime, perception management and psychological operations are, like the intelligence services, "an extremely forbidden necessity" for the strategist who wishes to succeed. The line drawn between these methods and general diplomacy and politics can become blurred, as can the same line drawn vis-à-vis economic contexts, in which false press releases, the spreading of rumors, etc. can have a manipulated speculative effect on stock markets. This fact means that it is as important to be aware of, and have the means to discover and check, the sources of such information as it is to have hacker-detection systems and firewalls in computer networks.

To sum up, information operations are characterized by the fact that there are no demarcation lines regarding their use in the scale of conflict made up by peace, crisis and war. These operations can be implemented in political, economic and military contexts. There is always a strategic purpose even if implemented at a low level within an organization. This means that an information operation must have the support of, and be synchronized at, the highest possible level: for a state at the highest security policy level, and for a company in the CEO's immediate circles. We cannot make a distinction between offensive and defensive expertise and capability: if you have one, then you have the other. The weapons can be cognitive, infological, electromagnetic and kinetic; it is the purpose that is the decisive factor.

The asymmetry which characterizes terrorism—it is no longer merely states which threaten states but also separate individuals/groups (e.g. bin Laden vs. the US)—is even more obvious in the field of information operations, since a single individual can theoretically cause serious IT attacks which affect important societal infrastructures. The media's role and what is known as "the CNN effect" reinforce this asymmetrical element. How much was the effect of the events of September 11 magnified by the fact that we could from early on and repeatedly with our own eyes follow the course of events and see the planes explode into the two towers of the World Trade Center? Would it have been equally traumatic if we had only heard about the event?

In the light of this, a crucial question has been how we can define the civil–military relationship in order to map out the relationships of responsibility within the civil services of different countries. One first important decision for the majority of defense forces in the mid-1990s was to define what was then called information warfare as either an operational- or an intelligence-oriented matter.

The Swedish armed forces, like those in the US and Germany, chose to regard this as an operational matter, i.e. as a weapons system, which a nation should be able to use in the same way as, for example, tanks and an air force to protect the country and ensure its survival. Others, including France and the UK, regarded the issue more as an intelligence matter. Consequently, completely different systems came to govern developments and influence what could be openly discussed. This has contributed to the fact that the EU has difficulty in addressing these issues. Even within NATO the discussion is limited because of differing national agendas.

To put it simply, based upon what the Swedish authorities have already published, we can talk about the following four dimensions of information warfare:

1 Defensive information warfare (IW-D), in which primarily the armed forces adopt measures even in peacetime to protect their own systems in the event of crisis and war. All countries talk about this.
2 Offensive information warfare (IW-O), in which the armed forces during crisis and war must have knowledge of such methods to uphold the

nation's sovereignty and survival. Only a few nations have spoken openly about this: the US, Germany and Switzerland.

3   Defensive information operations (IO-D) can be regarded as a "total defense in cyberspace." Since this can occur in peacetime, as well as in times of crisis and war, it is an issue for the national authorities in which the armed forces can only play a supporting role. It is perhaps only Switzerland and, in some cases, the US—though not when it comes to psychological defense—that openly declare their ambitions in this field.

4   Offensive information operations (IO-O) represent the most sensitive of these four dimensions, and state authorities are unwilling to comment publicly on them since these are best classified as skilled intelligence operations.

Since the threat against national infrastructures has nevertheless been observed in all nations, particularly after September 11, the term critical infrastructure protection (CIP) has become the concept which most closely corresponds to IO-D, and which is used to denote protection against this kind of civilian threat. In recent years most Western nations have seen the construction of new cross-sectoral management structures to better handle the necessary cross-sectoral problems.

Thanks to the legacy of its "total defense system," Sweden has an advantage which has not been fully exploited. This concept demands a highly holistic approach in the preparedness of both civil and military organizations in Sweden as the only way in which a small country might have any real opportunity of withstanding an attack from a much more powerful adversary. In addition to military defense, the system encompasses an integrated defense against economic, psychological, security and infrastructural threats. Even in peacetime, the different sectors of society, both government and private, have been required to look towards this overall defense goal.

The problem with this model is that it is not intended to operate during peacetime and only comes into effect in the event of national crisis or war. We are now faced with potential attacks which can occur at any time, and most likely in times of peace. This leaves room for conflicts of expertise (and interests) between different sectors—a situation which paralyzes and delays objectivity and the necessary organizational, structural and operational changes. Sweden is still investing approximately SEK40 billion in military defense, without knowing if it can, or is allowed to, use such vital resources for peacetime non-military threats. It is imperative that Sweden does not "throw the baby out with the bath-water," but rather tries to retain the holistic approach which characterized its highly agile and successful total defense legacy, by developing new peacetime structures.

Within the defense establishment, information warfare has caused some perplexity and anxiety, particularly when it comes to the fundamental axioms of military theory. Firstly, the boundaries between tactical, operational and strategic levels are becoming increasingly blurred.[1] If a leaflet, supplied by a military PSYOPS platoon within the KFOR force in Kosovo, comes under the cameras of CNN and ends up on the desk of the American President, it is definitely no longer a tactical issue. A tactical maneuver by units against an important network or telecommunications node must be synchronized at the highest level so that the effects do not exceed the intended ones ("cascading effects") and, at the same time, do not reveal or impede its own intelligence capability. Secondly, it has been discussed whether the Swedish philosophy of military leadership with its emphasis on assignment control has been a hindrance.[2] This leadership philosophy has been a hallmark of delegated decision-making (*auftragstaktik*) within most of the Swedish armed forces, but is less well suited to these contexts, since very strict command and control procedures are the only possible way of managing IO/IW during the early stages of conflict, which is when information warfare is most effective.

What, then, does the threat scenario look like today? The traditional formula of intelligence analysis, "Threat = Intentions × Resources," should in the Information Age be expanded to "Threat = Intentions × Resources × Vulnerability." A country with heavy IT dependency (like the United States) is naturally more vulnerable than an underdeveloped country that lacks societal IT structures (such as Somalia). On the side of the perpetrators, distinction should similarly be made between states, terrorists, criminals and individuals (hackers).

One basic difficulty for a defender is to reliably know the identity of the attacker in the event that this is concealed or that the address is false (spoofed). Despite bold rhetoric about hitting back, it can be hard to guarantee that one is not actually attacking a hospital and putting vital life-support systems out of action. Which state department takes responsibility is often confused and can lead to delays in launching an effective response: if the attacker is a nation-state, it would be the job of the armed forces; if, however, the attacker is a criminal or a terrorist, it would be the business of the police. But how do we know? Before we have figured it out, the question has probably become obsolete . . .

When it comes to individual countries, the US has the largest—and openly acknowledged—military capability in this area, but even nations like China are investing a lot in both doctrinal and structural development, setting up a special reservist organization for information warfare. A semi-official treatise, *Unrestricted Warfare* (1999), written by two Chinese colonels described the intention to use both "soft" PSYOPS methods and "hard" network attacks against, above all, the US. The aim was to particularly exploit the United States' Achilles heel in the form of the population's low tolerance for casualties: for example, it was believed that a terrorist attack against a military base with a resulting large number of dead soldiers would create pressure on the American government to withdraw from most conflicts which did not affect the American homeland. In many other countries, information operations are regarded as an intelligence matter and not as operational, which limits the amount of public knowledge.

We have not yet seen many cyber-attacks launched by terrorist groups. In a 1996 paper, Dr. Andrew Rathmell of King's College, London, compared the willingness of the Muslim organization Hamas and the IRA to use infra-structural attacks and IT weapons. He found that the IRA had sent people from their attack units on computer courses and had located crucial electricity nodes in London for a coordinated attack against commerce and the economy. Yet they had desisted. Why? His conclusion was that within traditional terrorist structures like the IRA, with its hierarchical organization and its blue-collar leadership, there existed a greater resistance to using these methods despite their effectiveness—they wanted things to go "bang." In contrast, Hamas, with its academic leadership and its network-oriented organization, might be more inclined to use cyber-weapons and infrastructure attacks.

Al Qaeda has begun to use IT methods to communicate secretly, but there are examples of more offensive use. The Aum sect in Japan, in addition to using poison gas in the Tokyo underground system, was also involved in developing both biological weapons and manipulating the software of government information systems. It turned out that a software company controlled by the sect had been responsible for programming the positioning system which the Japanese police used for their vehicles and police officers. The sect probably knew exactly where the police were at any given moment.

When it comes to serious crime involving IT elements, there are not many publicly acknowledged examples because it is in the very nature of this crime that the number of unrecorded cases is very high. There is only one known computer attack against a financial institution—against Citibank in 1994 when the Russian leader of a qualified hacker group, Vladimir Levin in St. Petersburg, succeeded in extracting $400,000—but he had been close to getting $70 million. Citibank reported the crime to the FBI, upon which Citibank's competitors announced to their own customers and the world: "We haven't had that problem." The immediate effect was that Citibank's four largest customers withdrew about $1 billion each. The incentive for companies to talk about similar events since then has not increased, even if rumors of successful computer-based coups and extortion against banks are occurring with much greater frequency.

When it comes to individual hackers, the most expensive attack to date was the "Love" virus in spring 2000. Originating in Manila in the Philippines, this virus caused damage worth an estimated $90 billion to information systems around the world. Sweden was helped by the time factor, even though SAS and others suffered. Asian companies discovered the virus first; American antivirus companies then had the night in which to find countermeasures, which Swedish companies could then use before booting up their systems in the morning. A number of IT security experts and administrators pointed out at the time that, if this was what two young people could do in five hours, what might a nation achieve with specially targeted viruses or by releasing some kind of mass virus close to the intended target?

How, then, should we view IT weapons? Can they be a force for good, in the hands of the democracies, or are they always likely to be the weapon of choice of the "bad guys"? As with all weapons, they are merely tools for the conduct of international affairs and these weapons will mirror the purposes for which they are launched. One major problem for the international community is the ability to intervene in international conflicts before they escalate to an unmanageable level. At the same time, more and more countries are concerned about their own losses in such conflicts. The American hesitation to send ground troops to the Balkans is one example of this. Thus, demand has recently arisen for a more flexible "toolbox" with more alternatives than the traditional military use of force using, for example, "smart sanctions" and conflict-prevention measures.

An article by an American military lawyer drew attention to Article 41 of the UN Charter, which proposes breaking off postal links and telecommunications with the aim of maintaining international sanctions. He constructed a scenario involving an application of IT weapons in accordance with Article 41 to maintain (what were in reality ineffective) sanctions against Rhodesia in the 1960s.

In this scenario, a unit would be able to identify, by means of a needs analysis, critical telecommunications nodes and knock them out. This would result in major communications blackouts, which would effectively maintain the sanctions. This action would be done within the broader concept of "use of force" and not within the narrower (and harder to decide on) "use of armed force." Even in the case of other international interventions, there would probably be a need for the UN-appointed military commander to have a more flexible toolbox.

Humanitarian aspects also indicate the need for an overhaul of international law. In a conflict involving an internationally sanctioned intervention such as the one in Kosovo in 1999, it is currently in accordance with international law to bomb a bridge on which there is a military truck even if 20 civilians also on the bridge are killed. In contrast, it is probably in conflict with the current interpretation of international law to cut off a civilian telephone line in the same area, even if that would have had a far greater effect on the war efforts of the Milosevic regime. International law and the laws of war are still based to a great extent on the legacy of experiences from the Napoleonic wars and are thus scarcely suited to the Information Age.

To sum up, in the light of possible conflicts and threats in the Information

Age, we can perceive changes in four important dimensions. Firstly, there is no longer any very clear difference between public and private dependencies. Reciprocal dependencies are at stake; and here the realization that the state bears a responsibility for the commercial infrastructure must have an effect. The state had no formal responsibility for the banks during the crisis of the Swedish krona in 1991, but there was no other authority that could take the responsibility, and the state's role as "insurer of last resort" then became obvious even in Sweden. Since such situations cannot be ruled out, we must also be able to plan for them.

From an information assurance standpoint, the role of the insurance and reassurance industry is crucial. If we can get their active involvement in developing insurances for the new risks—with low probability but with huge consequences and thus no actuary data—we could promote more sound risk management procedures within the private sector. When the costs for vital IA measures reach the CEO level instead of the CIO within a company, the foundation of the critical infrastructures would be much safer and more reliable. It would be a self-regulating mechanism based on market values and incentives. That also implies that the role of government—with the taxpayer's money—only needs to be support of measures for strengthening the private sector against effects that are beyond the business optimum, or where the knowledge of those threats primarily is out of bounds for the private sector and more of a government issue for the intelligence community.

The role of information sharing is also critical. When it comes to developing private–public partnerships and the necessity of creating information-sharing analysis centers (ISAC), there should be an impartial broker ("priest") to whom the companies can with trust perform their "confessions." This is in my view the reason why global integrity had such a success running the financial ISAC, compared to some others where different industrial associations were in charge. In the latter category there is often a sense that one's competitor would take advantage of vulnerabilities, which reduces the willingness to share.

Secondly, the relationship between civil and military authorities has changed. "During the Cold War, the civil defense was supposed to support the military defense—now it is the reverse." This statement was made by the former Norwegian Prime Minister Kåre Willoch, who in 2000 headed the Vulnerability Committee specially appointed by the Norwegian parliament. The final report contained demands for major structural changes to the Norwegian political establishment. Sweden has so far been hampered by what is known as "the Ådalen syndrome,"[3] which has maintained a very strict regulatory framework when it comes to the use of military resources. In 2001, however, a new committee of inquiry suggested that it should become easier to use these resources on the condition that the use of force was excepted. In summer 2006, Sweden at last got a law that permits law enforcement agencies to take advantage of military resources including the use of force under strict rules.

Thirdly, it is no longer possible to rely on any division of the conditions of peace, crisis and war when it comes to the relationships of responsibility for meeting the new threats. Flexible coordination between the police and the military must be established similar to that of the US, whose new Department of Homeland Security appears to have an increasingly strong mandate. In Sweden, for example, a civilian (police) command should be able to request the NBC unit which is currently being built up within the Swedish armed forces to handle this kind of terrorist event.[4]

Fourthly, there are no borders in cyberspace. Since the link between grand strategy and the economy has become increasingly strong, threats, in particular anonymous IT attacks from other nations, can occur against economic players in another country. International security measures must therefore be developed by means of collaboration between as many countries as possible and at all levels. Legal and technological regulations must be harmonized so that a cyber-attack can be traced and stopped almost in real time.

It is impossible to make any distinction between offensive and defensive "capability." It is only the hard-to-access "motives" which can provide guidance. Since technological equipment is extremely useful for both peaceful and antagonistic purposes ("dual use"), so in principle every young computer "geek" can, with completely legitimate motives, acquire the necessary equipment. This means that demands for arms control in the IT sphere are no longer applicable and that prospective enemies must be identified from a far larger arena than has previously been the case.

The realization that cross-sector threats demand cross-sector solutions must also influence the design of any national defense strategy in the field of information operations.

The dramatically increased need for a rapid connection between "threat" and "planning" can—if the will exists—be handled within the Swedish system. A first step has been taken with the establishment of the Swedish Emergency Management Agency, but further changes are also needed to overcome the stovepipe structure of government and achieve a more horizontal, layered structure. This can and should be developed through the collaboration of civilian, military and police authorities and, above all, in conjunction with the private sector.

Notes

This text was previously published in *Access*, no. 5, 2002.

1 Tactical measures involve direct battle planning in near time (hours/days); operational planning occurs at the higher levels of staff and concerns the entire geographical area of operations with a longer time perspective (days/weeks); whilst strategic planning occurs at the national headquarters and Ministry of Defense level (months/years).

2 Assignment control means that subordinate commanders can fairly freely solve the tasks given them by superior commanders with the allocated resources and with few other rules of conduct. Command and control involves control in detail.

3 In 1931 a strike in the district of Ådalen led to battles between the demonstrators and military forces in which five civilians were killed and five wounded. One result was the establishment of a national police force and a ban on the use of the armed forces against Swedish civilians.

4 The NBC unit is a military unit which will be established to handle (limited) attacks involving nuclear/radiological, biological and chemical weapons.

6

SUICIDE BOMBERS, SOFT TARGETS AND APPROPRIATE COUNTERMEASURES

Robert J. Bunker

Suicide bombings are receiving increased public attention now that they are taking place on an almost every other day basis against American and allied forces in the stability and support operation (SASO) environment of post-Operation Iraqi Freedom (OIF) Iraq. In November–December 2004 alone some 27 suicide bombings took place.[1]

This type of bombing is indicative of one of the changing dynamics of post-modern terrorism—terrorists purposefully engaging in suicide (martyrdom) in return for increased attack capability to support the greater vision of their cause. This form of individual and group sacrifice is analogous to the defense of a beehive undertaken by a swarm of liked-minded worker bees who can each deliver one sting before they die. In this case, however, the swarm is composed of a network of like-minded radical jihadi groups that compose al Qaeda.[2] Another swarm, based on emergent affinity Shia groups tied to Hezbollah or a central Asian "-Stan" entity, also has the potential to develop.

At some point in the future it is expected that suicide bombings will begin to strike the US homeland—at the very least on an irregular basis. The 9/11 bombings (using passenger airliners as the delivery mode and the aircraft's kinetic energy/fuel load as the warhead) represents the first of these suicide attacks.

With more, though probably less spectacular, attacks to follow, prudence suggests that we better understand suicide bombings at both the philosophical and the tactics and techniques levels. In addition, the relationship of suicide bombing to the soft targets within our society should be discussed and the appropriate countermeasures needed to respond to this threat identified. In doing this analysis, we must ask ourselves the question: Where do we draw the line when everything can't be protected 24/7?

Suicide bombers: Islamic tradition[3]

The Islamic martyrdom tradition is of specific interest because it is within this philosophical context that suicide bombings within Iraq and the United States have been and will be conducted. Raphael Israeli has written the best overview of suicide bombing's Islamic philosophical origins.[4] The conceptual basis is found with the legendary suffering and death of Hussein in Karbala in 680. This, along with the assassination of the son of Ali (first true Imam and successor of the Prophet), has had an extreme impact on the thinking of the Shia (Shi'ite) branch of Islam. The Shia cult of martyrdom is a tradition that originates with Hussein ibn Ali, grandson of the prophet Muhammad, who was killed by the army of Caliph Yazid at Karbala in 680. The idea of individual "selfless sacrifice" was then drawn upon during the Iran–Iraq War of the 1980s when units of Iranian children with the "keys of paradise" hanging on their necks cleared Iraqi minefields with their bodies. These Shia sacrifices were immortalized with the blood-red-colored water fountain of the martyrs in Tehran.

In 1982, the Iranian revolution under the Ayatollah Khomeini was then exported to Lebanon, where the "Islamic Resistance," the precursor of Hezbollah, launched a series of suicide attacks against US, French and Israeli targets. Thus, the creation of the Hezbollah (Party of God) in 1982 as a counter to the Israeli invasion provided the impetus for modern suicide operations. Hezbollah exploited the images of the cult of Hussein to inculcate self-sacrifice and "martyrdom" as an ideal for its fighters. This Shia group, which utilizes both terrorist and guerrilla techniques, conducted its first large suicide bombing in April 1983 against the US embassy in Beirut. That bombing was directly influenced by the first documented vehicular suicide bombing in December 1981 against the Iraqi embassy in Lebanon. The 1981 bombing was conducted by the Shia Amal group, which had cross-group linkages with Hezbollah upon the latter's formation.

Suicide bombings remained a Shia activity for a decade until Hamas (Islamic Resistance Movement), a Sunni terrorist group, conducted a suicide bombing within Israel in April 1993 against IDF soldiers. The reason for this transference from Shia to Sunni was based on two events. The first was the exile of over 400 Islamic activists, many of them Hamas members, by Israel to southern Lebanon in December 1992. Stranded and alone, these activists were befriended and protected by Hezbollah based on the simple rationale that "the enemy of my enemy is my friend." While in exile in Lebanon, the Hamas members were influenced by Hezbollah's suicide bombing CONOPS and brought these techniques back to the West Bank with them when they were repatriated. The second event was fundamentalist Sunni scholars who created fatwas (religious edicts) to rationalize how Shia concepts of "selfless sacrifice" could fit into Sunni thinking about martyrdom and ultimately the punishing of one's enemies. Suicide bombings spread to other fundamentalist

Sunni terrorist groups and then to more secular and nationalistic terrorist organizations such as the al-Aqsa Martyrs Brigades. The Brigades emerged in 2000 as an offshoot of Yasir Arafat's Fatah faction of the Palestinian Liberation Organization (PLO).

This migration of suicide bombings from the religious, initially with Shia groups and then to Sunni groups, and then to the secular, as with Arafat's Fatah, set the stage for Saddam Hussein's attempts in early to mid-2003 to draw upon this "criminal warfighting" technique against allied invasion forces in Operation Iraqi Freedom.[5] It also explains why suicide bombings have the potential to be conducted by any combination of former Iraqi Ba'ath party loyalists (to a limited extent) and fundamentalist Shia and Sunni terrorists now operating in Iraq.

From the perspective of individual and unit-level doctrinal employment, suicide bombers are advocated at both levels by radical Islamic elements. Suicide bombers look forward to death because, as martyrs (*shahid*), they expect to be rewarded by Allah in paradise while posthumously they and their families typically gain social status within their societies. Economic benefits, such as monetary payments, may also come to their family members as an additional bonus for the completion of a successful operation. For example, Saddam Hussein was noted for providing cash payments of $25,000 to the families of Palestinian insurgents killed in suicide attacks against Israeli targets during the Second Intifada.[6] Suicide operations range in organizational sophistication as well. For example, a single suicide bomber may act individually against a target, two or three may coordinate the bombings, or a larger number of suicide bombers may participate. This final scenario was seen with the 19 al Qaeda members who hijacked four US airliners on September 11, 2001, coordinating their activities as part of a larger strike force against multiple high-value targets.

Suicide bombers: tactics and techniques

Modern suicide bombings were first operationally employed in southern Lebanon by the terrorist Amal and Hezbollah groups in the early 1980s. This technique then spread to the Tamil Tigers in 1987 and to Hamas in 1993. Over the ensuing decade, an increasing number of terrorist groups engaged in suicide bombings: Palestinian Islamic Jihad in 1994, the Kurdistan Workers Party in 1996, al Qaeda in 1998, the Chechens in 2000, and the al-Aqsa Martyrs Brigades in 2002.[7] Since 1993, this pattern, with the exception of the Kurdistan Workers Party, is derived from radical Islamic groups netting together in a global insurgency against the United States and its allies.

Major groups engaging in suicide bombings can be analyzed by delivery modes (see Table 6.1) and target set (see Table 6.2). The Tamil Tigers and al Qaeda top the list in suicide bombing sophistication, followed by the Chechens and Hezbollah. Less sophisticated groups are Hamas, Palestinian

Table 6.1 Major groups by "suicide bomber" delivery mode

Group	Personnel (human)	Vehicular	Aircraft	Vessel
al-Aqsa Martyrs Brigades	Yes	Yes	No	No
al Qaeda	Yes	Yes	Yes	Yes
Chechens	Yes	Yes	No	No
Hamas	Yes	Yes	No	No
Hezbollah	Yes	Yes	No	No
Kurdistan Workers Party (PKK)	Yes	No	No	No
Palestine Islamic Jihad (PIJ)	Yes	Yes	No	Yes
Tamil Tigers (LTTE)	Yes	Yes	No	Yes

OSINT: Courtesy of Counter-OPFOR Program, NLECTC-West©2003, 2005.

Table 6.2 Major groups by "suicide bomber" target set

Group	Civilian (personnel)	Military/LE (personnel)	VIP	Transit	Aircraft	Vessel	Buildings/ infrastructure
al-Aqsa Martyrs Brigades	Yes	Yes	No	Yes	No	No	Yes
al Qaeda	Yes	Yes	Yes	Yes	Yes	Yes	Yes
Chechens	Yes	Yes	Yes	Yes	Yes	No	Yes
Hamas	Yes	Yes	No	Yes	No	No	Yes
Hezbollah	Yes	Yes	No	No	No	No	Yes
Kurdistan Workers Party (PKK)	Yes	Yes	Yes	No	No	No	Yes
Palestine Islamic Jihad (PIJ)	Yes	Yes	No	Yes	No	Yes	Yes
Tamil Tigers (LTTE)	Yes	Yes	Yes	Yes	Yes	Yes	Yes

OSINT: Courtesy of Counter-OPFOR Program, NLECTC-West©2003, 2005.

Islamic Jihad, and the al-Aqsa Martyrs Brigades—even though they have engaged in a greater number of suicide bombings than some of the other major groups. The Kurdistan Workers Party is at the bottom of the sophistication scale.

More sophisticated groups use larger and higher-order explosive devices. They engage in simultaneous (multiple suicide bombers/targets) and/or sequential attacks (secondary and tertiary suicide bombers at the same target) and often combine the attack with other weaponry. They have the ability to engage "hard" rather than solely "soft" targets (partially as a result of larger bombs and better explosives), and can draw upon more delivery methods. Triggering methods (fuses, pull cords and cell phones) also increase with sophistication, as does the lessened detection of explosive device by sensors (X-rays, metal detectors, dogs and soldiers).[8]

Operational advantages of suicide bombings over normal terrorist bombings include:

- *The device is precisely delivered to the target.* The suicide bomber functions as a "precision weapon," taking the explosive device right to the target. This is a dimensional stand-off attack in the sense that the terrorist is "invisible" (stealth-masked) until the device is detonated, which helps overcome the Western advantage of stand-off targeting based on physical distance.

- *Harder targets can be attacked.* Targets which cannot normally be attacked can now be reached. Heavily fortified compounds with proper stand-off distances will not be damaged by normal terrorist bombings, whereas suicide bombers can crash through the front gate of a fortified compound and reach the desired target. Such gate-crashing has taken place repeatedly in vehicular suicide bombings.

- *The device has no window of vulnerability.* The explosive device cannot be found and moved or rendered safe. No time period exists from when the device is left at the target and the terrorist escapes to safety prior to detonation.

- *No plan required for egress.* The explosive charge simply has to be delivered to the target.

- *No one left alive to interrogate.* Because suicide bombers are not typically captured, operational security (OPSEC) of the terrorist group is better maintained. The Tamil Tigers use poison capsules as a fail-safe method in this regard. Some of the Palestinian groups use a redundant, cell phone-activated detonator which can be set off by calling the cell phone in case the bomber attempts to back out of his or her mission.

- *No burden of wounded comrades.* Injured comrades create a logistical strain on a group.

- *Psychological factor.* Suicide bombers are blown to pieces, with the head (in the case of wearing a bomb vest) typically being separated from the body. Individuals also become concerned about other people close to them where suicide bombings take place with some frequency. This can create higher levels of anxiety for US troops when dealing with locals. Everyone in a crowd now has to be scanned for bulky clothing and unusual behavior.

- *Blood-borne pathogens delivery.* Suicide bombers infected with hepatitis and HIV can potentially create a "hazmat" incident by spreading disease to targeted personnel. Bone fragments and blood-covered bolts/nails may directly transmit pathogens from the bomber to nearby victims. While this is less commonly used and of questionable utility, infected bombers have been utilized by some Palestinian terrorist groups.

A strategic consideration must also be mentioned. Suicide bombings create martyrs for the group and the society it attempts to draw upon for recruitment. As more and more suicide bombers kill themselves and gain prestige and heavenly rewards (in the eyes of their society), the cycle of violence can continue to escalate. This can create a "religious movement" within the faithful. Already, Palestinian society is taking on characteristics of a death cult, with young children preferring to grow up to be suicide bombers rather than engineers and doctors. Recruitment of new suicide bombers is no longer difficult as the movement grows.

This should give us pause for concern and reflection, because radical Islamic networks, which include al Qaeda, are engaging in a global insurgency against the West. Martyrdom is one of the common bonds that hold this insurgency together, and it is increasing in strength as more terrorist groups engage in suicide bombings. The Roman Empire faced a similar strategic dilemma with Christian martyrs. We need to break the radical Islamic link to martyrdom, now over 20 years long, before it becomes too fully entrenched. Failure to do so has the potential to create a strategic dilemma for the United States.

Soft target environment

The United States is a stable and safe democratic state and, as a result, is full of "soft"—as opposed to "hard"—targets. Soft targets lack proper stand-off distances, access denial and/or blast protection that hard targets possess. The United States is now vulnerable because of a battlespace shift occurring in war and conflict. Older forms of homeland defense based upon conventional armies, air forces and navies (fourth-dimensional forces) are no longer able to protect us. The reason we are now defenseless is because these legacy forces are not able to stop the penetration of our country's borders from "stealth-masked" terrorist assault teams (fifth-dimensional forces).

Soft targets that can be attacked by jihadi suicide bombers range from low to high value. No hard-and-fast rule exists on how to determine whether one target is of greater or lesser value than another one. Typically low-value targets include individual and multi-tenant residences, small businesses and random groupings of individuals. Medium-value targets include schools, apartments, office buildings, hospitals, passenger ships and aircraft, and local and state government buildings.

High-value targets have a greater value placed on them than low- and medium-value targets because they are special for some reason. This is the case because they can be defined by one or more of the following attributes: they contain large numbers of people (e.g. a major sports arena), hold a high symbolic value to society (e.g. the Statue of Liberty) or are critical to the operating of the economy (e.g. Wall Street) or government (i.e. Congress or the President), or their destruction would result in catastrophic effects (e.g. a nuclear power plant or major dam).

These soft targets represent almost all of the public and private structures and infrastructure of the country and its entire populace, i.e. everything is basically the target of a potential suicide bomber and is threatened 24/7. Protecting everything is politically and economically impossible, even more so given the fact that our enemies currently possess a military (e.g. battlespace) advantage over us.

Taking this as a given, the question can then be asked: Where do we draw the defensive line? To answer this question, however, we need to better understand the targeting effects of suicide bombing (i.e. terrorism).

Targeting effects come in two forms. Conventional military force is based on thing targeting. This form of targeting relies upon fires and maneuver—things are destroyed (killed), damaged (injured) or seized (taken prisoner). Traditional "destructive firepower" based upon firearms, artillery, rockets/missiles and bombs falls squarely within this category. This is a form of targeting, and warfare, that the United States and its allies dominate.

Unconventional military force is based upon bond-relationship targeting (BRT). This form of targeting relies upon "disruptive firepower" that attacks the linkages between things. Terrorists rely upon this form of targeting because it is an asymmetric response to US (and other nation-state) domination of conventional military force. Rather than focusing on the point-of-impact concerns indicative of destructive targeting, this form of targeting focuses upon the shock waves generated by the targeting event. To use the pebble in the pond metaphor, it is the ever increasing shock waves generated by the impact of the pebble, rather than the impact of the pebble into the water itself, that is of significance to this form of targeting.

The effects of bond-relationship targeting can be better understood by viewing Table 6.3. Each year the United States loses about 2.4 million people to old age, disease, accidents, etc. This represents the annual fatalities of citizens in the society and is accepted as part of the human condition. The 9/11 suicide bombings resulted in about 2,800 deaths—0.00117 percent of the annual fatalities—a loss that, while tragic, is meaningless to the health and welfare of the society.[9] From a military firepower perspective, the 9/11 losses are also meaningless; losses of tens and even hundreds of thousands of individuals (remember the Somme?) are common in conventional nineteenth- and twentieth-century battles.

So then, why did 2,800 fatalities result in such widespread panic and societal disruption? While far more accidents, suicides and homicides take place

Table 6.3 Peacetime and suicide bombing fatalities

Fatalities	Number
Peacetime examples:	
Total US deaths (2001)	2,416,425
Accidents	85,964
Suicides	27,710
Homicides	11,328
Suicide bombing examples:	
September 11, 2001 attacks	*c.* 2,800
Operational Iraqi Freedom (first year)	813

Sources: U.S. National Vital Statistics, vol. 52, no. 9. Nov. 7, 2003; and Robert J. Bunker and John P. Sullivan, *Suicide Bombings in Operation Iraqi Freedom*, Land Warfare Paper 46W, Sept. (Arlington, VA: Institute of Land Warfare, Association of the United States Army, 2004), pp. 3–7.

in the United States each year than the 9/11 fatalities, they are spread out in time and space and occur individually and in small clusters and thus the effect is not the same. The second plane crashing into a World Trade Tower and the implosion of the Towers were viewed collectively by most of the US population. This allowed for the psychological shock waves generated by the incident—which are far more dangerous than the actual death and destruction created—to be effectively transmitted to the populace. As a result, the bonds that held society together were directly assaulted and frayed to an extent by the incident.

This brings us back once again to the question of drawing a defensive line. Such a line is not fatality-based—though fatalities may be a component in our considerations. Surely if 9/11 had resulted in 28 (1 percent) or even 280 (10 percent) fatalities the societal disruption generated by watching the incident unfold would not have been as great. Hence 9/11 represented a rock thrown into a pond, rather than a pebble, and rocks generate greater shock waves than smaller stones.

First and foremost, then, the defensive line against suicide bombers needs to be drawn at the disruptive firepower level. The effects of bond-relationship targeting need to be mitigated so that society does not suffer from the shock waves generated. Since shock waves are generated by more significant suicide bombing events rather than less significant ones, higher-value targets within society, especially ones with great symbolic value, should be protected first.

The problem that we face, however, is that any new suicide bombing or multiple suicide bombings that take place in America will be an immense disruptive firepower event. One plausible scenario is three or four indoor shopping malls being targeted by Iraqi insurgents who have infiltrated into the country in order to bring the war home to America.[10] Given that only 813 people were killed during the first year of all the suicide bombings in

Operation Iraqi Freedom, at best such a conventional domestic attack might yield 100 to 200 fatalities. After this event or events, the disruptive nature of future attacks should lessen—in contrast to the other view that American society could begin to turn in upon itself after successive waves of attacks. What mitigates the second view is the inability of OPFORs to sustain a suicide bombing campaign in our country; logistical and organizational restraints would be an inhibiting factor in such a terrorist campaign.

Appropriate countermeasures

At a minimum, the countermeasures needed are to defend against disruptive firepower (BRT) against society at the strategic level and to protect high-value targets to limit the BRT potentials generated. However, since we are probably looking at decades of war with jihadi insurgents, the public will not tolerate being undefended over the long term against the threat of suicide bombings within the country, even if the rates of fatality potential are well below the yearly national levels of suicides and homicide. As a result, a much broader suicide bombing countermeasures program will be required.[11] The outline of such a full-fledged program is presented in Table 6.4. Its evolution should be dependent on our perceptions of the intent and capability level of terrorist groups seeking to engage in suicide bombings and the resources that we are willing to spend to respond to the threats identified.

Greater dividends always result from proactive measures and programs at the strategic rather than the tactical and operational level. It is better to deter a suicide bombing or better yet shape an OPFOR (opposing force) in such a way that it will renounce martyrdom operations (such as suicide bombings) as one of its TTPs. Once we are forced to actually respond to an incident or, worse yet, deal with its consequences, we have gone from being proactive to being reactive. From a countermeasures perspective it is far smarter to retain the initiative and make an OPFOR contend with our actions rather than the other way around.

Full-fledged suicide bomber countermeasures would be broken up into four phases, with the first two ongoing and the second two incident-triggered. Ongoing programs focus on monitoring, analysis and coordination, and proactive measures. Tables 6.1 and 6.2 showing group delivery modes and target sets for suicide bombings are basic examples of intelligence collection and threat analysis—know thy enemy. Interagency networks, from the operational through the global level, would be based on the Los Angeles Terrorism Early Warning group model of intelligence fusion and network response capability. Incident-based programs focus on incident response and consequence management.

Each phase of the program could be broken down and discussed further, but such detail is not appropriate for this venue nor required to make the point that a very sophisticated countermeasures program can be created.

Table 6.4 Suicide bomber countermeasures program

	Tactical and operational	*Strategic*
Monitoring, analysis and coordination (ongoing)	Intelligence collection. Threat analysis. Interagency networks (operational area).	Intelligence collection. Threat analysis. Interagency networks (regional to global).
Proactive measures (ongoing)	Preemption capability/ hunter-capture teams (MIL—uniformed and plain-clothed). Facility design. Awareness training (responders and civilians). IPO: playbooks and target folders. RAM and playbook security measures (PSM). Electronic warfare. Information operations. Specialized equipment and training (LE/responders).	Redefine international and national law. OPFOR deterrence. BRT vs. OPFORs. BRT vs. martyrdom. Operations (religious). HUMINT/ group penetration and informers. ELINT/ group monitoring. Early warning notification system (cell/ pager).
Incident response (incident-based)	Force protection/ secondary devices. Bomb squads. SWAT teams. HAZMAT teams. Counter-surveillance teams.	Incident notification system (cell/ pager)
Consequence management (incident-based)	Medical treatment. Fire suppression and victim extraction. Hazmat clean-up/ decon post-blast forensics. Criminal investigation. Media liaison. Insurance claims. Restoration of services.	Societal disruption limitation (BR protection). TTP notification system (cell/ pager).

Courtesy of Counter-OPFOR Program, NLECTC-West©2005.

What is important is the question of the extent to which we actually implement such a program. This should be determined by the threats identified and our willingness to allocate resources against them in a proactive and rational manner. Unfortunately, what happens all too often is a reactive response made after a terrorist event or events have transpired. Given this perspective, we can expect that singular or multiple suicide bombings will have to take place domestically before we will know where the defensive line will be actually drawn.

Notes

This chapter was originally presented as a paper at the Law Enforcement—Intelligence Interactions: Problems and Opportunities Panel, Intelligence Studies Section, 46th International Studies Association Conference, Honolulu, Hawaii, March 3, 2005.

1 This does not include four unverified martyrdom operation claims made by the "Military Wing of Al-Qaiida's Jihad Committee in Mesopotamia." OSINT is derived from Reuters, CNN, BBC, AP and other news sources.

2 To better understand the new capabilities such groups possess, see Robert J. Bunker and Matt Begert, "Operational Combat Analysis of the Al Qaeda Network," *Low Intensity Conflict and Law Enforcement*, ed. Robert J. Bunker, Special issue "Networks, Terrorism and Global Insurgency," vol. 11, no. 2/3, Winter 2002 (published in Sept. 2004), pp. 316–39.

3 The suicide bombers sections are drawn from Robert J. Bunker and John P. Sullivan, *Suicide Bombings in Operation Iraqi Freedom*, Land Warfare Paper 46W, Sept. (Arlington, VA: Institute of Land Warfare, Association of the United States Army, 2004), pp. 3–7. The tables have been updated with more recent suicide bombing information.

4 Raphael Israeli, "A Manual of Islamic Fundamentalist Terrorism," *Terrorism and Political Violence*, vol. 14, no. 3, Winter 2002, pp. 23–40.

5 While suicide bombings spread to the secular socialist Kurdistan Workers Party years prior to Fatah's al-Aqsa Martyrs Brigades, it was probably too early to directly influence Iraqi thinking.

6 See "Iraq Continues Paying Palestinian Suicide Bombers' Families," *Iraqi Kurdistan Dispatch*, June 20, 2002, found at http://www.ikurd.info/news-20jun-p2.htm, and "Saddam Stokes War with Suicide Bomber Cash," *Sydney Morning Herald*, Mar. 26, 2002, found at http://www.smh.com.au/articles/2002/03/25/10174766310.html

7 These initial incident dates are drawn from open-source information (OSINT).

8 More specific information on tactics and techniques is outside the scope and venue of this work. Open-source documents that can be referenced are: International Institute for Counter-Terrorism at the Interdisciplinary Center, Herzliya, *Countering Suicide Terrorism*, Anti-Defamation League of B'nai, 2002; Human Rights Watch, *Erased in a Moment: Suicide Bombing Attacks against Israeli Civilians*, New York, 2002, access via www.hrw.org/reports/2002/isrl-pa/. US military and law enforcement should see the unclassified but restricted TSWG, *Suicide Bombing in World Terrorism*, June 26, 2003.

9 For more on misperceptions, see Clark R. Chapman and Alan W. Harris, "A Skeptical Look at September 11th: How We Can Defeat Terrorism by Reacting to It More Rationally," *Skeptical Inquirer*, Sept.–Oct. 2002, www.csicop.org

10 Ned Parker, "Iraqi Insurgents Threaten Attack inside the United States," Agence France Presse, Jan. 4, 2005.

11 Such a program is not stand-alone but would be integrated with other counter-terrorism programs dealing with other forms of threat weaponry and tactics, techniques and procedures (TTPs). Also, while conventional suicide terrorism currently generates only low levels of fatalities, CBRN (chemical biological radiological nuclear) terrorism (the E—explosive—purposefully excluded) has much greater fatality potentials.

7

TERRORIST USE OF NEW TECHNOLOGIES

Abraham R. Wagner

Introduction

While the religious and philosophic underpinnings of the current terrorist movements may be rooted in the Middle Ages, the various technologies that they employ certainly are not. Indeed, terrorist organizations operating in the Middle East, Asia and elsewhere have embraced a range of modern technologies to support their operations in areas including communications, targeting and recruitment, as well as in the design and use of weapons against military and civilian targets. At the same time, the intelligence services, military and law enforcement agencies engaged in counter-terrorism are operating in an era where a host of new technologies exist, and continue to evolve, that are of potential use in combating terrorism.

Technology has served as an "enabler" on both sides of the terrorism problem, in terms of making actual operations more difficult or the legal impediments presented by new technologies. While the term technology covers a very broad range indeed, and even the various technologies employed by terrorists and those engaged in counter-terrorism are also quite wide, the focus here is on four critical areas, including communications and IT, weapons and countermeasures, biological attack, and non-intrusive inspection.

Communications and information technology

What began as an MIT dissertation in 1962, and a DoD experiment after that, has evolved into a technological revolution, now known as "cyberspace" and the internet, that goes far beyond communications. Indeed, it is likely the most significant advance in media since printing and Gutenberg's invention of movable type in the sixteenth century. Internet use has exploded from a few scientists to a world where "net" access is almost universal, and terrorists are no exception; they have all become increasing users of the internet for a variety of functions. Where in earlier times they relied on other technologies such as telephone, radio, the mails and other systems, they cannot be barred from net access and will continue to use it for their purposes.

At the outset of the ARPAnet, e-mail and the web were not even a part of the vision. It began as a project at the DoD's Advanced Research Projects Agency (ARPA) utilizing a novel concept of "packet switching" as a more efficient use of a network than the "line switching" that had been used since the time of Morse in 1848.

The project did not lead immediately to the ARPAnet or the internet, since it was not a high priority in the 1960s, with the space race getting most of the funding. The program finally got off the ground in 1968, and the first elements of the ARPAnet were installed at UCLA and Stanford in 1969, connected by a 56 KB leased line between these first two nodes. Even after the prototype had been demonstrated, the net did not expand rapidly and received little notice outside the scientific and research community. Electronic mail (e-mail) was not even part of the initial ARPAnet concept. Likewise, media files and the "web" were never a part of the original concept, and it was years before these features were developed.

Exponential net growth resulted from other technology developments, such as the personal computer, or PC. By the late 1980s, millions of PCs were being sold, and means to connect this growing number of computers to the net were developed. Commercial network service providers (such as Prodigy and AOL) gave the broad population a means to access the net, heretofore limited to a few ARPA researchers, and in 1988 the ARPAnet transitioned to the internet for all to use.

A paradigm shift and exponential growth in cyberspace

The internet explosion of the last decade is largely the convergence of several related developments taking place at about the same time. It is possible to view this "explosion" in terms of four key technology developments:

- *Moore's Law—cheap computers for everybody.* This is now a world of increasingly cheap and powerful processors, making possible low-cost computers.
- *Packet switching.* Switched packet communications has enabled the internet and other communications networks.
- *Digital everything.* The world has moved rapidly from an analog world into a digital one where data, voice, video, text and all media are now in digital form.
- *Infinite/cheap bandwidth.* Fiber optic cable, advanced RF systems and other technologies have enabled order-of-magnitude increases in high-quality, low-cost bandwidth worldwide, at low marginal cost.

While the world has become digital, internet use has spread to even the most desolate areas, with users from all ages and walks of life—including terrorists. Indeed, as one analyst has written, "Cyberspace is not only a nascent forum for political extremists to propagate their messages but also a medium for strategic and tactical innovation in their campaigns against enemies."[1]

The internet is ideal for terrorists, providing communications and security at low cost. Recent evidence demonstrates widespread net use by various terrorist organizations worldwide in four major areas, including communications; access to information via the web for information on potential targets and technical data on weapons; websites for propaganda purposes and recruiting; and terrorist attacks on the internet, commonly known as cyberwarfare. In the Middle East, for example, Islamist terrorist organizations now employ over 300 websites.

Terrorist use of the internet for covert communications

The internet provides an ideal medium for terrorist communication, offering asynchronous worldwide service.[2] The sender and recipient of an e-mail or file transfer can be any place, at any time. Hamas, for example, has utilized the internet to send secure files and messages to operatives relative to attacks, including maps, photographs, directions, codes and technical details for various operations.

Terrorists, like other covert operatives, need to communicate while avoiding detection (location) or having their communications intercepted. Previously, terrorists' communications techniques ranged from messengers to a wide array of technologies.[3] Virtually all of these methods suffered from one or more serious problems:

- They were unreliable or not timely.
- Almost none were secure.[4] Most communications systems, particularly analog ones, were subject to location and intercept.[5]
- Specialized systems used by the government were not commercially available, and commercial systems were costly and cumbersome.

Compared to the early methods, or even costly systems, the internet is ideal, providing asynchronous global access at almost no cost. For most terrorist uses, dial-up access is sufficient and is available worldwide. Terrorists without computers or phones have been known to frequent internet cafés to access the net. Aside from the low cost of net access, the internet is probably more reliable than any other system, since it uses switched packet communications, and the net degrades by using alternate paths and moving more slowly, rather than simply failing as in the case of line-switched or point-to-point systems. Wireless internet access has proliferated rapidly, and access, using computers with 802.11 wireless capabilities, will certainly become increasingly popular with terrorists and may give new meaning to the term "T-Mobile hot spot." Such locations have all the advantages of the internet café, as well as greater bandwidth and anonymity.

Operational security (OPSEC) can be accomplished in several ways, such as the simple use of alias accounts among the vast number of servers around the world. Hiding among the millions of internet users is simple, and opening an account on a service such as yahoo.com or hotmail.com takes only a moment and is cost-free. Accounts can be used for only a few messages and abandoned in favor of other accounts and aliases. Without timely collateral intelligence on such accounts and names, it is virtually impossible to keep track of their use. Combined with the fact that the actual message content may be either encoded or employ cryptograms, the problem becomes even more difficult. Over the last decade, terrorist organizations have also set up internet servers of their own using commercial front organizations to purchase net bandwidth and register domain names.

Finding terrorist e-mail

Since the internet and e-mail have emerged as an ideal medium for terrorist communications, this raises the question as to how counter-terrorist intelligence services can locate and intercept these communications. Published reports indicated that the US and others were exceedingly slow to recognize this as a serious problem at all. Indeed, the potential for criminal or terrorist use of new technologies such as the internet and cellular telephones was largely ignored and greatly underfunded for much of the past decade. What exists now can likely be characterized as too little, very late.[6]

While it is possible to fault the intelligence community for inadequate attention to this area, as the 9/11 Commission has done, in the long run it may not be possible to accomplish a great deal. Here it may be close to impossible to find covert terrorist communications sent via e-mail if done "properly," where terrorists adhere to good OPSEC. Access to terrorist communications in the future will likely depend on their being stupid and employing sloppy procedures, as well as collateral intelligence. If e-mail account names are known only to a very few, and are changed with great regularity, it will be difficult to find the accounts and their communications.[7]

The impact of encryption

Few developments trouble the world's intelligence services more than the proliferation of commercial encryption. Cryptographers, using powerful computers, had some advantage in dealing with the problem, and until recently encryption has been limited to sensitive government communications. Commercial encryption has generally been a failure since, in the "analog era," high-grade systems were both costly and imposed an administrative and logistics burden on users. Encryption required hardware that included an analog-to-digital converter, as well as a digital encryption device, which was essentially a special-purpose computer.

The "digital revolution" changed this radically, with voice and data of all types now in digital form. With powerful processors, "mixing up the digits" wasn't a major problem, and good encryption algorithms have proliferated as well. User demands have also changed, with users now becoming more sophisticated while demanding privacy and security. Increased commercial use of the internet and growth in computer crimes have greatly heightened sensitivity in this area.

For intelligence services and law enforcement authorities seeking to find terrorists and others, this is a troublesome future. It is no longer a world where only a small number of encrypted communications will exist— everything will be encrypted. It remains to be seen whether encrypted communications can be located at all, and if it will be possible to decrypt terrorist communications in a timely and cost-effective manner so that they are of use.

In the future, finding important communications in a vast sea of encrypted digital bits and packets will become an ever daunting task, and even a brute force approach to searching data collected from the network is not likely to be highly productive. It will be necessary to have some "external" indication of the source or recipient of the data. On the other hand, the closer it is possible to get to the source (or recipient) through various means, the more it becomes possible to narrow the search considerably. In July 2004, for example, the seizure of several computers used by al Qaeda in Pakistan made it possible to examine the hard drives for stored messages and files.

Low-grade encryption is largely a thing of the past, with powerful processors and good encryption algorithms now available. Even where access is technically possible, the resources required for decryption, in terms of computer time and manpower, are significant and will severely limit the amount of access. For the world's intelligence services this is clearly not a happy thought, but the golden era of largely unbridled access is over.

Terrorist access to information

The internet has emerged as the world's greatest source of information, and it is clear that terrorists worldwide regularly use the internet to obtain operational information:

- *Potential targets for future attacks.* It is far more efficient and less costly for a terrorist organization to obtain target information from internet sources rather than costly and risky physical surveillance. Web-based information can be obtained on a very wide range of targets, with easy and anonymous access in most cases.[8] For many potential targets, essential data is often available, and frequently includes photographs, plans and information on hours of operation, as well as other geographic and operational data. Access to most websites is frequently not restricted and is generally free with access granted on an "anonymous" basis. While many sites do in fact record the IP address of incoming users, terrorists can do their web surfing from internet connections that cannot readily be traced to them.

- *Logistics for terrorist operations.* Terrorists increasingly use the internet for operational logistics. Travel tickets, for example, are purchased on the internet, as in the case of the 9/11 hijackers. Websites are now used for all aspects of travel, and these services provide great efficiencies. They also offer a means for terrorists to arrange for travel and other logistics, such as package shipping, in a way that provides substantial anonymity. The only time terrorists become "visible" is when they are screened at an airport, and this may be only a perfunctory check of their ID. False identification, such as driver's licenses, can easily be purchased, and while forged passports may be more difficult to obtain they do not pose a serious problem for terrorists today.[9]

   Since the world will not return to paper tickets, the question becomes one of what can be done to detect terrorists using the internet for such purposes. The "good news" is that such transactions are in digital form, records are generated, databases exist credit cards are generally used and authorities have access to this data. Here new "data mining" techniques may show signs of terrorist activity.[10] Ultimately the results are likely to be mixed, with "sloppy" operations easier to detect than ones where operational security has been well thought out.

- *Technical data for terrorist operations.* There is increasing concern over the amount of technical data freely available on the internet of use to terrorists. While public attention has largely focused on the extremes of nuclear technology and weapons systems development, enormous amounts of information are also available on conventional bombs design, chemical weapons, biological weapons, and other forms of radiological weapons. While a substantial amount of information has come into the public domain, and can be found on the internet, there is a large gulf between accessing information and constructing a working weapon. If a terrorist group were to obtain enough fissile material (enriched uranium or plutonium), the technologies required to actually fabricate a working nuclear weapon are very difficult and complex. Manufacture of chemical and biological agents is nowhere near as difficult as nuclear weapons, and internet data can be of substantial use.

Terrorist websites

The internet and the world wide web (www) protocol enable anyone with internet access to easily search and access a vast array of information.

Information, publications, records and data of every imaginable type are now "online," with massive amounts of additional information being added daily. There are in fact no good estimates of the amount of information on the web, although most users already find it is astounding.

- *Ease of access.* Anyone with a computer and internet access can access data, from any location. Research no longer requires going to a library or anywhere else.
- *Zero marginal cost to users.* The web can be searched endlessly for almost no additional cost.[11] The marginal economics here are truly compelling.
- *Zero marginal cost to publishers.* The economics for web publishers are also compelling, since the costs of establishing a website are nominal and content can be placed on a site for an infinite number of users at no marginal cost. No other media exist with such scale economies.

Terrorist organizations have established an increasing number of websites for the dissemination of information, recruitment of personnel, solicitation of funds, and other purposes:

- *Platform for terrorist propaganda.* Along with satellite television, the web is now a preferred medium for dissemination of news, propaganda, and other materials in the form of hypertext. Postings range from terrorist beheadings of captives to messages from Osama bin Laden.[12] Hezbollah, for example, maintains a multilingual website (www.hizballah.org), and its webmaster, Ali Ayoub, has stated "[Hezbollah] will never give up the internet. We successfully used it in the past when we showed video clips and pictures of the damage caused by Israeli bombings in Lebanon." Hamas has also established a net presence, and is possibly the most prolific of any online organization, with the website (www.palestine-info.org) providing both Arabic and English links to a wide range of Hamas resources, including the Hamas charter, official communiqués and statements of its military wing, the Izz al-Din al-Kassam Brigades.
  Other Islamic terrorist organizations dedicated to the destruction of Israel, such as the Palestinian Islamic Jihad (PIJ), have used the Hezbollah website, as well as other regional media, including the al-Jazeera satellite television network. The Palestinian Authority (PA) and its primary political party (Fatah), as well as the Al-Aqsa Martyrs Brigade, have increasingly used web-based media, such as the online edition of *Al-Hayat Al-Jadida* for statements in support of their activities, which in turn has been supported by the Palestinian Communications Ministry.
- *Platform for terrorist recruitment and fundraising.* The internet is also used to recruit terrorists and operatives. These sites still don't have online applications but have done everything else. The sites focus on Islamic youth and potential recruits, while Israelis, Jews and their supporters are

shown in the worst possible light, and frequently in terms of outrageous falsehoods, and the acts of Islamic martyrs are portrayed in heroic terms. Hezbollah, Hamas, PIJ and others have all used the internet to raise funds and transfer funds. Methods range from transfer of cash by electronic couriers to legitimate bank accounts then used by the terrorists, to financing of a wide range of front organizations, including Islamic charities, and similar institutions for supposed charitable or other humanitarian initiatives.[13] Hezbollah, for example, raises money in the US through the website of the Islamic Resistance Support Association, where users are given the opportunity to fund operations against the "Zionist Enemy."

While some terrorist websites have been shut down, others emerge and are frequently in places beyond the reach of the US and other friendly governments. This leads to the question as to whether the US can or should continue to try to stop such operations, since most web content is free speech, protected by the First Amendment. Aside from the legal issues, it is difficult to shut these sites down on any permanent basis, and their operators have shown they can quickly move websites to alternate locations. Efforts to impede terrorist web operations are likely to be a relatively short-term annoyance and not a permanent solution.

Terrorists and cyber-terrorism

Terrorist abuses of the internet include the potential for attacking the network itself, or "cyber-terrorism." While there is some debate about the cyber-terrorist threat, it often involves writers with no significant technical expertise in this area commenting on matters they don't understand. The discussion arises partly because some see two relatively disjoint phenomena (terrorism and cybernetics) and think they must logically intersect at some point.[14] Currently the major sources of cyber-attacks include:

- *Hackers.* Most attacks have come from "hackers" who are largely bored high-school kids who are malicious, but not terrorists. They don't seek money or destruction, and simply seem to get perverse pleasure from annoying others.
- *Criminals.* As commerce has increasingly moved to networked computers, criminals have moved here as well.
- *Disgruntled employees.* Many cyber-attacks come from former employees (e.g. system administrators) still angry with their employers, not the nation. Most cases involve current or former "insiders" with at least some technical skills, who still have passwords and access to the systems.

Thus far there is limited evidence of any significant terrorist effort to attack the internet. Captured terrorists show some computer skills, but they are mostly personnel supporting operations, and not involved in network attacks.[15] Looking at captured computers and hardware from various terrorists shows that they are avid internet users—not attackers.

There is also mixed evidence as to how serious this threat is, even from experienced attackers. In one US Department of Defense exercise, a team of 35 experts from the National Security Agency were given three months to plan and execute an attack on DoD systems using only public internet access and commonly available hardware and software. The team demonstrated that it could untraceably "bring down" the backbone of DoD command and control, as well as other key systems, although they only gained "root access" to 36 of the 40,000 Pentagon network servers. Evidence like this must be viewed in perspective. First, this exercise was undertaken in 1995—at least two net generations past—and, second, almost three dozen NSA technical experts were employed, a resource no terrorist group is likely to have. Finally, this group only gained access to a minute fraction of the DoD computers interrogated, and the vulnerabilities that permitted this type of access have long since been eliminated. More recently, a 2002 exercise using outside experts concluded that a terrorist cyber-attack was possible, but would require a $200 million investment and five years to accomplish. Again these experts were applying DoD concepts of program management and funding. Presumably a terrorist organization could undertake a less costly program, but it still suggests the magnitude of the problem for any terrorist group.

Weapons and countermeasures

Weapons technologies in use by terrorists range from "stone age" and improvised devices to the much-feared area of "weapons of mass destruction" (WMD). As a practical matter, terrorist groups are non-state actors and lack any significant industrial infrastructure. While they often have funding, they generally seek to acquire weapons and components on world markets.

Conventional arms

For terrorists today, acquisition of common conventional arms does not pose any economic or technical problem. Assault rifles (such as the AK-47 and others), hand grenades, small surface-to-air missiles, and a wide range of other "conventional" arms are widely available and have been acquired by terrorists and others for decades. Indeed, they are openly available in the markets of Afghanistan, Pakistan and elsewhere at low prices. Many in these nations purchase weapons for self-protection if nothing else. It is largely impossible to stem the flow of conventional arms, or control their sale in most Third World nations.

The question arises as to how to control the flow of such weapons into the

United States and other target states. While the US and others increasingly seek domestic gun control, the preferred option for terrorists would most likely be to smuggle them into the country, which raises the issue of inspection of cargo coming into the country, considered at greater length in the section "Non-intrusive inspection."

In Israel, for example, there is an ongoing arms flow to terrorists operating from Gaza and the West Bank. While the Israelis have sought to stem the flow of weapons from Egypt and elsewhere, success has been limited at best, despite substantial resources devoted to the problem, including well-trained forces and skilled intelligence operations. The flow continues through tunnels, smuggling, arrival by sea, and otherwise.

For terrorists in the 9/11 case, or more recently in the UK, conventional arms were not an issue. These groups sought large-scale destruction, and did not seek confrontation with local police or military. Acquiring assault weapons was not consistent with their operation. Indeed, obtaining such weapons could have been counterproductive and alerted the police unnecessarily. The al Qaeda operatives responsible for the 9/11 attacks in the US used no weapons at all, and employed simple knives ("box cutters") to hijack commercial aircraft which were then used as "weapons" to destroy the World Trade Center and part of the Pentagon.

Efforts following 9/11 have focused heavily on hijack prevention, with new systems being installed to detect metal and some explosives in airports. Here it may be worth noting that the current generation of metal detectors is useful against metal knives and similar objects, but does little to detect ceramic knives. So far no terrorist has been known to hijack an aircraft using such a knife, but these are kitchen knives readily available on the internet and in stores. Similarly, current detection systems for explosives are only capable of finding nitrate-based compounds, and are not effective against more modern high explosives, such as C-4 and others, which do not diffuse by-products of manufacture or decomposition—as in the case of older, nitrate-based explosives. New detection technologies are under development and may be deployed in the future, although this will be a very costly enterprise.

Improvised explosive devices

Terrorists frequently employ fairly crude technologies, such as the "improvised explosive devices" or IED—generally a weapon assembled using parts and explosives from other weapons. By one account, the classic IED is a "155 mm howitzer high explosive round with its detonator screwed off and replaced by a blasting cap and wires."[16] This is not always the case. More sophisticated IEDs have been constructed from scavenged arming devices or from purchased electronic components, many through internet sources. The degree of sophistication depends on the ingenuity of the designer, and the tools, as well as the materials available. IEDs are now extremely diverse and may contain various firing devices, plus commercial, military or contrived chemical or explosive fillers. Specific IEDs depend on the local context. Those found in Israel and the occupied territories, for example, differ from those found in Iraq. There are far fewer 155 mm howitzer rounds available to the Palestinians, and the form factor does not fit terrorist operations there. IEDs come in several forms, generally related to the delivery mechanism:

- *Personnel-borne explosives.* An increasingly common tactic has been the use of "suicide bombers," where an explosive charge, accompanied by crude metal shrapnel and a triggering mechanism, are hidden on the person of the bomber. The weight of the explosive charge is limited to a few kilograms, and the surrounding shrapnel has been only a kilogram or so of old screws, bolts and similar metal parts. Most of these attacks have employed using relatively crude devices, often constructed in covert shops, and in a number of cases have failed to explode as intended, leading to the capture of both the terrorists and the weapon. The most devastating results from this type of weapon come from the proximity of the suicide bomber to the proposed targets. Crowded venues, such as buses, cafés, markets, restaurants and other places easily accessed, have been the most common targets thus far, where a relatively small explosion will cause the largest number of casualties.

- *Explosive packages.* Terrorists have also placed explosive charges in parcels and luggage, either carried by "suicide bombers" and set off by the bomber, or left by the bomber and triggered later, with an internal mechanism such as a timer, or remotely through a cell phone or other technique. The form and size of these devices varies greatly, ranging from birdcages, with small amounts of explosive hidden in the bottom, to much larger devices contained in backpacks or other luggage which have been left on subways, buses and other places.[17] Here too the damage varies with the proximity to the target, and where packages can be left in a crowded venue the results can be substantial. Left packages are, of course, subject to detection but then, if the bomber has gone, he remains to leave more packages.

- *Explosives in vehicles.* It is also common for terrorists to place explosive charges in vehicles, such as cars, taxis and small trucks, which can be detonated by the driver as a "suicide bomber," or the vehicle can be left parked and then triggered later with an internal timing device or remotely. The major advantage of using a vehicle is that far more explosive can be employed than in any carried package. In Iraq and elsewhere now, exploding vehicles have been seen with hundreds of kilograms of explosives, and have caused substantial damage and death. The major disadvantage of such weapons is that they can generally not be brought as close to the target population as an individual suicide bomber or a package. In some cases vehicles have been left on streets, and in other cases they have been driven into targeted buildings, achieving greater proximity to the target.

Weapons of mass destruction (WMD)

Of greatest concern in the area of terrorist weapons continues to be the potential acquisition of the category of weapon termed "weapons of mass destruction" or WMD, which commonly refers to chemical, biological or nuclear weapons. The prospect of such weapons being developed in Iraq, for example, moved the US into war there even with faulty intelligence as to their actual development or existence. Current programs in Iran, North Korea and elsewhere are of major concern, as well as the prospects of weapons and fissile materials from existing nuclear powers falling into the hands of terrorists. Most experts agree that such weapons in the hands of terrorists would not be for deterrence, and would most likely be utilized. Since terrorists are generally seeking spectacular results with as much death and destruction as possible, such weapons are an ideal choice, and offer the prospect of far greater damage than the improvised explosive devices used by car bombers, suicide bombers and others. Table 7.1 illustrates the comparative effects of the three major WMD technologies:

- *Chemical weapons.* Military forces have employed chemical weapons to achieve leverage for over a century, with relatively mixed results. In general most chemical weapons lack the lethality of biological weapons, and pose a host of logistic problems. They are difficult to manufacture, transport and deploy. Even when used, atmospheric effects (largely wind) have often caused them to blow back on the user with disastrous results. For

Table 7.1 Comparative WMD effects

Using missile warheads	Area covered (sq. km)	Deaths assuming 3,000–10,000 people per sq. km
Chemical: 300 kg of sarin nerve gas with a density of 70 mg per cubic meter	0.22	60–200
Biological: 30 kg of anthrax spores with a density of 0.1 mg per cubic meter	10	30,000–100,000
Nuclear: 12.5 kt device achieving 5 lbs per in^3 over-pressure	10	23,000–80,000
1.0 megaton hydrogen bomb	190	570,000–1,900,000

Source: Peter Katona and Michael D. Intriligator.

these and other reasons, they are not optimal for terrorist use. This is not to say that some terrorists at some point won't use chemical weapons, but biological and nuclear weapons seem to hold greater promise for them.

- *Biological weapons.* Disease has not always been thought of as a military technology or weapon, although anthrax and several pandemics have probably caused more deaths over time than weapons of any type. Actual "weaponization" of biological toxins and agents has been undertaken by the major powers and others for decades now, and the potential lethality of such an approach has been fairly well identified. Since this presents the greatest practical potential for terrorists, it is considered at greater length in the section "Biological attack and bio-defense."

- *Nuclear and radiological weapons.* Certainly the greatest cause for fear from terrorist attack is the prospect that some terrorists actually come into possession of an operational nuclear weapon. Clearly it is the desire of various terrorist groups to obtain such a nuclear capability, and most analysts agree that they would most likely use any nuclear device for a "spectacular" strike against a target population.[18] The debate currently going on within the academic and intelligence communities largely focuses on the likelihood of a terrorist group actually obtaining an operational nuclear weapon, and to a lesser extent their ability to deploy and detonate such a device in the most destructive way. Among the most vocal here has been Graham Allison, who sees such a scenario as entirely likely, and calls on the US and other nations to react to this prospect.[19]

There are several parts to this scenario which are important to note. First, as Allison and others all agree, no terrorist group is capable of manufacturing fissile material—either weapons-grade uranium or plutonium. Only nations have the industrial base to do so, and thus far there are only nine of these. At the same time, there is a substantial amount of fissile material that has been produced, much of it in the former Soviet Union, that remains unaccounted for and could possibly fall into the hands of a terrorist organization. The prospect of a black market in fissile materials and even nuclear weapons cannot be discounted.[20]

Second, analysts are in some disagreement over how difficult it would be for a terrorist group to fabricate a nuclear weapon with fissile materials in hand. Beyond a critical mass of fissile material, construction of a nuclear weapon requires some other exotic materials, a substantial manufacturing capability, and knowledge about building such a device. Taking the last piece first, most analysts unfortunately agree that the knowledge about how to construct a working nuclear weapon is all too readily available. The other materials needed to construct such a weapon are also available on the world market. The greatest unknown is how sophisticated a facility is needed to do the construction. Pakistan and others have already demonstrated that a Los Alamos type of operation is no longer required. At the same time, a covert

machine shop of the type used to make improvised explosive devices is
ficient. This will remain a subject of debate for some time, and a
o which cannot be discounted.

An additional issue remains as to what size device could be constructed by
a terrorist group, and how it might be deployed against a target. The first
nuclear weapons built by the US were in the order of 10 tons each. During
the Cold War both the US and the Soviet Union developed much smaller
nuclear devices, including nuclear artillery shells, and what have been called
"suitcase" nuclear weapons.[21] While the US was able to shrink the size of
nuclear weapons, the technology needed to do so was not trivial, and is in
fact exceedingly complex. Whether a terrorist group could effectively duplicate this feat in the foreseeable future is a difficult question, about which
there continues to be substantial debate in the technical community.

The final aspect of the current debate revolves around whether terrorists
could bring either fissile materials or an entire weapon into a target nation
such as the US. At the present time, the answer is unfortunately "yes" in both
cases.[22] As discussed at greater length in the section "Non-intrusive inspection," the current generation of detectors will not find packages containing
nuclear materials that have even the most minimal amount of lead shielding.
More effective sensors are still under development. Ultimately, however, the
challenge in this area is along the lines that Allison suggests, that the existing
nuclear powers need to develop and exercise sufficient control over both
fissile materials and nuclear weapons themselves, to ensure these do not fall
into terrorist hands.[23]

One additional category of weapon which has a greater likelihood of being
acquired and used by terrorists—although with far less potential damage—is
a "radiological" or "dirty bomb." In practical terms this is simply a bomb
containing conventional explosives as well as some amount of radioactive
material, generally from a medical or industrial use. It is most unlikely that
this would be highly enriched uranium or plutonium (fissile material), and
would not produce a nuclear detonation. At best, the result would be some
damage, a "mess" and a substantial amount of shock and media attention.
The benefit of such an attack to terrorists lies largely in the spectacular nature
of the attack rather than the likely death toll. What the level of damage and
death toll would be depends on the exact nature of the device and how it was
utilized. Certainly there are no major technical bars to terrorists developing
and using such a device. Clearly the manufacture of IEDs presents no
problem, and a variety of nuclear materials are commercially available.
Combining these in a clandestine lab is not a major undertaking.

Biological attack and bio-defense

As already indicated, the use of pathogens and toxins as a weapon or military technology has taken on increasing appeal over the last several decades.

There is some evidence that such weapons were employed in the Iran–Iraq War, Africa and elsewhere in the Third World in recent years.[24] The appeal of using such bio-weapons is obvious, since the potential lethality of these weapons is enormous, and they can be manufactured or obtained with far less difficulty than nuclear weapons. Table 7.2 illustrates the major impact and lethality from biological attack utilizing three alternative bioagents. In each case the actual amount of bioagent involved is 100 kilograms or less, an amount easily carried in the trunk of a car, and the estimated number of deaths exceeds that experienced in the 9/11 attacks by an order of magnitude.

Table 7.3 again makes the important point that a biological attack holds the potential for far greater lethality than a chemical attack on a target population. The examples used in these scenarios compare 1,000 kilograms of a chemical agent (sarin nerve gas) with 100 kilograms of a bioagent (anthrax spores). Here ten times the physical weight of the chemical agent produces only a small fraction of the casualties estimated for one-tenth the weight of

Table 7.2 Comparative effects of biological agents

Biological agent	Amount released	Estimated damage/ lethality
Anthrax	100 kg spores released over a city the size of Washington, DC	130,000–3 million deaths
Plague	50 kg Y. pestis released over a city of 5 million people	150,000 infected 36,000 deaths
Tularemia	50 kg F. tularensis released over a city of 5 million people	250,000 incapacitated 19,000 deaths

Source: Peter Katona and Michael D. Intriligator.

Table 7.3 Comparative effects of chemical and biological agents

Using 1 aircraft dispensing 1,000 kg of sarin nerve gas or 100 kg of anthrax spores	Area covered (sq. km)	Deaths assuming 3,000–10,000 people per sq. km
Clear sunny day, light breeze		
Sarin nerve gas:	0.74	300–700
Anthrax spores:	46	130,000–460,000
Overcast day/night, moderate wind		
Sarin nerve gas:	0.8	400–800
Anthrax spores:	140	420,000–1,400,000
Clear calm night		
Sarin nerve gas:	7.8	3,000–8,000
Anthrax spores:	300	1–3 million

Source: Peter Katona and Michael D. Intriligator.

the bioagent. At the same time, the bioagent covers a far greater area. Results from both the chemical and biological attacks depend on the extant weather conditions—largely the wind, which influences the location and the extent of the damage.

Both scenarios postulate aircraft delivery of the toxic agent, since it is desirable that the agent not "blow back" on the perpetrator, as was often the case in World War I and others. Clearly terrorists are likely to have access to various small aircraft that can carry 100 kilograms of anything, and the 1,000 kilograms of sarin assumed in Table 7.3.[25]

The first major question here is therefore how difficult is it for a terrorist group to obtain or manufacture significant quantities of bioagents? Certainly it would be much easier for a terrorist group to accomplish this than manufacture an operational nuclear weapon. Indeed, the skills, materials and other resources necessary to manufacture biological agents are nowhere near as difficult as the nuclear weapons case. The various bio-toxins themselves, such as the three illustrated in Table 7.2, are well known. Anthrax, for example, has been around for centuries and is no state secret.[26] The bio-technology by which anthrax spores can be reproduced into the quantities needed for a bio-weapon is neither secret nor exceedingly difficult. In short, it would be possible for a dedicated terrorist group to manufacture several kilograms of anthrax spores for a weapon, either outside the target nation or at a covert facility inside the target nation. The specific laboratory technologies needed are common to the pharmaceutical industry and the dairy industry, and are not the subject of international controls and are in fact readily available on the world market. Common laboratory supplies can be easily obtained from commercial suppliers or from the internet, and are largely uncontrolled, unregulated and unknown. A terrorist group could easily construct a substantial laboratory with equipment and supplies purchased anonymously.

Second, it would be exceedingly difficult to locate and detect such a laboratory. If the equipment and supplies purchased were in fact purchased anonymously, and operational security maintained, there would be few (if any) "signatures" that could be detected by external means. A terrorist group seeking to develop the means to implement a biological attack and then implement such an attack on a target nation would be faced with the decision whether to make the agent "in country" or to manufacture the biological agent in another country and then ship it to the target nation. The answer to this question is not a serious technical or logistics one, and would really depend on the specific group and circumstances. Both are viable options, and need to be considered by counter-terrorist authorities.

Despite the various efforts of law enforcement and other authorities in the US, as well as other advanced target nations, it remains possible to transport almost anything into the country unimpeded. As far as biological agents are concerned, bringing 100 kilograms of a toxin through the border, by any one

of a number of means of transport, remains a relatively trivial m
Indeed, such materials do not emit radioactivity, as do nuclear materia
even the chemical signatures found with some explosives.

Non-intrusive inspection

The economies of the US and most other advanced nations have become
increasingly dependent on imports from abroad, and a transportation infrastructure has evolved by which massive amounts of goods of all types—
including raw materials, agricultural products, and finished goods of all
sorts—arrive daily. In the US, well over half of all imports arrive in cargo
containers, carried by ships, aircraft, rail and truck. Incoming containers are
handled by automated systems, and the relevant federal authorities are
largely dependent on the written declarations of the shippers as to their
contents. As a practical matter, only a small fraction of the cargo containers
that enter the US are inspected at all, and an even smaller number receive
anything that might be called a thorough inspection. By one recent estimate,
less than 10 percent of cargo containers coming into the US are inspected,
and less than 3 percent are "thoroughly" inspected.

For the criminal or the terrorist, the implications are obvious. Since the
likelihood of inspection and cargo interception is minimal, simply accept any
possible chance of loss and ship more. This painful reality has not been lost
on drug dealers. Despite occasional press reports of drug seizures, huge
quantities of illicit drugs continue to come into the country.[27] Seizures by law
enforcement have made no significant impact on the national supply, and
street prices, a good indicator of any supply shortage, have not risen and in
many cases have even fallen due market flooding. As a practical matter it is
simply not possible to physically open and inspect the vast number of containers arriving into the US daily, or inspect vehicles entering the country
through the many legitimate points of entry. Any inspection regime must be
"non-intrusive," employing an imaging or sensor technology of some sort to
detect contraband (be it drugs, explosives, fissile nuclear materials or some
other material) or identify some container, vehicle or individual for further
inspection.

- *Inspection of individuals.* Possibly the easiest inspection regime involves
  the non-intrusive inspection of individuals passing through critical entry
  points, such as those boarding commercial aircraft. Since 9/11, the US
  and most other advanced nations have instituted inspection regimes at
  airports in an effort to thwart further aircraft hijackings. This regime
  generally includes the non-intrusive inspection of luggage, hand luggage,
  other personal articles and some clothing such as shoes and jackets. It
  also includes the passage of individuals through metal detectors, and
  some additional external inspection of laptop computers and cameras.[28]

In addition, a subset of individual passengers are generally selected for a "more thorough" inspection of their luggage and a "pat-down" of their person.

The technologies employed here are fairly crude ones, seeking to image or detect metal objects in luggage and on the bodies of subjects. This approach is based on the fact that most guns and knives (hijack weapons) are made from metal. It assumes that no terrorist would ever acquire a ceramic knife or a gun made from composites. As previously stated, ceramic knives are widely available, and guns made from composites are on the market as well.[29] In addition, the detection programs used to identify explosives are largely focused on nitrate-based explosives, and exclude some of the more advanced highly energetic materials such as C-4 and others. In sum, these "inspection" programs assume that any terrorist will be stupid, and try bringing either metal weapons or old-style explosives through an inspection point. On balance, it is a large ongoing investment based on several assumptions that are highly suspect.

- *Inspection of cargo containers.* As indicated above, non-intrusive inspection of cargo containers is a most difficult problem. At a macro level, it is a critical issue of numbers. Any inspection regime that slows down the throughput at a major container facility threatens the national economy, and will simply not be tolerated.[30] By definition, any inspection regime here must be largely non-intrusive. To the extent that any inspections are performed, they are done by "imaging" with several types of sources, including magnetic and nuclear, with the object being to identify suspect cargo for further inspection. Current systems deployed include older X-ray, as well as more sophisticated radiation sources. Developmental systems, such as one employing pulse fast neutron activation (PFNA), have been tested. While the capabilities of these systems certainly vary, and some are better than others, it is generally the case that the "quality" of imaging depends on either the time available for imaging or the strength of the radiation source.[31] The net result is the same. Most contraband in cargo containers passes through the various ports of entry unimpeded. Systems, technologies and procedures currently in place are totally incapable of stopping this. Arrests and "busts" are little more than media events.
- *Vehicle inspection.* Aside from cargo containers, the US and other nations are highly dependent on cross-border vehicle traffic, for goods and cargo as well as a labor supply. At the US–Mexican border, for example, a large number of cars and trucks transit legitimate border crossings daily with goods and workers. Even cursory inspection of trunks has delayed the border traffic to the point of being unacceptable. The same is true at the US–Canadian border, where, for example, it is largely impossible to search truck traffic coming over the bridge from Canada into the US. To the extent any non-intrusive inspections take place, they are largely

superficial and employ antiquated and inadequate technologies—simple metal detectors, old X-ray and similar devices.

The reality of this situation is not secret, and has not been lost on drug dealers, smugglers or illegal aliens, or other criminals. Reports of this situation are increasingly found in the news media and on the internet. One can only assume that terrorists are reasonably well aware of this situation, and understand that it is possible for them to bring weapons, funds and materials of importance to terrorist operations into virtually any advanced nation. Technologies to combat this reality, either in place today or under development for future deployment, are either inadequate or grossly underfunded. It is entirely possible that target nations, such as the US and the UK, have simply given up and not been willing to admit that this is the case. In either event, counter-terrorist operations must be based on an acceptance of reality, and the fact that non-intrusive inspection is highly unlikely to prevent the entry of anything or anyone into a target nation.

Notes

1 Mark Last and Abraham Kandel (eds.), *Fighting Terror in Cyberspace* (Singapore: World Scientific, 2005).
2 Digital cell phones and PDAs operating on GSM systems are a useful adjunct to the internet, but do not offer the same level of security and capability as the net. Indeed, the increasing ability of the world's intelligence services to access cell phones has not been lost on the terrorists. They continue to use them, but now do so in ways that minimize their vulnerability. If nothing else, they change their SIM cards and related numbers with great frequency to avoid detection!
3 Some of these have actually been quite clever, such as using coded messages on the weather forecasts of local radio stations to pass operational signals.
4 As a practical matter, it has been known for centuries how to generate an unbreakable code for using a "one-time pad." Unfortunately, this process is administratively cumbersome, and not useful for long messages.
5 Following the attack on the La Belle Disco in Germany by Libyan terrorists, President Reagan authorized the release of communications intelligence information collected by the US against Libya as a part of his justification for retaliatory air strikes against Libya. While it served the political purposes of the time, it also served as a "wake-up call" to terrorists that the US was indeed paying considerable attention to their communications, and it was necessary for them to change their mode of operations.
6 Actual programs in this area are highly sensitive, and government officials are extremely reluctant to discuss them. The recent US House and Senate investigations of the intelligence community, as well as the 9/11 Commission Report, have shed a good bit of new light on the extent of these failings.
7 It would be technically, legally and politically difficult to comb, for example, all accounts on AOL, Yahoo or Hotmail seeking some that may have been used by terrorists.
8 It is possible for excessive data searching to set off some alarms. Recently a group of architecture students seeking data for a project involving a new building at Camp David, the US presidential retreat, made a sufficiently large number of web

searches for them to come to the attention of the secret service. This is likely a rare case, involving a large number of search requests against a very high-value target.

9 A senior EU official has recently stated that approximately 25,000 EU passports are listed as lost or stolen, and presumably some significant fraction of these are available on the black market.

10 At the time, the DARPA TIA Program received a significant amount of publicity in the press, much of which was critical and unfavorable to the program, largely on issues related to privacy. As databases grow in number and size, the tension between intelligence, law enforcement and the protection of individual rights to privacy will become one of increasing concern.

11 This is not entirely the case. There are a number of sites that do impose user fees, but these are still relatively limited.

12 The Lebanese terrorist organization Hezbollah, for example, is a model of modern media use. Over the past decade it has developed al-Manar as a source of information on regional events, broadcast by foreign news media and over the internet. See Reuven Paz, "Hizballah Considering Satellite Broadcasts," Mar. 22, 2000, at www.ict.org.il and www.hizballah.org

13 See Rachael Ehrenfeld, *Funding Evil: How Terrorism Is Financed—And How to Stop It* (Chicago, IL: Bonus Books, 2003). See also Reuven Paz, "Targeting Terrorist Financing in the Middle East," Oct. 23, 2000, at www.ict.org.il

14 See Abraham R. Wagner, "Cyber-Terrorism, Evolution and Trends: Relax Chicken Little, The Sky Isn't Falling, or Get a Grip on Reality," *Post Modern Terrorism, Trends, Scenarios and Future Threats* (Herzliya, Israel: ICT, 2003).

15 One good example is the July 2004 capture of an al Qaeda operative in Pakistan who was operating several internet servers, and transferring files brought to him on disk from Osama bin Laden to the servers for transmission.

16 Interview with Sgt. Maj. Willard Wynn (USA). Wynn continues: "The bad guy just hooks these two wires to a battery and bang. It's the big one. It will kill anything in the area."

17 The July 2005 terrorist attacks in London were of this type, with explosive packages left on three subways and one bus, in the first set of strikes. Reports on these strikes indicate that the actual bombs (packages) were constructed on a covert facility in Leeds, driven to London by the bombers by car in a picnic cooler to keep the explosives cool, and then taken to the subways and the bus.

18 There is little doubt in this regard. Terrorist websites, which now number over 300, are overflowing with "doctrine" and objectives about the aim of destroying infidel populations. Were al Qaeda or some other such organization to obtain a nuclear weapon, the same doctrine that guided the 9/11 attacks would likely apply again.

19 See Graham Allison, *Nuclear Terrorism: The Ultimate Preventable Catastrophe* (New York: Times Books, 2004), and Graham Allison, "How to Stop Nuclear Terror," *Foreign Affairs*, Jan./Feb. 2004.

20 At one point recently a terrorist group did purchase what it thought was fissile material from some Russian dealers, although it turned out to be a scam on the part of the dealers, and the shipment contained no real materials.

21 See here.

22 UCLA Chancellor Albert Carnasale, a nuclear expert and former member of the SALT delegation, has remarked that it's easy to smuggle a nuclear weapon into the US—just hide it in a bale of marijuana. Everybody knows it's easy to bring that into the country!

23 Allison summarizes this in terms of his three "NOs"—no loose nukes, no nascent nukes and no new nuclear states.

24 See Anthony H. Cordesman and Abraham R. Wagner, *Lessons of Modern War*, vol. III: *The Iran–Iraq War* (New York: Praeger Publishers, 1985).

25 Ownership, sale and rental of small aircraft, generally referred to in the US as "general aviation," is largely uncontrolled, unregulated and mostly unaccounted for. Getting a small plane is as easy as getting a car, and loading it with a chemical or biological agent would present little problem for any terrorist group.

26 Prior to the end of the Cold War, the Soviet Union was working with several strains of anthrax, about which not a great deal has been made known, but much is in the public domain about the more common strains of this disease.

27 Extensive research into this area has been conducted for close to two decades by the Office of National Drug Control Policy (ONDCP). The Counter-Drug Technology Assessment Center (CTAC) of ONDCP has undertaken research studies as well as several technology pilot programs in the area of non-intrusive inspection, and is probably the leading source of insight into this area.

28 Some facilities still ask laptop computer owners to turn the machines "on" and look through the viewfinders of cameras. Such procedures are apparently based on totally misguided notions that a computer containing a bomb would not turn on, or explosives in a camera would somehow block the viewfinder. Such an approach is, at best, comical.

29 See www.kyoceraadvancedceramics.com/Products/kitchen_faq.htm. Ceramic knives made by Kyocera and others are widely available from knife stores and kitchen supply stores, and over the internet.

30 Note, for example, that some 30 percent of all goods entering the US pass through two ports in the Los Angeles, CA area—the port of Los Angeles and the port of Long Beach. Anything slowing down the movement of goods here would have a rapid and massive impact on the entire US economy.

31 Several technical authorities have said that any system which met both the time and fidelity requirements would most likely melt the contents of the container, or "fry the luggage." Presumably consumers are not interested in either.

Part III

THE CHANGING DYNAMICS OF POST-MODERN TERRORISM

8

RESPONDING TO RELIGIOUS TERRORISM ON A GLOBAL SCALE

Mark Juergensmeyer

Terrorism has been described as "the greatest threat to international security in the twenty-first century."[1] The war against terrorism has been from the point of view of military and diplomatic leaders a kind of global anti-guerrilla war. It has been difficult to fight with weapons designed for warfare that is waged in a more conventional and technological way. Many supporters of al Qaeda and similar movements see these conflicts differently, however—not in military but in theological terms. For them, religious terrorism has been an aspect of cosmic war, one that need not be won in ordinary history, and one in which they are convinced eventually they will triumph.

This means that the usual way of viewing terrorism—as a political and military strategy—is insufficient for understanding and combating religious violence. In my recent book on the subject, *Terror in the Mind of God*, based on interviews with religious activists associated with incidents such as the bombing of the World Trade Center, suicide attacks in Jerusalem and Tel Aviv, and the Tokyo nerve gas assaults, I conclude that many of these acts were not so much strategic as symbolic. Even those that had specific goals were also to some extent forms of "performance violence"—dramatic acts meant to call attention to a vast, albeit hidden, war.[2]

Any government's response to religious terrorism, therefore, will have to take cognizance of this spiritual war and the human aspirations that often accompany it, including the desire for a renewed role for religion in public life. Faced with such motivations, how can authorities appropriately respond? The following is the range of approaches that governments have recently used in countering religious violence and my assessment of their varying degrees of effectiveness:

1 *Destroying violence with violence.* The first scenario is one of a solution forged by force. It encompasses instances in which terrorists are literally killed off or have been forcibly controlled. The destructive strategy can

work only in limited circumstances, when the opponents have been easily identified and—perhaps more important—contained within a specific region. The government of India was able to virtually obliterate the most militant of the Sikh separatists in 1992 in part because it embarked on a ruthless search-and-destroy mission against the activists within the confines of the state of Punjab. In other cases this strategy has backfired and produced more terrorism in response. In Algeria, attempts to eliminate Muslim militants had violent repercussions. In Israel, efforts to destroy Hamas leaders have led to an escalation of violent retaliation. After the US actions in Afghanistan and Iraq following the September 11 attacks, the worldwide incidents of al Qaeda-related terrorism increased.

The war-against-terrorism strategy can be dangerous, in that it can play into the scenario that religious terrorists themselves have constructed: the image of a world at war between secular and religious forces. A belligerent secular enemy has often been just what religious activists have hoped for. In some cases it makes recruitment to their cause easier, for it demonstrates that the secular side can be as brutal as they have portrayed it to be. In Algeria, for instance, when the military junta halted the elections and ran the country with an iron hand, popular support for the Islamic party and violent resistance against the junta escalated as well.

In order for the retaliatory strategy to work, a secular government must be willing to declare a total war against religious terrorism, and wage it over many years, as the Israeli government attempted to do against its terrorist opponents. Even then the prognosis for victory has been good only when the opponents were easily identified and contained within a specific region. Israel's attempts have had only varying degrees of success.

2   *Choking off terrorism through legal means.* Legal means of quelling a religious insurrection have been effective only when the government has had direct legal authority over the group. In Japan, for instance, the government was able not only to bring the Aum Shinrikyo leaders to trial and imprison them, but also to use its legislative and police powers to restrict the movement's activities. In 1999 China outlawed the Falun Gong movement, which it considered dangerous. But as the United States discovered in the case of the Libyan terrorists who allegedly destroyed Pan Am flight 103 over Lockerbie, Scotland, controlling activists in another country—especially an unfriendly one—can be a difficult matter.

3   *Infighting within violent groups.* Activist groups sometimes have destroyed themselves from within. Factions within groups have deemed each other satanic foes, and infighting within some movements has become so severe that they literally killed themselves off, or so weakened their military defenses that their government opponents could handily subdue them. The internal squabbling of various Sikh factions, for instance, made it easier for the movement to be conquered by the Indian government.

4 *Self-destruction*. In some cases, such as the non-terrorist but heavily armed Branch Davidian movement in Waco, the members of the movement have resorted to suicide when they perceived no viable options for the future. In the year before he instigated the nerve gas attack in the Tokyo subways, Shoko Asahara mentioned group suicide as a way out of what he thought was a government conspiracy against his movement.[3] Thus although it is difficult for a government's military power to obliterate a terrorist band, sometimes its own sense of desperation can accomplish the task by moving the cosmic war to the afterlife.

5 *Terrifying terrorists*. In this scenario the threat of violent reprisals or imprisonment so frightens religious activists that they hesitate to act. Though some fringe members of activist groups may have been sobered by such threats, it is doubtful that the "get tough with terrorists" strategy has had much of an effect on the more dedicated members. In the view of most of them, the world is already at war, and they have always expected the enemy to act harshly. They would be puzzled if it did not. In fact, a harsh response from the government might actually encourage the activists, since it helps to confirm their own perception of the world at war between secular and sacred forces.

The case that is sometimes offered as a successful instance of intimidating terrorists is the one involving Libya. In the mid-1980s, Libya was thought to harbor Muslim activists perpetrating a series of acts of international terrorism against the United State. In 1986, the US undertook an air strike against the leader of the country, Muammar Qaddafi, in reprisal. The missiles were aimed at one of his residences, and in fact a member of his family was killed in the attack, but Qaddafi himself survived. Over ten years later there were very few terrorist acts aimed at the United States that were attributed to Libya.

Were the air strikes effective? It is doubtful. Although it is possible that Libya was eventually intimidated by the strikes, the immediate response was quite different. According to the RAND–St. Andrews Chronology of International Terrorism, the number of terrorist incidents linked to Libya and directed against the United States rose in the two years following the US air strikes: 15 in 1987 and eight in 1988.[4] The most devastating terrorist attack against the US to which Libya has been implicated—the tragic explosion of Pan Am 103 over Lockerbie, Scotland, killing all 259 on board—occurred in December 1988.

It is not clear why terrorist attacks from Libya decreased in the years since then. Comments made by Qaddafi in 1998 indicated that the economic sanctions leveled against Libya were much on his mind.[5] He broke off relations with Arab states in a pique of anger after they failed to support the abandonment of the boycott. Perhaps he was eager to normalize relations with other governments for trade reasons as much as any other. In any event, there is no clear evidence that he or any other supporter of international terrorism has been intimidated by America's show of military might.

Libya's agreement to abandon its nuclear program and open its country to inspections in 2004 is sometimes offered as an example of an intimidating result of the US government's 2003 invasion of Iraq. Since the negotiations with Libya stretched back some years before the Iraq war, however, it is likely that Libya's decision was made for reasons of national self-interest that had little to do with the war in Iraq.

6 *Terrorists terrifying themselves.* In some cases terrorists have frightened themselves when the magnitude of their destructive acts has been so enormous that they were shaken into a realistic understanding of what their symbolic violence could in fact produce. After Timothy McVeigh destroyed the Oklahoma City federal building in 1995, the number of violent incidents from Christian militia members in the United States diminished. After Rev. Paul Hill killed abortion clinic staff in 1994, other members of the Army of God said there was no need for further action.[6] In other cases, activists have had an epiphany on their way to committing their deeds of destruction, as former Christian Identity activist Kerry Noble reported. When he was sent to destroy a gay church and its parishioners in Kansas City, the moments in which he sat in the pew before he was to trigger his bomb and depart was an occasion for him to seriously reflect on what his intended act would achieve. "All I could envision was torn bodies, limbs ripped from torsos," Noble recalled.[7] Sobered and shaken, he left the sanctuary with the briefcase containing the bomb still in his hand.

7 *Violence is transformed into political leverage.* This scenario is the reverse of those cases in which terrorism is defeated or defused: it is when the violence is used as leverage in political negotiation and the causes behind the struggle are met. Terrorism, in this way, wins. Yet such compromises are not always accepted gracefully by renegade members of activist movements, who insist on continuing their violent paramilitary campaigns. Public support for these compromise solutions is crucial, since the public can isolate perpetrators of acts of violence and undercut their public support.

Could this solution work in Palestine? When I asked the late Hamas leader, Dr. Abdul Aziz Rantisi, whether Jews and Muslims could live in harmony in the area he described as Palestine, he affirmed that they could—but not under the present arrangement. He could not accept "Israel's sovereignty over Palestinian land," he said.[8] But the two groups could live in peace if the situation were reversed and the land were controlled by Palestinian Arabs. "Jews would be welcomed in our nation," Rantisi explained, adding that he did not hate Jews as such. He pledged not to mistreat them "when we become strong." Needless to say, this has not been a solution enthusiastically embraced by Israel. Given that fact, and realizing that Israel holds a preponderance of military power in the region, could any part of his radical Islamic Palestinian objective be achieved?

The answer is yes, if—and this is a big "if"—Rantisi and his activist colleagues in Hamas could have accepted an incremental or compromise solution. In some cases elsewhere in the world the power accrued through terrorist acts has been converted into bargaining chips for negotiated settlements, and formerly terrorist organizations have been forged into political parties. An example of this process of what might be called the domesticization of violence was the negotiated peace settlement in Northern Ireland. Yet, as the bombing in Omagh in August 1998 revealed, such compromises have not always been accepted. The ideology of cosmic war does not easily submit to compromise.

The approach taken by the opponent—the government or some other enemy in a terrorist struggle—has sometimes made all the difference in a successful transition from violence to the politics of compromise. The attempted resolutions of the Northern Ireland and Palestine conflicts, respectively, were interesting cases in point. In the former case, the British did not blame Sinn Fein for the Omagh violence, and both British and Sinn Fein leaders formed a united front against it. Hence the public perception of Omagh was that of a senseless act, one that was peripheral and counterproductive to the political purposes of the North Irish Catholic community. In Israel, however, when Hamas terrorist activities were renewed after the Peace Accords, Benyamin Netanyahu and other Israeli leaders publicly blamed the Palestinian leader, Yasir Arafat, for the terrorism. Thus, perhaps inadvertently, the Hamas activists were given credibility by Netanyahu by equating them with Arafat, and the legitimacy of the secular Palestinian leader was undercut by blaming him for renegade activists whom he could hardly control. With Arafat weakened and Hamas emboldened by the effect of their acts of terrorism, the violence continued.

Thus a negotiated compromise with activists involved in terrorism is fraught with difficulties. It is a solution that does not always work. A few activists may be appeased, but others may be angered by what they regard as a sell-out of their principles. In the case of Arafat and Hamas, the case was complicated not only by the lack of cooperation from the Israeli side following the elections that brought Netanyahu into power, but also by the intractability of Hamas and its own fears of losing whatever leverage it had gained through its previous tactics. In 1996 some members of the movement advocated a shift of strategy and participation in Palestinian elections as a political party. It was a shift that the leadership of Hamas at that time rejected. One of their concerns was political: they knew that although they might have won in parts of Gaza their numerical support on the West Bank was not sufficient to rout Fatah and the other parties that supported Arafat's Palestinian Authority. Another concern of the Hamas leadership was ideological: once one has entered into the rhetoric of cosmic war, the struggle cannot easily be abandoned without forsaking the will of God. In 2006, however, their political and ideological calculations changed and Hamas roared to electoral victory.

8 *Separating religion from politics.* In this scenario the absolutism of the struggle is defused, and the religious aspects are taken out of politics and retired to moral and metaphysical planes. In some cases a more moderate view of the image of religious warfare has been conceived, one that is deflected away from political and social confrontation. The reforms proposed by Muslim leaders in Iran such as Abdolkarim Soroush exemplify this transition to what has been called the "privatization" of religion.[9] For such a transformation to come about two conditions must be met: members of the activists' religious community have to embrace this moderate form of social struggle as a legitimate representation of cosmic war, and the opponents have to accept it without being threatened by it. Secular authorities can do little about the first criterion, since it requires a transformation of thinking and leadership within the religion itself. But they can effect the second criterion by resisting the temptation to act like an enemy in a cosmic war and being open to a social role for religion on a less violently confrontational level.

Abdolkarim Soroush argued that political involvement was bad for Islam.[10] Soroush made a distinction between ideology and religion, and claimed that Muslim clergy had no business being in politics.[11] Similar statements were made by such moderate Islamic thinkers as Hassan Hanafi in Egypt, Rashid Ghannouchi in Tunisia, and Algeria's Mohammed Arkoun.[12] For them, the image of struggle was largely a spiritual battle, a contest between moral positions rather than between armed enemies. Positions such as these do not require the image of cosmic war to be removed from public life or abandoned altogether. Rather, it is redirected to the battlefield of ideas.

The extreme form of this solution—one in which religion returns to what Jose Casanova describes as its privatization in the post-Enlightenment world—is unlikely, however. Few religious activists are willing to retreat to the time when secular authorities ran the public arena and religion stayed safely contained within the confines of churches, mosques, temples and synagogues. Most religious activists have regarded the cosmic struggle as the very heart of their faith, and have dreamed of restoring religion to what they regard as its rightful position at the center of public consciousness.

9 *Removing the underlying sources of tension.* Many of the issues raised by Osama bin Laden in his 1998 fatwa against the United States are problems that could be alleviated—such as the presence of American

troops in Saudi Arabia and the perception that the US supports only the Israeli side of the Israeli–Palestinian dispute. Though it is unlikely that US policies would ever satisfy Osama bin Laden completely, there is no question that US support for Prime Minister Sharon's strong-handed position in Israel and the invasion and occupation of Iraq have exacerbated tensions between the US and the Middle East, and made bin Laden's extremist rhetoric more palatable to the population of Arab states.

The simple rule of thumb in responding to terrorist acts is to avoid acting like the evil enemy that the activists claim that you are. In other words, one should try to avoid inadvertently supporting their ideology by heavy-handed military tactics that buttress the image of an irrational and evil foe.

When one is treated like an enemy, of course, the temptation to act like an enemy is considerable. This has been especially so when the provocations have been savage. After the attack on the World Trade Center and the Pentagon on September 11, 2001, the US government faced enormous pressure to respond in a forceful way—to retaliate swiftly and strongly to appease its constituency. Understandably, governments cannot afford to let acts of terrorism go unnoticed. Governments must be vigilant in their surveillance of potentially terrorist groups, diligent in their attempts to apprehend those suspected of committing terrorist acts, and swift in bringing them to courts of law.

But the tit-for-tat approach to terrorism has usually failed if for no other reason than that few governments have been willing to sink to the savage levels and adopt the same means of gutter combat as the groups involved in terrorist acts. Moreover, any response to the perpetration of violent acts, even in the form of retaliatory strikes, will enhance the credibility of terrorists within their own community. Supporting moderate leadership within the communities, however, would diminish support for the extremists.

Examples of attempts to foster moderation may be found in the British reactions to the violence of the Irish Republican Army, and at least one moment in the Israeli response to Palestinian activism. When Britain's Prime Minister Tony Blair befriended Gerry Adams, the leader of the IRA's political wing, Sinn Fein, and when Israel's Prime Minister Yitzhak Rabin shook the hand of Palestine's Yasir Arafat, many in these Prime Ministers' respective countries were convinced that they had sold out to terrorists. Within Adams's and Arafat's camps there were those who felt that the Sinn Fein and Palestinian leaders had also abandoned their principles—and the Omagh tragedy and the Hamas suicide bombings were violent expressions of this displeasure. Yet the British and Israeli authorities persevered because they saw that they had the opportunity of supporting a peaceful solution over a violent one, and—for the most part—they continued on a path of reconciliation that rewarded those who favored a transformation from terror to cooperation.

10 *Taking the moral high road and co-opting the terrorists' claim to moral politics.* The most successful solutions are those that have been forged on a moral plane—those that have required the opponents in the conflict to summon at least a minimal level of mutual trust and respect. This respect has been enhanced and the possibilities of a compromise solution enlarged when religious activists perceive governmental authorities as having a moral integrity equal to, or accommodating of, religious values.

In some cases where religious violence has been quelled, religion has literally been subsumed under the aegis of governmental authorities. In Sri Lanka, for instance, the efforts of the government to destroy the Janatha Vimukthi Peramuna (JVP)—the People's Liberation Front, supported by many radical Buddhist monks—were double-pronged. The harsh measures involved tracking down and killing the most dedicated members of the radical movement. The more accommodating measures included efforts to win the support of militant religious leaders. President Ranasinghe Premadasa provided a fund for the financial support of Buddhist schools and social services, and in 1990 he created a Ministry of Buddhist Affairs, naming himself the first Minister. Premadasa formed a council of Buddhist advisors, including Buddhist monks who had been quite critical of the secular government. One of these told me that, after Premadasa's pro-religious measures, the government was finally beginning to "reflect Buddhist values."[13]

In other cases, such as the British response to Irish terrorism, the government's stance in following the rule of law and not overreacting to terrorist provocations demonstrated to both its friends and its foes the government's subscription to moral values. This made it difficult for religious activists—with the possible exception of the Rev. Ian Paisley—to portray it as a satanic enemy. It also increased the possibility of some sort of accommodation with religious activists on both sides of the Northern Ireland dispute—leading to the signing of a peace accord in 1998.

Such measures have not erased all sources of opposition to a government, of course, but at least they have greatly reduced the terrorists' basis of support within their own communities. Since violent religious activists have relied on this support to carry out their ventures and receive approval for them, a diminution of community support has been tantamount to cutting off terrorism's lifeblood.

Governments that chose the other route—abandoning their own democratic principles in response to terrorism—have embarked on perilous journeys. The violence that erupted in Algeria after the military junta annulled elections in 1992 was in part due to the perception that the government had discredited itself. In the eyes of many supporters of the

Islamic Salvation Front the secular leaders had demonstrated that they could not meet the mundane moral standards of secular democracies, much less the presumably higher standards suggested by religion.

It is poignant that the governments of modern nations have so often been accused by religious activists of being morally corrupt. After all, the Enlightenment concepts that launched the modern nation-state did so with a fair amount of moralistic fervor. Jean-Jacques Rousseau coined the term "civil religion" to describe what he regarded as the moral and spiritual foundation based on "the sanctity of the social contract" that was essential for any modern society.[14] As the historian Darrin McMahon has pointed out, even then religious critics of such secular thinkers accused them, perhaps unfairly, of cloaking self-interest in the garb of high-minded abstractions.[15] It is their ability to label secular leaders as hypocrites—and charge secular society with spiritual vacuousness—that has animated religious activists from the time of the Enlightenment down to the present day.

This point was brought home to me in a direct way in a peculiar place—the Federal Penitentiary in Lompoc, California—where a convicted terrorist lectured me on my lack of moral and spiritual purpose. Mahmud Abouhalima, imprisoned for his role in the World Trade Center bombing, accused me and all secularists as being hypocritical. He challenged our dedication to the virtue of tolerance when we could not tolerate religious activists such as himself. He insisted that he knew what people like me lacked: "the soul of religion," he said, "that's what's missing." He went on to say that people in the secular world "are just living day by day, looking for jobs, for money to live." They were living, he said, "like sheep."[16]

Several thoughtful observers of Western society have agreed with Abouhalima's latter point. Perhaps, they suggest, the time has come for religion to re-enter the public arena. A French theorist, Marcel Gauchet, has argued that Western society needs to recover the spiritual roots that it abandoned early in the Enlightenment when it transferred the sense of sacredness from God to the nation.[17] An American theologian, Reinhold Niebuhr, put forward a similar argument many years ago, even though Niebuhr was wary of religion's intrusion into politics. Niebuhr was suspicious of religion's ability to absolutize and moralize political calculations that were done for reasons of self-interest. Yet he could see a political role for what he called the "illusions" of religion in providing the ties that bound people together "in spite of social conflict." He described these as "the peculiar gifts of religion to the human spirit."[18]

I agree with Niebuhr that what religion provides society is not just high-mindedness, but a concern with the quality of life, a goal more ennobling than the simple accretion of power and possessions. It is for this reason that religious rhetoric has entered into political discourse at times like these—the turbulence of an increasingly global post-Cold War world—when the moral and spiritual roots of traditional communities have been challenged or have been in danger of being severed.

At such times, groups have seized on religious ideas to give a profundity and ideological clarity to what in many cases have been real experiences of economic destitution, social oppression, political corruption, and a desperate need for rising above the limitations of modern existence. The image of cosmic struggle has given these bitter experiences meaning, and the involvement in a grand conflict has been for some participants exhilarating, even empowering. Persons and social movements engaged in such a conflict have gained a sense of their own destinies. In such situations, acts of violence, even what appears to us as terrorism, have been viewed by insiders in some cultures of violence as both appropriate and justified.

In responding to religious terrorism, then, those of us who oppose it have to find a way of rejecting the violence without also rejecting the religion. The goal is to affirm the public validity of religious values, even those to which acts of violence have been occasionally and deviously attached. This is not a goal that will be achieved easily or quickly. There are turbulent years ahead in which religious activists will continue to attempt to stake their claims in vicious ways, and in which we will be called upon to exercise a great deal of patience, reason and a continuing subscription to our own moral values in response.

Notes

1 Press conference with U.S. Secretary of State Madeleine Albright, reported on ABC Nightline, Aug. 21, 1998.
2 Mark Juergensmeyer, *Terror in the Mind of God: The Global Rise of Religious Violence* (Berkeley, CA: University of California Press, 2000). This chapter incorporates revised excerpts from the last chapter of the book.
3 Shoko Asahara, in a speech given in April 1994, cited in Ian Reader, *A Poisonous Cocktail? Aum Shinrikyo's Path to Violence* (Copenhagen: Nordic Institute of Asian Studies, 1996), p. 69.
4 Cited in Bruce Hoffman, *Inside Terrorism* (New York: Columbia University Press, 1998), p. 192.
5 Douglas Jehl, "Despite Bluster Qaddafi Weighs Deal," *New York Times*, Nov. 1, 1998, p. 8.
6 Interview with Rev. Michael Bray, Reformation Lutheran Church, Bowie, MD, Apr. 26, 1996.
7 Kerry Noble, *Tabernacle of Hate: Why They Bombed Oklahoma City* (Prescott, Ontario: Voyageur Publishing, 1998), p. 146.
8 Interview with Dr. Abdul Aziz Rantisi, MD, co-founder and political leader of Hamas, Khan Yunis, Gaza, Mar. 1, 1998.
9 Jose Casanova, *Public Religions in the Modern World* (Chicago, IL: University of Chicago Press, 1994), pp. 40ff.

10 Robin Wright, "Islamist's Theory of Relativity," *Los Angeles Times*, Jan. 27, 1995, p. A1.
11 Behrooz Ghamari-Tabrizi, "From Liberation Theology to State Ideology, Modern Conceptions of Islam in Revolutionary Iran: Ali Shari'ati and Abdolkarim Soroush," unpublished article, 1997. See also an article by Robin Wright, "Iran Moves to Stifle Exchange of Re-formist Views," *Los Angeles Times*, Dec. 30, 1995, p. A6.
12 For English translations of some of these writings, see Charles Kurzman (ed.), *Liberal Islam: A Sourcebook* (New York: Oxford University Press, 1998).
13 Interview with the Venerable Palipana Chandananda, Mahanayake, Asigiriya chapter, Sinhalese Buddhist Sangha, in Kandy, Sri Lanka, Jan. 4, 1991.
14 Jean-Jacques Rousseau, *On the Social Contract*, Ch. 8 of Book IV: "On Civil Religion."
15 Darrin McMahon, *Enemies of the Enlightenment: The French Counter-Enlightenment and the Making of Modernity* (New York: Oxford University Press, 2001), Ch. 1.
16 Interview with Mahmud Abouhalima, Sept. 30, 1997.
17 Marcel Gauchet, *The Disenchantment of the World: A Political History of Religion*, trans. from the French by Oscar Burge (Princeton, NJ: Princeton University Press, 1998).
18 Reinhold Niebuhr, *Moral Man and Immoral Society* (New York: Charles Scribner's Sons, 1932), p. 255.

9

TERRORISM IN ALGERIA

The role of the community in combating terrorism

Anneli Botha

Introduction

September 11, 2001 confronted the world with the question: How could this happen to the only remaining superpower, given its advanced technical infrastructure? After a period of retrospection, investigations and an official commission, the American public and the world had to accept that the lack of human intelligence had made the inexpensive and low-technology attack possible. The guidance provided by the United Nations and regional and sub-regional institutions since 9/11 focused the attention of the international community on formal counter-terrorism structures, with specific reference to counter-terrorism legislation and bilateral and multilateral agreements. The primary focus of these initiatives is to criminalize acts of terrorism and to enable governments and their security forces to share information. Despite being a step towards effective regional and international counter-terrorism strategy, a very important national element is excluded from these initiatives: the average citizen whom these formal initiatives seek to protect. Most governments and their security forces consider the "war against terrorism" as their sole mandate. The need for secrecy in intelligence-driven counter-terrorism operations is recognized, but a balanced approach is also necessary. However, the lack of knowledge of the threat and reality of terrorism could equally lead to a sense of exclusion, with subsequent frustration and instability. Important lessons could be learned in ways to implement a more effective strategy in preventing and combating terrorism by appreciating the difference between the Algerian experience, with its confined, immediate threat to life, and the current US-led "war against terrorism." While the many complexities of the Algerian conflict are beyond the scope of this chapter, it is important to recognize the devastating effect of the conflict on the Algerian population as a whole. Community members became actively involved as a result of a feeling of desperation. The Algerian inclusion of its population faced obstacles, as will be explained, but a culture of participation and responsibility that developed during the war for liberation was nurtured.

Historical background

Simplistic analyses date the onset of Islamic-motivated terrorism to the interruption of elections in January 1992 with the banning of the Islamic Salvation Front (FIS). The FIS and its armed wing, the Army of Islamic Salvation (AIS), were initially the primary vehicles for opposition to the military-controlled government, but more extreme splinter groups later emerged in the 1990s. Since then, civilians, government officials and foreigners in the country have been targeted in attacks, indiscriminate bomb explosions and fake roadblocks.

An overview of the conflict is necessary to explain and understand the need, role and function of the "groupes d'autodéfense" (self-defense groups) or "patriotes" (Patriots). The conflict in Algeria developed in four stages:

- *First stage.* Attacks were aimed at security forces and government employees. Islamic rebel groups first targeted military bases and specific individuals who opposed their cause. In 1992 Mohammed Boudiaf, the State Council President, was assassinated while delivering a speech.
- *Second stage.* Attacks were directed at intellectuals, journalists, lawyers, artists and foreigners. Increased violence across the country indicated in 1993 that there had been a major split in the Islamist camp between the FIS and the GIA (Armed Islamic Group). The GIA issued a warning to all foreigners to leave the country or become legitimate targets.
- *Third stage.* Attacks were directed at the general infrastructure of the country, e.g. bridges, schools, railways and the electricity supply.
- *Fourth stage.* Attacks were directed at the entire population.

The fourth stage can be subdivided into two phases, the apex phase and the declining phase.

Apex phase (1995–1997)

This phase, the bloodiest of the conflict, was characterized by collective slaughters that targeted rural and isolated communities. The aim of these attacks was to terrorize and punish the population hostile to the attackers, or those who formerly supported them but had since withdrawn their support, or relatives and current supporters of rival armed groups.[1]

The GIA has been one of the most deadly organizations in its willingness to target civilians. Abu Selman (Farid Hamani) advocated a policy of massacring villagers in December 1996. Since 1996 the "eradicateur faction" of the GIA has undertaken massacres of villages aimed at terrifying the rural population. The GIA undertook its most controversial strategy under the leadership of Antar Zouabri: the slaughtering of villagers by cutting their throats. Mustapha Kamel (Abu Hamza), a veteran of the jihad in Afghanistan who at the time preached at the Grand Mosque in Finsbury Park, London, religiously backed Zitouni and his successor Zouabri. Abu Hamza, under the head of the Religious Affairs Committee, requested that Abu Moundher write the 60-page manifesto entitled "The Sharp Sword." The author explained in Chapter 8 of the manifesto that the GIA had never accused Algerian society as a whole of impiety but that this changed when the lack of commitment to the jihad against the unlawful Algerian government became apparent.[2] This was followed by a fatwa declaring non-members as infidels and thereby permitting their murder. In an underground bulletin distributed in September 1996, Zouabri issued a fatwa entitled "The Great Demarcation" (al-mufassala al-kubra), in which he called the self-defense committees "Harki militias" and "Zeroual's dogs." The religious judgments contained in the fatwa included "application of divine law" to enemies, i.e. putting them to death, the expropriation of their belongings (as ghanima or war booty) and the abduction of their women, who could be treated as sabaya (captives). In June 1997, the same bulletin contained a communiqué from the group's religious interpreter, Assouli Mahfoudh, declaring that the murder of women and children consorting with "enemies of Islam" was lawful and that the innocent among them would be admitted to paradise. His readers were informed that those who had their throats cut in towns and villages were "supporters of the tyrant" (taghout).[3] These texts seem to have set off the massacres, but they also appear to have triggered the defection of several Islamist sections from the central GIA.

The darkest period in the chronology of massacres in Algeria began in the summer of 1997 and lasted until the middle of 1998. For example, during the period between August and September 1997 a series of massacres that claimed the lives of hundreds of people took place in dramatic succession:

21 August	63 villagers were killed in Souhane, south of Algiers.
24 August	29 people were killed in a village near Medea, south of Algiers.
26 August	64 people were killed in Beni Ali, south of Algiers.
28 August	Five children were killed in attacks in west Algeria and 16 people were killed in attacks on villages in the south-west.
29 August	300 people were killed in the Sidi Moussa district south of Algiers.
30 August	In two attacks south of Algiers, 47 people were slashed to death.
4 September	The throats of 22 people were slit in a village in northern Algeria.

5 September	63 people were killed in the Algiers suburb of Beni Messous.
6 September	Nine people were killed in the Saida and Miliana regions of Algeria.
19 September	Seven people were killed in Abouyene, a village on the Algerian–Moroccan border.
21 September	53 people were killed in Beni-Slimane, south of Algiers.
23 September	85 people were killed in an attack by Islamic militants in Baraki.
26 September	15 people were killed in a village south of Algiers.
29 September	48 people were killed in the village of Sidi Serhane, south of Algiers.
29 September	19 people were killed in three separate incidents in suburbs of Algiers.
5 October	The throats of 16 people were slit in Sekmouna, a village south of Algiers.
5 October	A school bus south of Algiers was attacked, which resulted in the killing of 16 schoolchildren and their driver.
11 October	The throats of a family of 11 were slit in a village south of Algiers.
12 October	43 people were killed at a fake roadblock in Sig, west of Algiers.
14 October	54 people were killed on a bus near Sig, west of Algiers.

Massacres continued and the government and military forces seemed unable to protect civilians against the GIA. Subsequently local villagers have been provided with arms to defend themselves and their families against the GIA. The strategy of arming civilians was designed to preserve the extensive urban centers by involving a large part of the population in the war. In the areas around Algiers and Oran, the self-defense groups served as a buffer between the National People's Army (ANP) and the Islamist units. But the strategy backfired: by turning terrorism away from conventional military targets, it unleashed a wave of bloodshed in the villages.[4]

Declining phase (1998–present)

By 1997, the GIA was seriously divided. Support from foreign Islamist groups expected to be sympathetic to their cause had diminished amid accusations that the group was guilty of either un-Islamic slaughter of innocent civilians or conspiracy with secularists in the security services. As a result a number of factions broke away from the GIA, including the Salafist Group for Combat and Preaching (GSPC), which focused their attacks against members of the security forces.

This phase was characterized by the disintegration of terrorist organizations, owing to:

- civil society mobilizing against the terrorists;
- the focusing and enhancement of the democratic process, especially through the setting up of elected institutions;
- a strengthening of counter-terrorist strategy through the creation of Patriot and self-defense groups;
- acquisition of experience in counter-terrorism initiatives from other countries, which facilitated the neutralization of several terrorist groups;
- increase in the possibility of conflict within terrorist groups;
- promulgation of the (Rahma) Clemency Bill under President Liamine Zeroual during 1995–1998 (under which 4,000 repented) followed by Civil Concord in 1999 initiated by President Bouteflika resulting in the repentance of 6,000 terrorists.[5]

A decline in support resulted in less hazardous tactics (for the terrorists) that included car bombs and fake roadblocks. The damages and losses caused by terrorist operations were destructive and had harmful effects. Besides human losses, terrorist action caused important material damage through the destruction of economic structures and school and health facilities under construction.

Civilian rule was ostensibly restored after the election of President Abdelaziz Bouteflika in 1999. Despite a decline in terrorist-related activities as compared to the late 1990s, the conflict in Algeria still has a remarkable impact on its population. During 2003, fewer than 900 people, including 420 radical Islamic extremists, were killed in violence led by or directed against extremists. This is in comparison with a total of 1,400 deaths in 2002 and 1,900 in 2001. Previous official figures recorded 9,418 bomb attacks from 1995 to 2001. Bombings peaked in 1998 with the detonation of 2,864 explosive devices; 245 explosive devices were detonated during 2001, claiming 72 lives.[6]

Patriots

The role of the Patriots during the liberation war

The active role of the community through the Patriots in counter-terrorism operations might be attributed to a very bloody liberation struggle against France. The older generation knew the value of sacrifice, while the younger generation grew up on stories of martyrdom. Algeria was built on an ideology of martyrdom in which the Patriots (also referred to as the war heroes) played a central role. Remarkably, the structuring and conduct of the liberation forces had a direct influence in the activities of the later "terrorists."

Between March and October 1954, the Revolutionary Committee of Unity and Action (Comité Révolutionnaire d'Unité et d'Action, CRUA) organized a military network in Algeria consisting of six military regions. In October 1954, the CRUA renamed itself the National Liberation Front

(Front de Libération Nationale, FLN), which assumed responsibility for the political direction of the revolution,[7] while its military arm, the National Liberation Army (Armée de Libération Nationale, ALN) was responsible for the war of independence within Algeria. The ALN was organized in three separate elements:[8]

1   uniformed fighters known as *mujahedines*;
2   paramilitary auxiliaries known as *moussebilines* who functioned on a part- time basis: the *moussebilines* acted as guides and intelligence agents and served as a reserve which the *mujahedines* could call on in an emergency; and
3   part-time helpers and fighters known as *fedayines*: less organized than the *moussebilines*, they were available to provide strategic support when necessary.

The ALN was ruthless against collaborators or traitors: "A common method of execution was the 'Kabylie Smile', a euphemism for a split throat,"[9] a tactic that was later adopted by the GIA.

Formation of the Patriots

Prior to the decree, since 1993, civilian "self-defense" groups had also been taking part in the conflict based on the principles of the *moussebilines* and *fedayines* during the liberation war. On September 22, 1993, a decree (Décret executif 97–04 fixant les conditions d'exercice de l'action de légitime défense dans un cadre organisé) established the communal guards to participate in missions for the maintenance or restoration of law and order. Legislation authorized each household to have a firearm and ammunition. The self-defense groups operated under the control of the army or the gendarmerie, whichever was nearer. Within the village, Patriots established a committee to be responsible for its activities. Ammunition provided to a household had to be accounted for and was intended to be used only for self-defense. With the introduction of the Patriots, the National People's Army, whose strength was estimated at 140,000 men at the beginning of the conflict, nearly doubled its strength by setting up village militias. These Patriots used by the Ministry of Defense, the Groupes de Légitime Défense (GLD), or legitimate defense groups, and the communal guard companies controlled by the Ministry of the Interior, were organized into 5,500 sections. Together they numbered over 80,000 men, mostly from the 1,541 communities directly affected by violence. Apart from defending villages, some of these units also assisted in the protection of strategic points in the countryside such as dams, power stations and gas pipelines.[10] Those who became members of the Patriots mainly came from the former Mujahedin (former fighters in the war of national liberation), their families and friends, as well as relatives of victims of the terrorism. At the height of the conflict, local communities were encouraged to become part of and participate in Patriot-related activities: state television and the printed press gave coverage of the activities of militias, praising their role in "combating and eradicating terrorism," and even ran "advertising spots" encouraging men to form militias; the slogan was *rijal khuliqu li-l-watan* (Men born for their motherland).[11]

According to Algerian officials, its broad counter-terrorism strategy was aimed at achieving four main objectives:

1 mobilizing the population against terrorist groups through sensitization and information;
2 countering the formation of formal as well as informal support networks;
3 getting terrorists away from cities and urban centers;
4 confining terrorists to uninhabited areas to neutralize them.

Tactical implementation consisted of:

1 securing the urban area and its surroundings by intensifying inquiry activities with the assistance of the Patriots;
2 enhancing cooperation between the four basic units: security forces, gendarmerie, military units of the People's National Army (ANP), and Patriots;
3 protecting social, economic and cultural structures by setting up specific regulations, including the possibility of resorting to private security and watching bodies;
4 securing rural and isolated populations through the creation of self-defense groups working in coordination with proximity police and military deployed units;
5 exerting constant pressure on terrorist groups by carrying out intelligence and search-and-destroy operations.

Since the implementation of this strategy, terrorist activities have substantially decreased in urban areas, where the infiltration attempts of terrorist groups have been eliminated. The terrorist activities only remain in rural zones and are directed against remote farmers or isolated nomads in places where there is no basic terrorism fighting unit. Although the tactical capacity of terrorist groups has declined, its impact is still felt in rural areas.

Defensive role of the Patriots

Since "most of the massacres took place near the capital, Algiers, and in the Blida and Medea regions, in the most heavily militarized part of the country" members of the Patriots began to protect their respective villages,[12] and a number of successes were recorded, especially when comparing the number of casualties in villages with and without Patriots. On July 27, 1997, for example, 53 people were killed and 23 wounded in a massacre near Benimessous (outside Algiers). In contrast, an attack on July 25, 1997 in Ain Khalil (near Tlemcen), killed 12 people and wounded three. The difference between the two attacks was that the latter village had been armed for self-defense: Patriots in the village had been instructed on how to respond if attacked so as to provide time for the security forces to come to their aid.[13] Patriots also provided specific protection: In one instance, four members of local militia were killed while escorting a minibus in a separate car on October 5, 1997. The minibus was transporting children from their village of Sidi Selhane to their school in Bouinan (near Blida). In addition to the four Patriots, 16 children between ages 12 and 15 were also killed in the attack.[14]

Offensive role of the Patriots

Offensive operations included setting up roadblocks and checkpoints, organizing and initiating ambushes, and other anti-terrorist operations.[15] These measures resulted in a plethora of first-hand reports accusing members of the Patriots of being involved in the execution of individuals suspected to belong to or support armed Islamic groups outside the framework of the law: "In interviews with Amnesty International delegates, foreign journalists and on Algerian television, members of the Patriots gave details of how they had ambushed, pursued, tracked down and killed 'terrorists', and of their determination to kill as many 'terrorists' as they could find, so as to 'clean-up' the areas."[16] Given the absence of the rule of law and conduct and an example set by government forces, one might expect that individuals would act outside the parameters of best international practice.

As a result of the Patriots taking the law into their own hands, the United Nations Commission on Human Rights began to question the legitimacy of a transfer of power by the state to private groups and to emphasize the risk to human life and security the exercise of that power entailed.[17] Originally the decree stated that members of the "groups of legitimate defense" could use force and firearms "in case of aggression, or attempted aggression, or in case of duty to assist persons in danger." Based on this provision, human rights organizations and other international observers began to question the legality of offensive operations.

The relationship between security forces and the Patriots

Patriots particularly assisted security forces in identifying suspected members and supporters of terrorists, therefore enhancing the human intelligence capacity of the military. Later in the conflict, this advantage turned out to be a disadvantage—reinforcing the need to keep the identity of individuals providing information secret. The knowledge of those who provided information led to isolation and even physical attacks. It was therefore of concern that those involved in Patriot-related activities became the target of terror operations. Knowing the terrain, Patriots also acted as guides for security forces in counter-terrorism operations.

Amnesty International in its 1997 report complained that, in addition to receiving arms and ammunition from the army and security forces, the Patriots did not appear to be subject to any chain-of-command control or accountability to authorities.[18] Patriots also made the news in a few instances, when accused of being responsible for massacres.[19] This resulted in recommendations that the Algerian authorities need "to bring to justice those responsible; and to take concrete measures to prevent further violations, including disbanding all government-backed militias."[20] The organization also called on armed opposition groups to end the killing and abduction of civilians.

Lessons learned from the Algeria conflict

A distinction exists between civilian participation in initiatives to prevent and combat terrorism and vigilantism. The observation that a more immediate and direct threat of terrorism has a greater possibility of the community taking matters into their own hands could be addressed if there was respect for the rule of law—recognized and honored by both the government and its populace. Oversight by international governmental and non-governmental organizations is essential to ensure that actions from either government or the population involved in counter-terrorism initiatives do not fall outside the framework of the law and that, if either overstepped this boundary, those responsible would be held accountable. Although there is an understandable need to enable the community to protect themselves from acts of terrorism in a post-9/11 strategy, this strategy need not involve physical protection through the use of firearms, but could rather be achieved by equipping the community with knowledge.

It is beyond the focus of this chapter to evaluate the Algerian government's legal response to terrorism, but it should be noted that the example set by the government and its security forces had a direct bearing on the role and conduct of the population in counter-terrorism initiatives. In the search for an effective counter-terrorism strategy, it is important to acknowledge that the Algerian government adopted legislation from 1992 onwards that has contributed to large-scale human rights violations. These violations include arbitrary arrests, secret detention in unofficial centers, widespread use of torture, summary executions, "disappearances," non-observance of the timeframes established for police custody and pre-trial detention, violations of the right to a fair trial, and infringement of the right of association and the right to demonstrate and freedom of the press. In the period before

9/11, these actions were criticized as inhumane and illegitimate under any circumstances. Unfortunately, the same countries that judged the Algerian government for adopting these initiatives often make use of the same tactics in the post-9/11 world.

Acknowledging that security forces will be the primary role-players in counter-terrorism activities, citizens need to be encouraged to support counter-terrorism initiatives, and to accept dual responsibility for their safety and security. This can only be achieved under the following conditions:

- *A culture of participation* is often created and nurtured in the aftermath of a conflict period, or after an act of terrorism, which had not been anticipated, had had an immense impact on society. For Algerians, the concept of the "patriot" is synonymous with patriotism, founded on the participation of ordinary people during the liberation war with France which ended colonialism in that country. As with Algeria, a number of conflicts in Africa resulted in the formal or informal establishment of civilian self-defense units, especially as one or both parties of the conflict began to target civilians and non-combatants. Some observers have welcomed the creation of militias as the sole means of protection in the face of the inability or unwillingness of the security forces to protect the civilian population. Others, however, have opposed it because they believe that the presence of militias contributes to the complexity of the conflict. For example, in Uganda, Local Defense Units were formed to assist government forces in border control against the Lord's Resistance Army (LRA) and the Allied Democratic Forces (ADF). In contrast to Uganda and Algeria, where civilian forces are under the direct or indirect control of the government or its security forces, the independent creation of self-defense militias could make the security situation in this region more precarious. It is an unfortunate reality in a large number of African countries that governments and/or their security forces are unable to protect and ensure the safety of all their citizens. Recognizing the arguments both for and against the formation of civilian self-defense groups, the right to live and to protect prove to be the determining elements. Instead of dismissing this reality, governments could take advantage under specific control measures in the prevention and combating of terrorism. Neither government nor individuals or groups, however legitimate the cause, could utilize a strategy based on the indiscriminate killing of civilians (terrorism). The Algerian experience illustrates the possibility for an escalation should one react in applying the same tactics in the name of countering terrorism when confronted with terrorism.
- *Education.* Being willing and able is not enough. It is essential to educate the populace at all levels from a very young age, not only to participate but also in the way to participate. Since the primary role of the populace will be to act as eyes and ears, they need to be educated on their role,

without overreacting or taking matters into their own hands. The African Union acknowledged the role of citizens in counter-terrorism efforts in the OAU/AU Convention on the Prevention and Combating of Terrorism of 1999. With reference to Article 4(i):

> State Parties . . . shall establish effective co-operation between relevant domestic security officials and services and the citizens of the State Parties in a bid to enhance public awareness of the scourge of terrorist acts and the need to combat such acts, by providing guarantees and incentives that will encourage the population to give information on terrorist acts or other acts which may help to uncover such acts and arrest their perpetrators.

Primary lessons to be learned will be on the nature of terrorism, underlying causes and what to be aware of, without falling into the trap of racial profiling. It is therefore important that the education process be impartial.

> Organize educational programs against acts of terrorism through the educational system and the mass media. As mentioned, the focus of this process should be on education and not sensation.
>
> Encourage civil society to support the effort made by the state in the fight against terrorism and denounce the political forces favorable or neutral to terrorism.
>
> Provoke the upsurge and promotion of new civilization referents to avoid leaving the ideological field free for terrorists.
>
> Develop and conduct repentance programs toward terrorists, taking into account disarmament and social rehabilitation in order to reduce the potentialities of terrorist groups and encourage defection inside them.

All possible measures need to be incorporated (differing from country to country) to ensure that members of the community do not act outside the framework of the law, therefore contributing to an existing sense of instability. Community education is essential to prevent this from occurring.

- *Communication.* Trust in government and its security forces is essential. An open line of communication can only be built on this principle. Coordinated initiatives on the local level would also imply that the government and its security forces improve the way they collect, analyze and utilize information relating to the threat of terrorism, including the current classification system. The over-classifying of information not only hampers interagency and cross-country cooperation, but it also prevents the sharing of credible information with the public.
- *Independent media.* The absence of independent media coverage within

Algeria resulted in the inability to verify the number and identity of those killed or the circumstances in which they were killed. In the absence of accurate and verified information, rumors and speculation have thrived, adding to the confusion and insecurity. Within authoritarian political systems, interest groups and the media are often harassed and even silenced under the auspices of a "campaign against terrorism." Under the renewed emphasis to counter and prevent terrorism, the media in particular have a very important role to play—first as a source of information but also as a "watchdog" against the misuse of power by government and its security forces.

The Algerian experience also presents another important lesson, namely that the decision to adopt a specific counter-terrorism strategy should correspond to, or address, the nature of the threat. A distinction is therefore needed between a possible direct and indirect threat depending on specific information on the nature and target of the threat. Heightened alert is important in the prevention of terrorism, but if used extensively it could easily lead to the opposite effect from that originally intended, namely a sense of insecurity and fear.

The question left is how to stimulate or encourage participation. Returning to a comment made earlier in this chapter: Communities that experienced the physical threat of terrorism will be more receptive to information and be willing to participate and cooperate in counter-terrorism initiatives. In the aftermath of 9/11, a number of programs have been launched where participants were taught how to recognize the "signs of terrorism" and what and how to report to the police and/or other law enforcement officers. In other examples, the Michigan State Police distributed a video called "The Seven Signs of Terrorism" that contained information about terrorists conducting surveillance, acquiring supplies and preparations for an attack.[21] In addition to this broad "education strategy," more specific or organized initiatives were launched, for example the decision to make use of truckers through the Alexandria Virginia-based Highway Watch, initiated by the American Trucking Association in 1998. Funded by Homeland Security since 2002, members have proved to be a valuable source of information on unusual or suspicious activities. The program teaches truckers to spot potential threats and how to prevent terrorists from using their cargo (especially hazardous materials) in terrorist attacks.[22] Within the academic community, colleges and universities throughout the United States have added homeland security programs to their curriculum since 9/11.

Conclusion

Many of us considered the conflict in Algeria, as with so many other conflicts on the continent, to be remote, domestic or even a consequence of their own making. In retrospect, after the senseless killing of thousands of civilians in the most gruesome manner, 9/11 first had to break the barriers for people throughout the world to take note of the plight of so many victims of terrorism in Africa.

A holistic approach that incorporates community involvement and participation proved to be an important element in preventing and combating terrorism in Algeria. While the involvement of local communities is almost common practice in African countries as a result of war, insurgency or terrorism, communities in the United States and other Western countries increasingly recognize the role of the ordinary man on the street in their counter-terrorism strategies. The distribution of information should not only be reactive in informing the public, but also proactive in the equipment of the ordinary citizen to prevent terrorism. In addition, better understanding is needed on how people become involved in terrorism, what motivates them and how governments, law enforcement agencies and ordinary people should react and address these underlying issues. In other words, an inclusive preventive approach is encouraged.

In addition to providing compelling arguments in favor of community involvement and participation in counter-terrorism initiatives, the Algerian conflict also provides examples of the worst-case scenario of community involvement: community members became "legitimate" targets, which resulted in them taking matters into their own hands, therefore becoming "part of the problem." Despite the need to participate, it is equally important to set the parameters for involvement. Community involvement should always remain supportive, though secondary, to the functioning of law enforcement agencies as the primary agents in safety and security. In addition, all initiatives should fall within the framework of the law.

Terrorism in all its manifestations is a direct threat to human security that compels people not only to learn more about the "enemy," but also to participate in ensuring a more secure environment. Ordinary citizens can be a valuable resource to governments throughout the world. In addition to governments' responsibility to protect, the need to know on the part of the public should be coordinated in structured initiatives to prevent and combat terrorism. Despite the initiatives briefly mentioned, much more is still needed.

Notes

1 Amnesty International, "Civilian Population Caught in a Spiral of Violence," Nov. 18, 1997, pp. 7–8.
2 G. Kepel, *Jihad: The Trail of Political Islam* (London: I.B. Tauris, 2003), pp. 268, 411.
3 Ibid.
4 D. Benramdane, "Election Shrouded in Confusion: Algeria Accepts the Unacceptable," Mar. 1999, http://mondediplo.com/1999/03/03algeria

5 Agence France Presse, "Armed Islamists Still Active in Algeria," Sept. 8, 2005, http://www.dailystar.com.lb/
   article.asp?edition_id=10&categ_id=2&article_id=18303
6 Associated Press, "Algerian Study Shows 10,000 Bombs Exploded in the Last Seven Years," Feb. 20, 2002.
7 H.S. Suliman, *The Nationalist Movements in the Maghrib: A Comparative Approach*, Scandinavia Institute for African Studies, Research Report No. 78, 1987, pp. 56–7.
8 E. O'Ballance, *The Algerian Insurrection* (Connecticut: Archon Books, 1967), p. 74.
9 J.J. McGrath, *Comparative Analysis of the National Liberation Movements in Northern Ireland, South Africa and Algeria* (Michigan: Dissertation Information Service, 1991), p. 87.
10 Benramdane, "Election Shrouded in Confusion."
11 Amnesty International, "Civilian Population Caught in a Spiral of Violence," pp. 20–1.
12 Amnesty International, Algeria Report 1998, http://www.amnesty.it
13 United Nations, "Report of the Panel Appointed by the Secretary-General of the United Nations to Gather Information on the Situation in Algeria in Order to Provide the International Community with Greater Clarity on that Situation," http://www.un.org/NewLinks/dpi2007/contents.htm, pp. 16–17.
14 Amnesty International, "Civilian Population Caught in a Spiral of Violence," pp. 17–18.
15 Amnesty International, Algeria Report: UNCHR 50-Years Anniversary, http://www.amnesty.org/ailib/intcam/unchr50/algeria.htm
16 Amnesty International, "Civilian Population Caught in a Spiral of Violence," p. 21.
17 United Nations Commission on Human Rights: Sub-Commission on Prevention of Discrimination and Protection of Minorities, "Question of the Violation of Human Rights and Fundamental Freedoms, including Policies of Racial Discrimination and Segregation and of Apartheid, in all Countries, with Particular Reference to Colonial and Other Dependent Countries and Territories: Report of the Sub-Commission under Commission on Human Rights Resolution 8 (XXIII)," E/CN.4/Sub.2/1998/NGO/28, Aug. 4, 1998.
18 Amnesty International, Algeria Report 1997, http://www.amnesty.it
19 B. Madani, "Algeria: Stronghold of the Pouvior," *Middle East Intelligence Bulletin*, vol. 3, no. 5, May 2001, http://www.meib.org/articles/0105_me1.htm
20 "Truth and Justice Obscured by the Shadow of Impunity," *Algeria Watch*, Nov. 28, 2000, http://www.algeria-watch.de/mrv/mrvrap/aireport2000.htm
21 "Michigan State Police Trains Citizens with Video to Spot Terrorists," Jan. 19, 2005, http://www.wzzm13.com/printfulstory.aspx?storyid=35456
22 "Homeland Security Wants More Truckers," *Tahlequah Daily Press*, Jan. 31, 2005, http://www.tahlequahdailypress.com/articles/2005/01/31/news/top_stories/homeland.prt

10

COOPERATION ISSUES IN THE GLOBAL WAR ON TERROR

Barry Desker and Arabinda Acharya

Introduction

September 11, 2001 was a day of unprecedented shock and suffering in the history of the United States.[1] As the authorities in New York and Washington scrambled together to put the traumatized nation back on its track, the US launched a global war against terrorism. The rest of the world, which had watched the death and destruction at the World Trade Center and the Pentagon with consternation, awe and horror, pledged support to the US to rid the world of terrorism. In fact, in the immediate aftermath of the September 11 terrorist attacks, the sense of vulnerability from the threats of transnational terrorism was almost universal and unprecedented. So also was the desire on the part of the international community to delegitimize terrorism in any of its manifestations. Outlining the strategy for what he called "the first global war of the twenty-first century," President Bush said that this "war on terrorism would be fought on a variety of fronts, in different ways," and that "every means of diplomacy, every tool of intelligence, every instrument of law enforcement, every financial influence would be committed to bring the terrorists to justice."[2] In a similar vein, Paul Wolfowitz, US Deputy Secretary of Defense, said that America's latest war is not about initiating a single military strike or "just simply a matter of capturing people and holding them accountable, but removing the sanctuaries, removing the support systems, ending states who sponsor terrorism. And that's why it has to be a broad and sustained campaign."[3] Simultaneously, the United States also launched a strike on the financial foundation of the global terror networks to starve the terrorists of funding.[4] While leading the coalition in the military front against the terrorists, Washington also took the lead in orchestrating a broad-based coalition of nations to target terrorist finances.

It would be reasonable to assess the outcomes of the US-led coalition's assaults on the terrorist haven in Afghanistan and their state sponsor, the Taliban, as remarkably successful. The Taliban's ouster was accompanied by the disruption of al Qaeda bases, training facilities and other logistical networks not only in Afghanistan but also in many other parts of the world. Many top-ranking leaders of al Qaeda and its associate groups were either killed or captured in almost 102 countries across the globe. It is estimated that about 3,200 out of about 4,000 of the core al Qaeda cadre have been effectively neutralized by the coalition actions.[5] Unprecedented coordination among various countries was successful in rooting the organization out of its stronghold in Afghanistan. Intelligence and information sharing among law enforcement and counter-terrorism agencies prevented a series of planned attacks in many parts of the world. Counter-terrorism cooperation also became more institutionalized with the participation of the United Nations and other regional and multilateral agencies and institutions. According to a US estimate, over 170 nations have participated in the war on terrorism by providing military forces and other support and interdicting terrorist finances. "International organizations are becoming more agile, adapting their structures to meet changing threats."[6]

However, recent incidents suggest that these successes have not been commensurate with the regenerative and adaptive capabilities of al Qaeda. Al Qaeda and associated groups remain resilient enough to continue with their campaign of terror, targeting not only the interests of the United States, but those of its allies and supporters worldwide. They have mutated into new forms and continued to engage in a "jihad against the Jews and the crusaders." On the other hand, however, many obstacles have emerged in counter-terrorism efforts, and the international coalition against terrorism seems to be weakening.

Several factors explain this apparent ineffectiveness and emerging setbacks. At the strategic level, the spirit of cooperation has been undermined by some of the policies of the United States. At a more tactical level, the failure can be attributed to two major factors. One is the failure to understand the nature of the threat, especially the "al Qaeda phenomenon," in its entirety, including the vision, capabilities and acumen and the organizational skills of Osama bin Laden, "terrorism's CEO."[7] Second and more important is the failure to address the core issues that have brought transnational Islamist groups into the center-stage of conflict against the West in the first place, and helped sustain their campaign.

Understanding the threat: the new face of terror

Today, a variety of transnational terrorist groups threaten an unusual range of regimes and interests.[8] Al Qaeda additionally demonstrated how it is possible to use terrorism as a "global instrument" to "compete with and challenge" traditionally organized state power and mobilize new global conflicts.[9] The extraordinarily coordinated and synchronized attacks on the World Trade Center and the Pentagon in the United States on September 11, 2001 demonstrated how the terrorist threats cross national borders and geographical delimitations. The attacks, conducted by 19 suicide hijackers, killed about 3,000 citizens from 78 countries.[10] Suspects in the attack have been rounded up in about 60 nations. Most of the planning and coordination for the attacks took place in countries in Europe and South, Southeast and Central Asia. Similarly, the money for the planning, preparation and execution of the attacks was procured from different sources and moved through diverse means. The terrorists were able to use civilian technology (civilian commercial airliners) to destructive use, which was a manifestation of the degree of their sophistication and professionalism. As Osama bin Laden asserted, September 11 demonstrated how the "West had become the 'weak horse' that could be defied with impunity."[11]

The international strategic environment of the post-Cold War era has transformed the trends and patterns of terrorism. The changes encompass the character, the motivations, the organizational patterns and the support structures as well as the nature and quality of weaponry and training facilities. The organizational pattern of "new terrorism," for example, has become that of leaderless and flattened hierarchies.[12] Within the groups the terrorist cells have grown smaller and hence more amorphous and autonomous. In recent years also, the terrorists have evolved an international matrix of operational, logistical and financial networks. Such networking has been quite significant among the militant Islamist groups, such as al Qaeda, Hamas and Jemaah Islamiyah. The groups have established elaborate transnational affiliations based on religious or ideological affinity and a common hatred of the enemy, mainly identified as the United States. With the gradual demise of state sponsorship, the contemporary terrorist groups have learned to take advantage of the prevailing political and economic conditions, especially opportunities provided by trans-border mobility, advances in communications technologies, etc., to reposition their operational structure. Even the most anti-modernist puritanical movements have embraced modern technology, jet travel, global trading, and finance and instant communications networks to conduct their campaigns.[13] With a combination of decentralized cells operating across the globe, united by religion and ideology, the relatively cheap cost of carrying out attacks and the ability to disguise fund transactions, the "new terrorists" have now emerged as the "harbingers of a new and vastly more threatening terrorism, one that aims to produce casualties on a mass scale."[14]

In recent years, al Qaeda reshuffled the entire understanding and assessment on terrorism by creating a complex "confederation" of militant groups and "aggregating support networks."[15] Al Qaeda and its leader, Osama bin Laden, brought disparate Islamist groups from the Middle East, Asia and the Horn of Africa together by creating a common platform and a common agenda.[16] Al Qaeda's rallying point revolves around the call for universal jihad against the United States, its allies and regimes, including moderate Muslim governments, accused by the group of imposing dysfunctional and immoral ways of life across the globe.[17] They manipulate Islam as a tool of mass mobilization, by extracting prophetic truths from the Koran to show "the inherent incompatibility of modern day concerns with the sacred texts."[18] One of the biggest accomplishments of Osama bin Laden was the effective "melding of the strands of religious fervor, Muslim piety and a profound sense of grievance into a powerful ideological force,"[19] and turning it into an "effective weapon with the technological munificence of modernity."[20] As Peter Bergen puts it, "this grafting of entirely modern sensibilities and techniques to the most radical interpretation of holy war is the hallmark of Bin Laden's network."[21] With a robust and techno-savvy propaganda and communication network, al Qaeda continues to disseminate its campaign of hatred against the West among its supporters and sympathizers. In the context of Iraq especially, this campaign appears to have struck a responsive chord throughout the Muslim world and possibly helped rejuvenate the movement in a more virulent manner.[22]

Thus, even though al Qaeda suffered significant attrition in terms of leadership, facilities for training and finance, it would, as Bruce Hoffman put it, be imprudent to write its obituary just yet.[23] In fact, the optimism expressed after the arrest of Khalid Sheik Mohammad, al Qaeda's chief of operations ("we have got them nailed, we are close to dismantling them"[24]), has now proved to be short-lived, especially in the context of events rapidly unfolding in Iraq.

After it was rudely disrupted from its safe haven in Afghanistan, al Qaeda demonstrated remarkable dexterity in adapting to an environment of borderless existence. In a recent statement, Chilean ambassador Heraldo Munoz, the head of the Security Council committee monitoring United Nations sanctions on al Qaeda, remarked how "Al-Qaeda had been going through a major change since 2001, shifting from a centralised network with a strong hierarchy to a decentralised movement even as it kept growing."[25] Attacks in Tunisia, Pakistan, Bali, Yemen, Mombasa, Riyadh, Istanbul, Madrid and now in Iraq tell us how the movement's strength is not in its location in a defined geographical territory, but in its fluidity and flexibility.[26] It still retains the means and the methods to prosecute its campaign throughout the world. Al Qaeda and Osama bin Laden were able to franchise the cause of Islamic jihad, linking home-grown movements to a global agenda.[27] Besides, the terrorist threats seem to have shifted from groups to a cadre of highly motivated and resourceful individuals such as, for example, Abu Mus'ab al-Zarqawi, now active in Iraq. This was facilitated by a network of sophisticated training facilities that al Qaeda established and maintained. These camps produced most of the world's contemporary high-caliber terrorists. Most of the camp veterans still remain unaccounted for, thus providing a ready pool of well-trained and battle-hardened fighters for al Qaeda to draw its cadre from. Similarly, despite a plethora of measures taken against terrorist financing, the group continues to be robust in managing and manipulating its resources. As the Report of an Independent Task Force Sponsored by the Council of Foreign Relations noted, building al Qaeda's financial support network was Osama bin Laden's foremost accomplishment and the primary source of his personal influence.[28] The group has proven itself more adroit than states at adapting to various control regimes in global financial systems, turning to alternative methods and diversifying financial means to avoid detection. Unfortunately, international efforts to interdict terrorist financing remain insufficiently coordinated.[29]

Thus, the post-September 11 al Qaeda has shown itself to be a remarkably nimble, flexible and adaptive entity.[30] It is too soon to write off either bin Laden or his faithful lieutenants spread across the globe. "Because of what Al Qaeda sees as America's global 'war on Islam,' the movement's sense of commitment and purpose today is arguably greater than ever."[31] Significantly, by forcing national governments and the international community into a reactionary mode of response, bin Laden continues to dominate and dictate the rules of engagement.[32] "Al Qaedaism" has become a movement on its own, and it would not probably much alter the dynamics of militant Islamic terrorist threat even if Osama bin Laden were killed or captured or if al Qaeda were completely decimated.

Indeed, al Qaeda has tapped a radical strain in Islamist ideology, which has been a notable feature of challenges to colonial and post-colonial authorities. In states with Muslim majorities and significant Muslim minorities, there will be resurgent challenges to governments from radical Islamists. The American notion of a "war on terror" is therefore misleading, unless we are thinking of an unending war akin to medieval religious wars with periodic ebbs and flows.[33]

Understanding ideology

President Bush equated the challenges of transnational terrorism to the ideological challenges of the past, namely fascism and totalitarianism trying to impose their radical views through threats and violence. "We see the same intolerance of dissent; the same mad, global ambitions; the same brutal determination to control every life and all of life."[34] In a similar vein, former Prime Minister of Singapore Goh Chok Tong remarked that "militant Islamic terrorism is to the 21st century what communism was to the 20th—a global ideological battle."[35] He urged the international community to fight terrorism with "ideas, not just armies." In a way, the former Prime Minister was reflecting the sentiment of an increasing number of scholars and analysts who take the position that the threat is just not a new form of asymmetric warfare explainable in terms of military or quasi-military doctrine and manageable through the use of force. In many ways the threat encompasses ideological radicalism underlying belief systems among the Muslim community all over the world. Within the ideological spectrum, however, the tendency to treat the threat more as a civilizational conflict and the equating of contemporary terrorism with Islam have become the predominant discourse, especially after September 11. For example, in *The Roots of Muslim Rage* Bernard Lewis wrote how "we are facing a mood and a movement far transcending the level of issues and policies and governments that pursue them."[36] Similarly, Samuel Huntington argued that the present conflict could turn into a "clash of civilizations," one of the cultural conflicts he predicted in his now famous work, *The Clash of Civilizations and the Remaking of World Order* several years ago.[37] While Nicholas Kristof, using Cicero's dictum *oderint, dum, metuant*, urged "let them hate, as long as they fear,"[38] Ralf Peters advised the United States not to "waste an inordinate amount of effort trying to win un-winnable hearts and minds."[39] The temptation to stereotype the conflict in this manner has also taken its strength from the fact that the game the terrorists are in today is almost zero-sum. "There is no room for compromise except as a tactical expedient. America may be the main enemy but it is not the only one. What Osama bin Laden offered Europe (in his April 2004 message) was only a 'truce', not a lasting peace."[40]

However, it a fallacy to accept an uncritical perspective of the inevitability of conflict between Islam and the West in the parameters set by Huntington, Peters, Lewis *et al.* At the same time, though, it will be dangerous to discount the potency of the Islamic religious discourse in fueling the contemporary wave of terrorism. It is rather far too easy to assume what Ann Coutler said: "not all Muslims may be terrorists, but all terrorists are Muslims."[41] As Farish A. Noor puts it, the "Islamic threat must not be taken seriously as an epistemic category"[42] or, as put by Edward Said, "many Muslims cannot be simplified."[43] It is critically important to deconstruct the ideology and discover wherein lies its universalistic appeal, which creates and sustains the psychic tensions that demand a "purge through a spasm of violence."[44] This needs to be part of the overall strategy to mitigate the problems of global terrorism. An overemphasized militaristic approach risks further marginalizing the disaffected and increasing the ranks of the jihadis.

Rather than being a "clash of civilizations," the present conflict is more about a struggle for the soul of Islam within the global Muslim community today. Extremist statements like those calling for jihad against the West have more to do with a bitter struggle now unfolding between moderates and radicals for the hearts and minds of the Muslim community.[45] Many Islamic scholars point to the "moral and ideological crisis" that has beset "the collective Muslim mind."[46] A category of self-appointed defenders of orthodoxy seems to have hijacked some of the key instruments of the ideology, i.e. jihad, fatwa and shariah, to make them serve their politically utilitarian and instrumental purposes.[47]

The current form of radical Islamic thought has got firmly entrenched in the hearts and minds of sizeable pockets of ideologically exclusionist and politically repressed young Muslims throughout the world.[48] This has been made more complicated and difficult by the conditions imposed on the community as it makes the painful transition to modernity. Muslim communities are finding themselves threatened, disadvantaged and marginalized by the processes of globalization, which they see as benefiting the West and harming vast segments of the Muslim world. Political Islam has exacerbated the conflict by transforming economic grievances into a mistrust of Westernization and even into an antagonism to modernity.[49] A result of failed and incomplete modernization, radical Islam has festered in societies where contact with the West has produced more chaos than growth and more uncertainty than wealth.[50] Radical Islam has manipulated the inherent tension between "secularizing, homogenizing and avaricious capitalism and ethnic and religious fundamentalism"[51] to construct and nurture its campaign of hatred against the West. In the context of a growing disillusionment with the conventional modes of political dialogue and negotiation, the radicals have misused the Islamic religious discourse as a framework for a moral or ethical critique of power and recreated a religiocentric viewpoint from which, as they claim, an idealistic pan-Islamic Muslim society can be reconstructed.[52] This form of radical Islam has become immensely appealing, because it purports to explain the loss of values and cultural disorientation facing Muslim societies confronting the challenges of globalization and modernization.[53]

The strategic center of gravity of radical Islam rests in the living, vibrant Muslim ummah or the global Muslim community. The stress on Islamic orthodoxy is bound together with a desire to acquire political space and a resentment of the perceived inferior position of Muslims in a globalized world. As Samuel Huntington wrote, for Islam the problem is about the people who "are convinced of the superiority of their culture and are obsessed with the inferiority of their power."[54] They know "their knowledge is superior, and have the capacity to affect change,"[55] but are overwhelmingly frustrated and angered by the "inadequacy, by the sense of being left out and the sense of being done injustice, to the point of desperation."[56]

Paradoxically, globalization has geared together "modern supranational networks and traditional, even archaic, infra-state forms of relationships (for instance, tribalism or religious schools' networks),"[57] which are now being manipulated by the ummah to keep recreating the hatred against the West and increasing the spectrum of Islam's global identity. In this, al Qaeda has been playing the role of some sort of vanguard of Islamic forces, the revolutionary catalyst. With remarkable sophistication it has managed to harness the Muslim extremist forces to coincide with the zeitgeist of increasing religious orthodoxy and the politicization of the ummah. More than any other leader before him, Osama bin Laden has been able to unify radical Islam and to focus its rage. Osama has always depicted the US as the main Western power—the head of the poisonous snake—threatening the very existence of Islam and the Muslim ummah. This rhetoric of revenge and hate focusing on the US and its allies has found resonance with sizeable pockets of disaffected youth across the globe.[58] Bound together by an increasing hatred against the West, these groups are able to continue with their domestic struggles, but, additionally, are reflecting it through the prism of a global cause and a global purpose, namely the defense of Islam.[59]

Factored into this equation is Osama bin Laden's call for revenge. In an audiotape released on April 7, 2003, bin Laden urged his followers to mount suicide attacks "to avenge the innocent children . . . assassinated in Iraq." He has projected the ruling to kill Americans and their allies as a sacramental obligation and a duty against the enemy that is corrupting the life and the religion.[60] Revenge as a motivation to terror is nothing new. In 1911, Leon Trotsky wrote how terrorism, before it is elevated to the level of a method of political struggles, makes its appearance in the form of individual acts of revenge. Trotsky cited how the flogging of political prisoners "impelled Vera Zasulich to give expression to the general feeling of indignation by an assassination attempt on General Trepov. Her example was imitated in the circles of the revolutionary intelligentsia, who lacked any mass support. What began as an act of unthinking revenge was developed into an entire system in 1879–81."[61] The desire for revenge has also provoked the periodic rounds of "tit-for-tat" killing that characterized much of the terrorism in Northern Ireland. Similarly, the 1985 terrorist bombing of Air India flight 182 was driven by a desire to avenge the honor of Sikhism following the Indian Army's 1984 storming of the Golden Temple in Amritsar in India. The perpetrators were so driven by the need for revenge that they were willing to kill "hundreds of innocent people."[62] Viewed in this context, the September 11 attacks on the United States also appear to have been at least partly motivated by revenge, a desire to kill large numbers of Americans for US actions against the Muslims everywhere.[63] Referring to "more than 80 years" of dispossession, Osama bin Laden, in his first broadcast after the September 11 attacks, said: "What America is tasting now is something insignificant compared to what we have tasted for scores of years. Our nation [the Islamic world] has been tasting this humiliation and this degradation for more than 80 years. Its sons are killed, its blood is shed, and its sanctuaries are attacked."[64] For the likes of Osama bin Laden, the dispossession and humiliation must be redressed through revenge. The enactments occurring in Iraq now follow this trend. As one of the Iraqis himself expressed it:

> For Fallujans it is a shame to have foreigners break down their doors. It is a shame for them to have foreigners stop and search their women. This is a great shame for the whole tribe. It is the duty of that man, and of that tribe, to get revenge on this soldier—to kill that man. Their duty is to attack them, to wash the shame. The shame is a stain, a dirty thing; they have to wash it . . . we cannot sleep until we have revenge. They have to kill soldiers.[65]

The protagonists of radical Islam have come to see the world as satanic, dominated by the forces of imperialism and decadence. The rage and a sense of injustice leading to disappointment, disillusionment and frustration are breeding revengeful followers with extreme positions.[66] This rage is projected on to scapegoats—hence the need to have enemies—and results in violence.[67]

This differentiates today's radical Islamic terrorists from the stereotypes of yesteryear's particularistic and largely secular breed. Today's archetypical Islamic radicals clearly place themselves and their enemies in a theological context. They understand themselves to be fighting on behalf of Islam against the enemies of God, epitomized by the US—the "Hubal of this age" and literally "Satanic," being in league with the Devil.[68] They are not necessarily the marginalized elements in society—ill educated, impoverished, destitute or disenfranchised.[69] Ironically, these new brands of supranational neo-fundamentalists are more a product of contemporary globalization than of the Islamic past.[70] What motivates them is not material deprivation but an all-consuming ideology. They are not just Muslims but also Islamists pursuing goals they consider higher than life itself. More than reacting against Westernization, which they believe "masquerades as globalization and whose chief instruments are the military, cultural, and economic powers of the United States," Muslim anger is being propelled by a vision that treats Islam as the answer to every conceivable problem.[71] And it is not just about releasing built-up frustrations, but also about seeking spiritual answers through violence.[72] Beset by alienation and loneliness and consummated by an intense search for identity, these people have fallen prey to a formalistic understanding of Islam that breeds violent radicalism.[73] While Islam has always been the faith of "the very rich and the very poor," the radical orthodoxy has united "the very angry and the very worried (and eager to channel this anger away towards conflicts with what they perceive the 'Hegemonic' powers)."[74]

The current conflict therefore is not against Islam as a religion or a civilization but rather with what Francis Fukuyama calls Islamo-fascism, that is against a radically intolerant and anti-modernist doctrine.[75] From the point of view of the coalition fighting terrorism, various civilizations identified by Huntington are all on the same side against the forces of terror. From al Qaeda's perspective, this is a conflict between the true followers of God and God's enemies including Muslims who align themselves with the Jews and the crusaders to let them defile the holy lands and places.[76] This conflict nests in the realm of politics and ideology.

Responding to the conflict involves changing minds and winning hearts by addressing the grievances that underlie the call for jihad.[77] In the political spectrum, it behooves the West, especially the US, to convince Muslims that the West is a friend of Islam and prove it through concrete action. There is a need to persuade Muslims that the West harbors no ulterior motive, no desire to subjugate them, as the radical Islamic movement suggests.[78] At the same time the West must assist moderate, progressive Muslim leaders and intellectuals who want Islam to make a successful transition to modernity.[79] For the ideological combat against radical Islamism to be effective, "Muslims must conduct it."[80] It is, therefore, imperative that "moderate Muslims ... reclaim center stage"[81] to undercut the appeal of radical Islam. The Muslim community must rescue key concepts of the ideological discourse from the rigid pedagogical structures that have kept it in static mode and infused it with inherent conservatism. For it is not Islam which obstructs its progress, but its "wrong and rigid interpretations."[82]

The United States, as Singapore's former Prime Minister said, cannot "lead the ideological battle." It has little credibility. The sources of Muslim distrust of the US are complex, and its anger greater today than ever before. Washington has also failed in public diplomacy directed at the Muslim world. "In many instances," American Defense Secretary Donald Rumsfeld admitted, "we are not the best messengers."[83] This has much to do with what many perceive as America's double standards—taking action against Iraq but not against Israel for non-compliance with UN Security Council resolutions. Washington's acquiescence in Israel's disproportionate use of force against the Palestinians and, most recently, Tel Aviv's policy of "targeted assassinations" has furthered the anger and the disappointment.[84] On its own part, according to some observers, the US, obsessed with its own security, has unleashed the passions of its own nationalism in a wave of redemptive global violence abroad, such as in Iraq.[85] Media reports and revelations of mistreatment of Iraqi detainees at the Abu Ghraib prison in Baghdad and the accompanying global revulsion have undermined US claims to the high moral ground.

Understanding the response

Threats posed by the contemporary terrorist groups are global in nature. Hence these threats can only be countered through a global response and strategy, not simply by any single power. Until the September 11 incidents brought the magnitude of the threat to the fore, counter-terrorism had almost always been looked at as a law enforcement problem and left to the initiatives of the individual states. This is, however, not to discount various international initiatives especially at the behest of the United Nations, taken against terrorism. However, the effectiveness of these measures was limited owing to the moral ambivalence among the international community about the general issue of terrorism. This ambivalence was translated into euphemisms such as "One man's terrorist is another's freedom fighter." Until September 11, the attitude of the governments, especially in the United States and the West, was one of indifference to the conflicts in Asia, the Middle East, Africa and Latin America, which had been the primary generators of terrorism.[86] As terrorist groups targeted public places, killing civilians including children in the global South, the West looked the other way, granted asylum to many terrorists and refused to interdict their financial infrastructure and support base abroad, citing human rights concerns, lack of evidence and incompatibility of criminal justice and prison systems.[87] September 11 marked a threshold between "good" and "bad" terrorists. As subsequent investigations revealed, among the perpetrators involved in the entire planning and execution of the attacks were a disturbing number of individuals who were not monitored by their respective countries because they were treated as merely "terrorist supporters," not actual "terrorist operatives." This lack of priority facilitated a permissive operating environment that enabled the terrorists to maintain elaborate support structures. As September 11 decisively demonstrated, the international neglect of conflicts returned to haunt the West with a vengeance.[88]

However, the September 11 incidents significantly changed the interests of actors in international politics, especially involving those who were willing to make exceptions for the "freedom fighters" and those who did not give their opposition to terrorism priority over other foreign policy issues since many did not see themselves as targets of terrorism.[89] New coalitions and alignments emerged against terrorism based not only on American power but on the perceived self-interest of other states as well. There was now an understanding that terrorism affects all and none can achieve its own anti-terrorist objectives without supporting a global effort against terrorism.[90] Much of the rest of the world sought to work more closely with the US against terrorism.[91] Increased coordination in surveillance and intelligence operations and attempts to break terrorist webs and their support structures along with solicitation of active military help (the Philippines, for instance) are the results of shifts in perceptions of interest arising from what Robert Keohane termed "public delegitimation of terrorism."[92]

At the same time, support for international institutions and regimes, especially the United Nations, also grew substantially. There was at least an expectation that UN involvement could elevate the actions from the policy of one country or limited set of countries to a policy endorsed on a global basis. This shift was more pronounced in the case of the US, as, from its perspective, it was both important and necessary for the UN to be the source of collective legitimization for Washington's global campaign against terrorism.[93] Various measures adopted within the span of a few weeks after the September 11 attacks, by both the General Assembly and the Security Council, underscore this depth of shared international commitment to an effective, sustained and multilateral response to the problem of terrorism.[94] The UN mechanism, together with other regional initiatives and the engagement of specialist institutions and agencies, proved invaluable to effectively internationalize counter-terrorist initiatives, especially from the US perspective. It spared Washington prodding its counterparts across the globe to take actions against specific terrorists, and ensured that nations would not have to employ any unilateral, "US-only" model of counter-terrorist strategy.[95]

However, despite an array of measures, a fully articulated and sustained response against transnational terrorism still eludes the international community. Though building coalitions against a common threat has never been new to the world community, the dynamics of the collective response to the new terrorist threats were based largely on the experience of the first Gulf War and, more recently, the Balkans. Unfortunately, however, these were poor guides to the evolution of effective cooperation against terrorism. In both scenarios, the United States played leadership roles. But its allies contributed substantial forces and allowed the use of their territory for force deployments such as in the Gulf, thereby accepting significant political and physical risks. In Bosnia, Kosovo and Macedonia, the coalition activity had a strong institutional basis, through the UN and NATO.[96] But more than four and a half years into the US "war on terror," the international coalition against terrorism remains fragmented with a patchwork of domestic, bilateral and regional efforts.

Many structural factors—lack of capacity, domestic politics, pressures of diplomacy and lack of substantive enforcement capability by the international institutions—could explain why an effective counter-terrorism regime has been slow to evolve. What has been increasingly missing now is the political will on the part of the international community.

One way to explain this inconsistency in the intent and action on the part of the members of the international community is to take a look at the theories of international cooperation. In economic and security affairs, cooperation among states has proved to be "as elusive to realize as to analyze."[97] Cooperation here is described as the "coordinated mutual adjustment of state's policies" undertaken to yield benefits to participants.[98] In the international arena, the need for cooperation is seldom contested irrespective of the participants' world view, as for example cooperating against a common threat even under an anarchic, self-help system. International cooperation is not necessarily based on assumptions of altruism on the part of the individual states. Often, the values and priorities underlying specific-issue areas and expectations of mutual benefit set the context of cooperation. In a system of sovereign nation-states, however, cooperation is "organized horizontally rather than vertically, through the practice of reciprocity."[99]

Possibilities of "cheating"—breach of promise, implying unobserved non-compliance and defection—often inhibit cooperation. Concern for maximizing one's benefits may predispose one to cheat. Similarly, common interest on an issue or against a threat may exist side by side with conflicting interests such as pressures of domestic politics. Each state's pay-offs, its perceptions of balance, its time horizon and expectations about the future are heavily conditioned by its domestic situation.[100] When there are domestic consequences of external conflict that are positively valued, the net cost of conflict is lowered, reducing the appeal of cooperation.[101]

Importantly, unilateral behavior in which actors do not take into account the effects of their actions on others also impedes cooperation. Though such actions may not be directed at reducing the gains of others, these tend not to address the negative consequences for others of one's policies.[102] This is problematic, especially if—as with the US in the "war on terror"—such actors' involvements in a cooperative arrangement are significant. Many analysts would point to the United States' "proclivities toward unilateral multilateralism"[103] and very "aggressive go-it alone" attitude[104] as being responsible for declining support from the rest of the world community in the global "war on terror." This attitude was evident when President Bush said "You are either with us or with the terrorists" and "If you are with the terrorists, you will face the consequences."[105] Similarly, on the eve of the Afghanistan attacks, Condoleezza Rice, the US National Security Advisor, commented that the US did not need a mandate from the UN Security Council to launch reprisals in response to the September 11 attacks. "We will see what further we need to do with the UN. But I do not believe the President . . . needs further authority to act in self-defense."[106] This is reflective of the predominance of the "neo-conservative" discourse in the Bush administration. The so-called "Neocons" disdain the UN and believe that all of Washington's foreign policy objectives—regime changes in rogue states and the democratic transformation of the Middle East—can be achieved by the US alone, with no need for assistance from, or involvement of, the UN.[107]

Linked to the unilateralist tendencies is the perception of the Washington policy makers that the war on terror can be a prescription for global order. This somewhat stems from the legacy of vulnerability that September 11 incidents set up for the Bush administration. Under threat from almost invisible non-state actors with global reach, the Bush administration shifted to a policy of preemptive defense which culminated in Washington's controversial engagement in Iraq. It went on the offensive to forestall or prevent hostile acts by its adversaries and, if necessary, to strike terrorists abroad so as to keep the homeland safe. The overall strategic goals for its war on terror are woven around the key concepts of preventing a "nuclear Pearl Harbor," "forestall[ing] or prevent[ing] hostile acts by our adversaries [and] . . . if necessary, act[ing] preemptively."[108] Translated into strategy, this means aggressive unilateralism, less importance to multilateralism and almost total neglect of international institutions. This strategy nevertheless has put too many issues at stake—nuclear non-proliferation, democratic transformation, regime change and the like.[109] In the words of Colin Powell,

> We fight terrorism because we must, but we seek a better world because we can—because it is our desire and our destiny to do so. This is why we commit ourselves to democracy, development, global public health, and human rights as well as to the pre-requisite of a solid structure for global peace . . . They are our interests, the purpose our power serves.[110]

Another issue involves how the US sought to define the threat of trans-national terrorism from the prism of its own vulnerability. Arguably, for the United States, the epicenter of the threat lies not on its own soil but elsewhere. It therefore seeks to pursue the terrorists "across the geographic spectrum" spanning the entire globe.[111] But its treatment of the regional conflicts does not take into account the political, cultural and historical contexts in which the conflicts themselves are embedded. In the Asia-Pacific context, for example, the problem of terrorism is one of internal domestic challenge which has put the defense of territorial sovereignty and regime survival to severe test.[112] But the purported universality of the campaign against terrorism since September 11 has significantly changed the communal dynamics in many countries where age-old contentions have resurfaced along with new tensions specifically over the role of religion in the public sphere. This has made it difficult for other countries to contribute to the coalition against terrorism on an equal footing.

The logical implications of this tendency could be that Washington was seeking its own security at any cost, even at the risk of making the rest of the world more insecure.[113] It probably believed that "America is so strong, it can safely ignore other nations' national interests and 'go it alone.'"[114] The perceived manifestation of this unilateralism was that, several months into the war on terror, it became clear that "the 'coalition' was an inch deep"; in essence, "most of its members were not being asked to do anything much beyond lining up behind American military action."[115] As dust settled down on the ruins of the World Trade Center and the Pentagon, this attitude was proving to be counterproductive for the United States.

The US engagement in Iraq has seriously jeopardized the global terrorism campaign, creating a major diversion from and a major division among its allies fighting the "war on terror."[116] The war in Iraq had also further radicalized the Islamic world against America. As recent incidents indicate, the magnitude of the resistance in Iraq against US-led forces has completely overturned Washington's strategic calculus for the Middle East, which saw regime change in Iraq as a precursor for a strategic transformation of the Middle East. This has emboldened the terrorists, especially al Qaeda and its leader, Osama bin Laden, to gloat that "The enemies have been stunned by the ferocity of the resistance and they did not enjoy a plain sailing."[117] The Iraqi prison abuse exposé only added to Washington's woes. For this, the United States, in the words of Thomas Friedman, was in "danger of losing America as an instrument of moral authority and inspiration in the world."[118]

While Asian, African and Latin American states have maintained close ties to the United States, there has been a sharpening division in Europe. To the Europeans, Americans appear to be besotted with power and becoming increasingly "overbearing, jingoistic and rash."[119] According to a survey by the Pew Research Center, skepticism about the US's motive in the global anti-terror campaign has led to a growing popular support for disengagement with Washington in foreign and security policy, including on terrorism.[120]

The basic ambivalence to the general issue of terrorism is now beginning to resurface. Debates are emerging about how much of the Islamist terrorism is anti-American and how much is against the West as a whole. Osama bin Laden probably correctly diagnosed the emerging rift when, in a taped message, he sought to isolate the rest of the West, especially the Europeans, from the United States. The growing evidence of disagreement at the political level is having its obvious repercussions on the campaign on various fronts against terrorism, and is making the global counter-terrorism strategy less and less effective.

Conclusion

Iraq is clearly emerging as the new epicenter of transnational terrorism, in much the same way as Afghanistan following the Soviet occupation in 1979. Al Qaeda is propagating the view that the US occupation of Iraq is the manifestation of its evil scheme to "dissolve the Islamic identity in the whole of the Islamic world." Osama bin Laden, in his latest audio message, urged that "this is a rare opportunity ... and a priceless one in its essence, to sharpen the faculties of the Ummah and to break its shackles, in order to storm forward towards the battlefields of Jihad in Iraq, to bury the head of international infidelity."[121] Unfortunately, however, as the ranks and the resolve of the terrorists seem to be on the rise, the well-established mainstays of the global order—the Western alliance, European unity and the United Nations—seem to be cracking under stress.[122]

In a "war on terror," the cardinal question is not necessarily about who wins. The problem of terrorism is likely to persist and is not easily eradicated completely. There is much unfinished business in conceptualizing appropriate responses to the challenge of international terrorism; one might even argue that this work has barely started.[123] The means one uses today, therefore, will shape "the ends one might perhaps reach tomorrow."[124] The war on terror has given the United States a core security interest in the stability of societies. But because of the complexity of the challenges and the changing nature of transnational relations, the US cannot "go it alone." "America must mobilize international coalitions to address shared threats and challenges."[125] The United States must rebuild its relations with the world by matching its military build-up with diplomatic efforts that demonstrate its interest and engagement in the world's problems and making the world comfortable with its power by leading through consensus.[126]

The world, as Singapore's former Prime Minister Goh Chok Tong warned, "stands at a 'turning point' in the war against terror, and a wrong move now can cause a turn for the worse."[127] The war in Iraq demonstrates that a strategy based on armed might alone will not work. Although the use of force may be successful in the short term as a counter-terrorism strategy, it is unlikely that a transnational terrorist network rooted in a millenarian religious ideology can be destroyed through armed combat alone. It is incumbent upon the international community to work together to roll back the threat of radical Islamic terrorism by helping the "the collective Muslim mind" recover from the "moral and ideological crisis" and ensuring the conditions under which Muslims can achieve a balance between personal piety and peace, freedom and prosperity.[128]

Notes

1 The 9/11 Commission Report, Final Report of the National Commission on Terrorist Attacks upon the United States, Official Government Edition (Washington, DC: US Government Printing Office, 2004), p. xv.

2 White House, Office of the Press Secretary, "President Freezes Terrorists' Assets," Sept. 24, 2001, available at http://www.whitehouse.gov/news/releases/2001/09/20010924-4.html

3 United States Department of Defense, "DoD News Briefing: Deputy Secretary Wolfowitz," News Transcript, Sept. 13, 2001, available at http://www.defenselink.mil/news/Sep2001/t09132001_t0913dsd.html

4 White House, Office of the Press Secretary, "President Freezes Terrorists' Assets."

5 Rohan Gunaratna, "The Post Madrid Face of Al Qaeda," Washington Quarterly, vol. 27, no. 3, Summer 2004, p. 93. The figure of 4,000 members comes from al Qaeda detainee debriefs, including the FBI interrogation of Mohommad Mansour Jabarah, Canadian operative of Kuwaiti–Iraqi origin in USA custody since 2002.

6 White House, "Progress Report on the Global War on Terrorism," Sept. 2003, p. 7, available at http://www.whitehouse.gov/homeland/progress/

7 Bruce Hoffman, "The Emergence of New Terrorism," in Andrew Tan and Kumar Ramakrishna (eds.), The New Terrorism, Anatomy, Trends and Counter Strategies (Singapore: Eastern University Press, 2002), p. 35.

8 Ian O. Lesser, "Coalition Dynamics in the War against Terrorism," International Spectator, 2, Oct. 2002, p. 46, available at http://www.pacificcouncil.org/pdfs/lesser.pdf

9 Rohan Gunaratna, Inside al Qaeda: Global Network of Terror (New York: Columbia University Press, 2002), p. 1.

10 US Department of State, "Patterns of Global Terrorism 2001," Washington, DC, May 21, 2002, p. 1, available at http://www.state.gov/s/ct/rls/pgtrpt/

11 Cited in Graham Allison, "Nuclear Terrorism Poses the Gravest Threat Today," Wall Street Journal, July 15, 2003, available at http://www.frontpagemag.com/articles/ReadArticle.asp?ID=8926

12 Kevin A. O'Brien, "Networks, Netwar and Information Age Terrorism," in Andrew Tan and Kumar Ramakrishna (eds.), The New Terrorism, Anatomy, Trends and Counter Strategies (Singapore: Eastern University Press, 2002), p. 90.

13 "Responding to Terrorism: What Role for the United Nations?," International Peace Academy, Oct. 2002, p. 19, available at http://www.ipacademy.org/PDF_Reports/Conference_Report_Terr.pdf

14 Steven Simon and Daniel Benjamin, "America and New Terrorism," Survival, vol. 42, no. 1, Spring 2000, p. 59.

15 Jean-Charles Brisard, "Terrorism Financing: Roots and Trends of Saudi Terrorism Financing," Report prepared for the President of the Security Council, Dec. 19 (Paris: JCB Consulting, 2002), p. 6, available at http://www.nationalreview.com/document/document-un122002.pdf

16 Rohan Gunaratna (ed.), *The Changing Face of Terrorism* (Singapore: Eastern University Press, 2004), p. 14.

17 Peter Chalk, "Al Qaeda and Its Links to Terrorist Groups in Asia," in Andrew Tan and Kumar Ramakrishna (eds.), *The New Terrorism, Anatomy, Trends and Counter Strategies* (Singapore: Eastern University Press, 2002), p. 109.

18 Karim Raslan, "Now a Historic Chance to Welcome Muslims into the System," *International Herald Tribune*, Nov. 27, 2001, available at http://www.asiasource.org/asip/raslan.cfm

19 Bruce Hoffman, "Emergence of New Terrorism," p. 38.

20 Ibid., p. 35.

21 Peter Bergen, *Holy War Inc.: Inside the Secret World of Osama bin Laden* (New York: Free Press, 2001), p. 28.

22 Bruce Hoffman, "Al Qaeda, Trends in Terrorism and Future Potentialities: An Assessment," Paper presented at the RAND Center for Middle East Public Policy and Geneva Center for Security Policy, 3rd Annual Conference: The Middle East after Afghanistan and Iraq, Geneva, Switzerland, May 5, 2003, p. 6.

23 Ibid., p. 3.

24 Unnamed intelligence expert, cited in Dana Priest and Susan Schmidt, "Al Qaeda's Top Primed to Collapse, US Says," *Washington Post*, Mar. 16, 2003.

25 "Al-Qaeda Turning to Crime to Raise Funds," *Straits Times* (Singapore), May 27, 2004.

26 Hoffman, "Al Qaeda, Trends in Terrorism," pp. 15–16.

27 Barry Desker, "Islam in Southeast Asia: The Challenge of Radical Interpretations," *Cambridge Review of International Affairs*, vol. 16, no. 3, Oct. 2003, p. 421.

28 Council on Foreign Relations, "Terrorist Financing: Report of an Independent Task Force Sponsored by the Council on Foreign Relations," Oct. 2002, p. 6, available at http://www.cfr.org/pdf/Terrorist_Financing_TF.pdf

29 Matthew Levitt, "Stemming the Flow of Terrorist Financing: Practical and Conceptual Challenges," *Fletcher Forum of World Affairs*, vol. 27, no. 1, p. 61.

30 Hoffman, "Al Qaeda, Trends in Terrorism," p. 15.

31 Ibid., p. 22.

32 Elena Pavlova, "Terrorism after September 11," in Rohan Gunaratna (ed.), *The Changing Face of Terrorism* (Singapore: Eastern University Press), p. 51.

33 Barry Desker, "Taking the Long View in Countering Terrorism," *Straits Times*, Feb. 20, 2004.

34 Remarks by US President George W. Bush to the Warsaw Conference on Combating Terrorism, White House, Nov. 6, 2001, http://www.whitehouse.gov/news/releases/2001/11/20011106–2.html

35 Goh Chok Tong, "Fight Terror with Ideas, not just Armies," Speech of the Prime Minister of Singapore at the Council on Foreign Relations, Washington, DC, on May 6, 2004, as reproduced by the *Straits Times*, May 7, 2004.

36 Cited in Samuel P. Huntington, *The Clash of Civilizations and the Remaking of World Order* (London: Simon & Schuster, 1996), p. 213.

37 Francis Fukuyama, "History and September 11," in *Worlds in Collision: Terror and the Future of Global Order*, ed. Ken Booth and Tim Dunne (New York: Palgrave Macmillan, 2002), p. 28.

38 "Cicero Was Wrong," *New York Times*, Mar. 12, 2002, available at http://www.nytimes.com/2002/03/12/opinion/12KRIS.html

39 Ralph Peters, *Beyond Terror: Strategy in a Changing World* (Mechanicsburg, PA: Stackpole, 2002), p. 59.

40 Goh Chok Tong, "Fight Terror with Ideas, not just Armies."

41 Cited in Bret T. Saalwaechter, "Militarism, Fear, and New Communism," *Democratic Underground.com*, May 7, 2004, available at http://www.democraticunderground.com/articles/04/05/p/07_fear.html

42 Farish A. Noor, "Globalization, Resistance and the Discursive Politics of Terror," in Andrew Tan and Kumar Ramakrishna (eds.), *The New Terrorism, Anatomy, Trends and Counter Strategies* (Singapore: Eastern University Press, 2002), p. 159.

43 Edward W. Said, "Impossible Histories: Why the Many Islams Cannot Be Simplified," *Harper's*, July 2002, pp. 69–70.

44 Kumar Ramakrishna and Andrew Tan, "The New Terrorism: Diagnosis and Prescriptions," in Andrew Tan and Kumar Ramakrishna (eds.), *The New Terrorism, Anatomy, Trends and Counter Strategies* (Singapore: Eastern University Press, 2002), p. 4.

45 Robert W. Hefner, "September 11 and the Struggle for Islam," Social Science Research Council, available at http://www.ssrc.org/sept11/essays/hefner.htm

46 See Farish A. Noor, *New Voices of Islam* (Leiden: Institute for the Study of Islam in the Modern World, 2002).

47 Farish A. Noor, "The Evolution of 'Jihad' in Islamist Political Discourse: How a Plastic Concept Became Harder," Social Science Research Council, available at http://www.ssrc.org/sept11/essays/noor.htm

48 Barry Desker and Kumar Ramakrishna, "Forging an Indirect Strategy in Southeast Asia," *Washington Quarterly*, vol. 25, no. 2, Spring 2002, p. 166.

49 Timur Kuran, "The Religious Undercurrents of Muslim Economic Grievances," Social Science Research Council, available at http://www.ssrc.org/sept11/essays/kuran.htm

50 Fareed Zakaria, "The Return of History: What September 11 Hath Wrought," in *How Did This Happen?*, ed. James F. Hoge and Giden Rose (New York: Public Affairs, 2001), p. 316.

51 See Benjamin R. Barber, *Jihad vs Mcworld: How Globalism and Tribalism are Reshaping the World* (New York: Ballantine Books, 1996).

52 Noor, "Globalization, Resistance and the Discursive Politics of Terror," p. 161.

53 Fukuyama, "History and September 11," p. 32.

54 Huntington, *Clash of Civilizations*, p. 217.

55 Shibley Telhami, "It's Not about Faith: A Battle for the Soul of the Middle East," *Current History*, vol. 100, no. 650, Dec. 2001, p. 415.

56 Surin Pitsuwan, "Strategic Challenges Facing Islam in Southeast Asia," Lecture delivered at a forum organized by the Institute of Defence and Strategic Studies and the Centre for Contemporary Islamic Studies, Singapore, Nov. 5, 2001.

57 Olivier Roy, "Neo-Fundamentalism," Social Science Research Council, available at http://www.ssrc.org/sept11/essays/roy.htm

58 Barry Desker, "The Jemaah Islamiyah Phenomenon in Singapore," *Contemporary Southeast Asia*, vol. 25, no. 3, 2003, pp. 489–90.

59 Ibid., p. 493.

60 Marc Sageman, *Understanding Terror Networks* (Philadelphia, PA: University of Pennsylvania Press, 2004), p. 19.

61 Leon Trotsky, "Terrorism (1911)," *Education for Socialists*, 6, Socialist Workers Party (Britain), March 1987, http://www.marxists.de/theory/whatis/terror2.htm

62 "Revenge behind A-I Bombing," *Rediff.com*, Apr. 29, 2003, http://www.rediff.com/us/2003/apr/29can.htm

63 "Terrorism: The Problems of Definition," Centre for Defence Information, Aug. 1, 2003, http://www.cdi.org/friendlyversion/printversion.cfm?document ID=1564

64 Cited in Farhang Rajaee, "The Challenges of the Rage of Empowered Dispossessed: The Case of the Muslim World," in *Responding to Terrorism: What Role for the United Nations?*, International Peace Academy, Oct. 2002, http://www.ipacademy.org/PDF_Reports/Conference_Report_Terr.pdf, p. 35.

65 Cited in M. Danner, "Torture and Truth," *New York Review of Books*, June 10, 2004, p. 46.

66 Rajaee, "Challenges of the Rage of Empowered Dispossessed," p. 37.

67 See Vamik Volkan, *The Need to Have Enemies and Allies: From Clinical Practice to International Relationships* (Northvale, NJ: Jason Aronson, 1994).

68 Chris Brown, "Narratives of Religion, Civilization and Modernity," in *Worlds in Collision: Terror and the Future of Global Order*, ed. Ken Booth and Tim Dunne (New York: Palgrave Macmillan, 2002), p. 293.

69 Desker, "The Jemaah Islamiyah Phenomenon in Singapore," p. 502.

70 Roy, "Neo-Fundamentalism."

71 Timur Kuran, "The Religious Undercurrents of Muslim Economic Grievances," Social Science Research Council, available at http://www.ssrc.org/sept11/essays/kuran.htm

72 Desker, "The Jemaah Islamiyah Phenomenon in Singapore," p. 502.

73 Abdurrahaman Wahid, "Best Way to Fight Islamic Extremism," *Sunday Times* (Singapore), Apr. 14, 2002.

74 Pavel K. Baev, "Examining the Terrorism–War Dichotomy in the 'Russia–Chechen' Case," *Contemporary Security Policy*, vol. 24, no. 2, Aug. 2003, p. 21.

75 Fukuyama, "History and September 11," p. 32.

76 Brown, "Narratives of Religion, Civilization and Modernity," p. 296.

77 Michael Mandelbaum, "Diplomacy in Wartime: New Priorities and Alignments," in *How Did This Happen?*, ed. James F. Hoge and Giden Rose (New York: Public Affairs, 2001), p. 263.

78 Desker and Ramakrishna, "Forging an Indirect Strategy in Southeast Asia," p. 168.

79 Ibid., p. 167.

80 Derk Kinnane, "Winning over the Muslim Mind," *National Interest*, 75, Spring 2004, p. 98.

81 Raslan, "Now a Historic Chance to Welcome Muslims into the System."

82 Mahathir Mohamad, "Breaking the Muslim Mindset," *Sunday Times* (Singapore), July 28, 2002.

83 Cited in Kinnane, "Wining over the Muslim Mind," p. 94.

84 Goh Chok Tong, "Fight Terror with Ideas, not just Armies."

85 Neil Smith, "Global Executioner: Scales of Terror," Social Science Research Council, available at http://www.ssrc.org/sept11/essays/nsmith.htm

86 "Terrorist Financing: Report of an Independent Task Force Sponsored by the Council on Foreign Relations," p. 32.

87 Rohan Gunaratna, "Terrorism in Asia before and after 9/11," in *Responding to Terrorism: What Role for the United Nations?*, pp. 32–3.

88 Ibid., p. 32.

89 Robert Keohane, "The Public Delegitimation of Terrorism and Coalition Politics," in *Worlds in Collision: Terror and the Future of Global Order*, ed. Ken Booth and Tim Dunne (New York: Palgrave Macmillan, 2002), p. 141.

90 Ibid., p. 142.
91 Fred Halliday, "A New Global Configuration," in *Worlds in Collision: Terror and the Future of Global Order*, ed. Ken Booth and Tim Dunne (New York: Palgrave Macmillan, 2002), p. 236.
92 Keohane, "Public Delegitimation of Terrorism," p. 144.
93 Ibid., p. 143.
94 "Report of the Policy Working Group on the United Nations and Terrorism," Annex to A/57/273, United Nations, New York, available at http://www.un.org/terrorism/a57273.htm#top
95 E. Anthony Wayne, "International Dimension of Combating the Financing of Terrorism," Testimony to House Committee on International Relations, Subcommittee on International Terrorism, Nonproliferation and Human Rights, US Department of State, Washington, DC, Mar. 26, 2003, http://www.state.gov/e/eb/rls/rm/2003/19113.htm
96 Lesser, "Coalition Dynamics in the War against Terrorism," pp. 44–5.
97 Helen Milner, "International Theories of Cooperation among Nations," *World Politics*, vol. 44, no. 3, Apr. 1992, p. 466.
98 Robert Keohane, "The Analysis of International Regimes: Towards a Europe–American Research Programme," in Volker Rittberger (ed.), *Regime Theory and International Relations* (Oxford: Clarendon Press, 1993), p. 23.
99 Ibid.
100 Milner, "International Theories of Cooperation among Nations," p. 489.
101 Robert Jervis, "Cooperation under Security Dilemma," *World Politics*, vol. 30, no. 2, Jan. 1978, p. 177.
102 Milner, "International Theories of Cooperation among Nations," p. 468.
103 Thomas J. Biersteker, "Targeting Terrorist Finances," in Ken Booth and Tim Dunne (eds.), *Worlds in Collision: Terror and the Future of Global Order* (New York: Palgrave Macmillan, 2002), p. 84.
104 Mwesiga Baregu, "Beyond September 11: Structural Causes of Behavioral Consequences of International Terrorism," in *Responding to Terrorism: What Role for the United Nations?*, p. 42.
105 "Fact Sheet: White House on Halting Financial Flows to Terrorists," *Washington File*, Washington, DC, US Department of State, Nov. 7, 2001, http://usinfo.state.gov/topical/pol/terror/01110711.htm
106 "US Sees No Need for Mandate from UN," *Dispatch Online*, Sept. 24, 2001, http://www.dispatch.co.za/2001/09/24/foreign/AABINLAD.HTM
107 "Neo-Conservatives: What and Who They Are," *Americans against World Empire*, http://www.iconservative.com/neoconservatives.htm
108 "Disarming Iraq: Prospects for Disarmament of Iraq: The UN Route, and UN Weapons Inspections," Second Report of the House of Commons Select Committee on Foreign Affairs, Dec. 19, 2002.
109 James Thomson, "US Interests and the Fate of the Alliance," *Survival*, vol. 45, no. 4, Winter 2003/04, p. 212.
110 Colin Powell, "A Strategy of Partnerships," p. 34.
111 "The National Strategy for Combating Terrorism," White House, Washington, DC, Feb. 2003.
112 Amitav Acharya, "Southeast Asian Security after September 11," *Asia Pacific Foundation of Canada*, Foreign Policy Dialogue Series, 2003.
113 Baregu, "Beyond September 11," p. 42.
114 "Neo-Conservatives: What and Who They Are," *Americans against World Empire*, available at http://www.iraqwar.org/point3.htm

115 Tony Judt, "The War on Terror," *New York Review of Books*, vol. XLVIII, no. 20, Dec. 20, 2001, pp. 102–3.
116 "Fighting a New Cold War," *Business Week*, Mar. 29, 2004, available at http://www.businessweek.com/magazine/content/04_13/b3876020.htm
117 "Transcript of Osama bin Laden Audio Taped Message," CBS News (London Desk), May 6, 2004.
118 Thomas L. Friedman, "Bush Team Must Eat Crow and Do What's Right," *Straits Times*, May 8, 2004.
119 Madeleine Albright, "Farce to Tragedy in One Act of US Folly," *In Review*, Jan. 19, 2004, available at http://www.inreview.com/showthread.php?s=&threadid=15250
120 "A Year after Iraq War: Summary of Findings," Pew Research Center for the People and the Press, Mar. 16, 2004, available at http://people-press.org/reports/display.php3?ReportID=206
121 "Transcript of Osama bin Laden Audio Taped Message."
122 Fareed Zakaria, "The Arrogant Empire," *Newsweek*, Mar. 24, 2003, available at http://www.fareedzakaria.com/articles/newsweek/032403.html
123 Baev, "Examining the Terrorism–War Dichotomy," p. 29.
124 Stanley Hoffman, *Duties beyond Borders: On the Limits and Possibilities of Ethical International Politics* (New York: Syracuse University Press, 1981), p. 197.
125 See Joseph Nye, *The Paradox of American Power: Why the World's Only Superpower Can't Go It Alone* (Oxford: Oxford University Press, 2002).
126 Zakaria, "Arrogant Empire."
127 Cited in "World at Turning Point in Terror War," *Straits Times*, May 7, 2004.
128 Pitsuwan, "Strategic Challenges Facing Islam in Southeast Asia."

Part IV

FUSING TERRORISM PREPAREDNESS AND RESPONSE INTO A GLOBAL NETWORK

11

FROM COMBATING TERRORISM TO THE GLOBAL WAR ON TERROR

Brian M. Jenkins

In 1972, President Nixon ordered the creation of the Cabinet Committee to Combat Terrorism. It was the US government's first formal organizational response to the growing phenomenon of international terrorism. Chaired by the Secretary of State, the committee was charged with ensuring the coordination of intelligence, security, law enforcement and diplomatic efforts. Although the committee itself met only once—other issues, including the Cold War, the escalation of military action against North Vietnam to encourage progress in the stalled negotiations, and a presidential election in November, crowded the agenda—it appointed a working group at the assistant secretary level that, with several changes of name, continued its work over several administrations.

The choice of the term "combat" reflected limited expectations. It implied an enduring task rather than a final victory. Again, the historical context is important. The United States was in the third decade of the Cold War, and it was looking for an acceptable way to end American participation in the Vietnam War—hardly a propitious moment to be starting new wars or implying military victories over amorphous foes. International terrorism was a threat to be contained.

To be sure, the inclusion of the Secretary of Defense on the committee implied the potential use of military force, depending on the specific circumstances. Military analysts spoke about terrorism as "a new mode of conflict," and the term "war" regularly appeared in political speeches of the 1970s. Even the terrorists themselves boasted that their attacks were the beginning of World War III. It was my view that combating terrorism crossed into the domain of war in 1984 when the US Secretary of State spoke of the need for an "active defense," and a new National Security Directive ended debate over whether the United States would use military force in response to terrorism. Military force was used overtly in response to terrorism in 1983, 1986, 1987, 1993 and 1998. It was, however, not until after the terrorist attacks of September 11, 2001 that continuous warfare, as opposed to the occasional use of military force, became a reality.

We are today engaged in a "global war on terror," a war with many objectives: It is a campaign to destroy al Qaeda's terrorist enterprise. It has become inextricably intertwined with the struggle to suppress an insurgency in Iraq. It is a selective effort to defeat other terrorist organizations that threaten the United States and its allies. It is a continuation of ongoing efforts to combat terrorism as a mode of conflict. And it has become conflated with efforts to prevent the proliferation of weapons of mass destruction on the presumption that their development by states will lead inevitably to their acquisition by terrorists.

The evolution of counter-terrorism from 1972 to its current state is the subject of this chapter. However, that development reflects the dynamic nature of the terrorist threat itself, so we begin with a brief review of the trends in terrorism.

The evolution of contemporary terrorism

Terrorism has evolved considerably since its emergence as a global problem in the late 1960s. Most dramatic has been the escalation in terrorist violence. In the 1970s and 1980s, the primary objective of terrorist tactics was to advance political agendas, and this imposed constraints on those who worried about maintaining group cohesion and not alienating perceived constituents. In the 1980s, however, a more ferocious form of terrorism began to appear. Driven increasingly by religious fanaticism, this "new terrorism" became increasingly bloody.

The worst incidents of terrorism in the 1970s caused fatalities in the tens. In the 1980s, fatalities from the worst incidents were measured in the hundreds. By the 1990s, attacks on this scale had become more frequent. On September 11, 2001, fatalities ascended to the thousands—and the toll easily could have been higher. This is an order-of-magnitude increase almost every decade. We now look ahead to plausible scenarios in which tens of thousands could die.

Still, self-imposed constraints have not entirely disappeared. Some terrorists continue to operate below their capacity, and even fanatic jihadists, who believe that God mandates slaughter, debate the utility of indiscriminate violence and the morality of killing women and children.

At the same time, our worst fears about what terrorists might do have not been realized. While the release of nerve gas on Tokyo's subways in 1995 and the anthrax letters sent in 2001 seemed to confirm long-held fears of chemical and biological terrorism, the consequences were less than had been imagined. Twelve persons died in the Tokyo attack, and the anthrax letters killed five. Since then, a number of foiled terrorist plots have involved ricin, which, while deadly, is not useful as a weapon of mass destruction.

Fears of terrorists armed with nuclear weapons attracted official concern in the early 1970s and were heightened by the collapse of the Soviet Union and the exposure of its vast nuclear arsenal to corruption and organized crime. Nuclear terrorism remains a potential threat.

Although precision-guided surface-to-air missiles are widely available and have been in some terrorists' arsenals for years, they have not been used against commercial aircraft outside of conflict zones. Terrorists, insofar as we know, have not attacked agriculture, nor have they attempted to seize or sabotage operating nuclear reactors. No cities have been held hostage.

The ways in which terrorism is financed have also changed. When contemporary terrorism emerged in the late 1960s, rival superpowers and their local allies were willing to support their surrogates in the field, but this support declined with the end of the Cold War, and combatants were obliged to find new ways to finance their operations. They relied more heavily on criminal activities—ransom kidnapping, extortion, protection rackets, petty crime and credit card fraud. The drug traffic offered large-scale returns, which benefited groups in South America, Central Asia and the Middle East. And ethnic diasporas, émigré communities and co-religionists could be tapped for contributions, especially when the practice of charity was "ordered" by a religion.

Some terrorist organizations have acquired considerable skill in moving money through informal banking systems, money order and cash wire services, and regular banks. Although recent efforts by authorities to impede terrorist financing have reduced high-volume transactions, it is not clear that these measures have seriously interrupted terrorist cash flows.

The problem of state-sponsored terrorism has gradually diminished. The end of the Cold War removed strategic interests from what had always been local conflicts. This facilitated the resolution of a few armed struggles, as some guerrillas and some governments made peace. Other groups degenerated into criminal gangs. Still other groups, notably the Afghan veterans of al Qaeda, survived and created new, more autonomous enterprises.

The end of the Cold War also altered the calculations of the handful of states identified by the United States as sponsors of terrorism. For example, it removed Soviet protection from Syria. Iraq, locked in a costly war with Iran, had already sought Western assistance. This was reversed when Iraq invaded Kuwait in 1990, but the muscular US-led response demonstrated a new reality, and the 2003 invasion brought down the regime. Libya continued its support of terrorist operations, but it became more circumspect and eventually sought rapprochement with the West. The Taliban government in Afghanistan was removed.

The decline of state sponsorship, however, made it more difficult to monitor the activities of some inherently dangerous actors. The reduction of material assistance, and in some cases outright abandonment of clandestine combatants, resulted in the loss of influence and intelligence sources within these movements.

Terrorists have evolved new models of organization. They have moved away from hierarchical organizations—miniature armies with little general staffs—toward flatter, more fluid networks. Al Qaeda seems to be one of the first groups to have patterned itself on a lean international business model, hierarchical but not pyramidal, decentralized but linked, able to assemble and allocate resources and coordinate operations, but hard to depict organizationally or to penetrate.

Networks provide numerous operational benefits, because they are adaptive and remarkably resilient. In order to work well, however, networks require strong, commonly held beliefs, a collective vision, some original basis for trust, and excellent communications. Networks have become a subject of intense analysis in the intelligence community. Whether the global jihadist network created by al Qaeda is unique or represents organizational innovations that can be replicated by future groups remains a question.

The creation of online manuals to exhort and instruct would-be terrorists brings us closer to the concept of "leaderless resistance" suggested years ago. In a leaderless-resistance model, self-proclaimed combatants, linked by common beliefs and goals, operate autonomously to wage a common campaign of terrorism. The leaderless-resistance model may be possible for isolated actions, but major operations still require structure, and that in turn requires some basis for trust, which is difficult to establish on the internet.

One of the principal features of contemporary terrorism is its transcendence of national frontiers. Many terrorists have depicted their movements in global terms. In the 1970s, there were terrorist alliances, including those between European and Japanese terrorists and Palestinian groups, which cultivated foreign recruits and relationships to increase their operational capabilities. The brief and ephemeral coalescence of Europe's left-wing terrorist groups led to concerns about "Euro-terrorism," while the IRA and Spain's ETA exchanged technical know-how.

The jihadists who are inspired and guided by al Qaeda's ideology represent a further development. Al Qaeda may properly be called a global insurgency— a phrase that implies both scale and reach. The jihadist terrorist enterprise, with organizational connections in 60 countries, has attracted recruits and funding from all over the world. Al Qaeda has never been a centrally directed, disciplined organization, even when it was operating training camps in Afghanistan. It was capable of centrally directed action and able to assemble resources for specific operations, but it always remained more of a network than a hierarchy. And this capability has proved to be the strength of the jihadist movement, at a time when al Qaeda's historic center is under pressure.

The most significant technological development for terrorists has been the ability to communicate directly with their chosen audiences. By the late 1960s, the spread of television throughout the world, communications satellites, more-portable television cameras, uplinks that connect remote crews with newsrooms and living-room screens, and global news networks made it possible for terrorists to reach audiences worldwide almost instantaneously. By carrying out visually dramatic acts of violence, terrorists could virtually guarantee coverage, thereby inflating both the terror they sought to create and their own importance.

Access to the media brought the kind of coverage the terrorists sought. But the same human drama that drew the news media to acts of terrorism— death, destruction, the suspense of lives hanging in the balance, tragedy and pathos—often obscured the terrorists' political message. Whatever they had to say was often lost in the anguish caused by their attacks.

Today, unedited propaganda can be disseminated on the internet, allowing direct communications between terrorists and their audiences: recruits, sympathizers, broader constituencies, enemy states and groups of citizens who disagree with their own governments' policies. A terrorist incident no longer consists of a bombing followed by a phone call to a wire service. Today's terrorists operate websites, publish online magazines, explain their causes, debate doctrine, and provide instruction in bomb-building, and they can use the same channels to clandestinely communicate with operatives.

Webcasts of "executions" are disgusting, but their appeal to violence-prone young men as fulfillment of revenge fantasies, as vicarious "blooding," and as encouragement to violence should not be underestimated. The terrorists' exploitation of electronic communication points to the democratization of violence—a supply-side push toward individual extremism in which "buyers" can shop for belief systems that will submerge them in a virtual group that encourages and approves their violent behavior.

Of all the tactical adaptations and innovations over the years—from hijackings to ransom kidnappings to seizures of buildings and barricade-and-hostage situations—suicide attacks have had the greatest psychological effect. These first appeared in the 1970s as isolated incidents, but by the 1990s they had become a frequent occurrence, employed by some groups as a strategic capability. Their increase corresponded to an increase in large-scale indiscriminate attacks, both reflections of growing secular and religious fanaticism. Suicide attacks obviate many security measures, but perhaps more significantly they create greater terror.

While we tend to focus on the motivations of the individual suicide attacker, it is the terrorist organization's ability to recruit, persuade, equip and deploy a steady stream of suicidal volunteers that gives the tactic its strength. Suicide attacks have now become the benchmark of commitment.

Terrorist actions have long been effective in achieving tactical results. Through dramatic acts of violence, terrorists have been able to attract attention to themselves, create alarm, cause disruption, provoke crises and oblige governments to divert resources to security and even, occasionally, to make concessions.

Terrorist actions have been less effective at the strategic level. Terrorists have rarely created powerful political movements, nor have they been able to fundamentally alter national policies. Terrorists themselves have brought down no governments. They have, however, been able to upset negotiations and impede the resolution of conflicts. And in countries where democracy was fragile to begin with, terrorists in some cases have provoked the overthrow of government by elements, usually the armed forces, determined to take a stronger line against the terrorists themselves.

The larger-scale attacks of recent years have produced significant economic and political results. The 9/11 attacks not only killed nearly 3,000 people, but they also caused between $40 billion and $80 billion in insured damages, and further business losses are estimated to be in the high hundreds of billions. Moreover, the attacks had a profound effect on US policies, as the "global war on terror" became the framework for American foreign policy. The 9/11 attacks provoked two invasions and led to significant reorganizations in the US government.

But terrorists have yet to achieve their own stated long-range goals anywhere. The South American urban guerrilla groups that initiated a wave of kidnappings and bombings in the 1970s were wiped out in a few years, having achieved no political result beyond provoking brutal repression. The movements launched by their counterparts in Europe, Japan and North America were suppressed years ago. The IRA has made its peace and is no longer a fighting force. It now pursues its objectives politically. In Europe, only Spain's ETA, now well into its fourth decade, fights on, no closer to its goal of an independent Basque state.

It cannot be denied that acts of terrorism kept the hopes of a Palestinian resistance alive when Arab governments were defeated on the battlefield, that terrorism galvanized the Palestinian population and contributed to the concept of a Palestinian state, and that terrorism helped persuade Israel to withdraw from Gaza. But a final verdict must await history.

Today's jihadists speak of driving the infidels out of the Middle East, of then toppling the weakened apostate regimes that now depend on Western support, and of going on to destroy Israel and ultimately re-establish the Caliphate. Assaults on the scale of the 9/11 attacks achieved significant results, and the jihadists are determined to carry out even more-ambitious attacks. Destroying their terrorist enterprise will take years, during which there will inevitably be surprises. Further terrorist attacks are likely, but, again, the verdict on their success must await the judgment of future historians.

Terrorism's future course

Prophets in the realm of terrorism are likely to suffer Cassandra's fate: Their predictions may be accurate, but they will not be listened to. In the early 1970s, any predictions of suicide truck bombs, sarin on the subways, hijacked airliners crashing into skyscrapers—all of which actually occurred in the

186

following three decades—would have been dismissed as the stuff of novels. This has now changed. The attacks of 9/11 redefined plausibility. Now, virtually no threat can be dismissed. Worst-case scenarios have become presumptions.

Terrorists fuel our fears with their own fantasies. Nearly 30 years ago, a German terrorist observed that, with a nuclear weapon, terrorists could make the Chancellor of Germany dance atop his desk on national television. His adolescent boast was appropriately dismissed, but al Qaeda's documented quest for chemical, biological or nuclear weapons cannot so easily be ignored. Fortunately, the jihadists' capabilities still trail their ambitions.

In part, the shift in perceptions also reflects a fundamental shift in the way we assess threats. Traditional threat-based analysis, based upon an estimate of the enemy's capabilities and intentions, has been replaced by vulnerability-based analysis, which starts with a vulnerability, postulates a hypothetical terrorist foe, and produces an invariably worst-case scenario. This type of analysis is appropriate for assessing consequences and evaluating preparedness, but it is no substitute for an actual threat. Nevertheless, what begins as a theoretical possibility in these assessments frequently becomes seen as probable, inevitable and, by the bottom of the page, an imminent threat.

Getting it right is difficult. To some extent, we may be focusing too much on worst-case scenarios and not enough on the most likely scenarios. Moreover, some of the worst-case scenarios are the products of threat advocacy in an environment where threats must compete for attention and resources; upon close analysis, they may offer gripping persuasive narratives yet make little sense.

We also have to keep in mind that many of things we have worried about for decades have not occurred. While this leads us toward making conservative estimates, we cannot argue that, because something has not happened, it will not happen in the future. We do not want another "failure of imagination" of the kind that prevented us from anticipating the 9/11 attacks before they occurred. We cannot dismiss the possibility of a truly catastrophic event.

The distance between the edge of our fears and actual events has narrowed. Another 9/11-scale event, perhaps with a different scenario but with casualties in the thousands, would create the impression that such events will be a regular feature of the landscape. That expectation would have profound economic, societal and political effects.

One serious terrorist chemical attack—the release of nerve gas on Tokyo's subways—has occurred. With better chemistry or better dispersal, it easily could have killed not 12 people but ten or a hundred times as many. It is a scenario that cannot be disregarded.

We know that some terrorists are interested in the dispersal of radioactive material. We tend to envision this scenario as some kind of terrorist Chernobyl, but that would require enormous quantities of radioactive material. More likely is a conventional bomb seeded with a modest amount of such material. Most of the fatalities would result from the explosion itself, not exposure to radiation, but it would be difficult to combat the immediate and continuing terror that such an incident would generate. Absent persuasive public education, we could expect spontaneous evacuation, reluctance to return, and the blaming of all subsequent health problems on exposure.

There is considerable concern about significant disruptions of oil supplies. Sabotage of oil terminals or blocking the Strait of Malacca with a hijacked tanker are the most often cited scenarios. Oil terminals, however, are large, hard targets. To seriously interrupt loading from Saudi Arabia would require putting at least two of its three terminals out of action for weeks. The terrorist threat to the Strait of Malacca—a perfect example of threat advocacy—crumbles upon close examination. A hijacked tanker would not block the strait. And, anyway, going around the strait adds a mere one or two days of transit time.

Of greater concern would be a sustained campaign of terrorism in Saudi Arabia accompanied by continuing sabotage of Saudi oil facilities. The Saudi government has contained the round of terrorist violence that opened with the attacks in Riyadh in 2003, but the outcome of the conflict in Iraq, the dispersal of skilled Iraqi insurgents (including Saudi volunteers) to other fronts, and the Saudi kingdom's restive Shi'ite population are sources of concern.

Some fear that tomorrow's terrorists will move beyond big bangs and high body counts to the realm of cyberspace, not merely to propagandize their causes as they do now, but to sabotage vital infrastructure, disrupt financial systems or crater the internet itself.

There is also concern that, through the internet or by means of coordinated conventional attacks, terrorists will take out vital nodes in the power grid, causing widespread, long-lasting blackouts, with all the attendant economic and social disruptions seen in the wake of natural disasters such as Hurricane Katrina.

Biological terrorism became a reality with the deliberate dissemination of salmonella in Oregon in 1984 and the anthrax letters in 2001. This was comparatively small stuff. The real nightmare would be the deliberate dissemination of a contagious disease such as smallpox. That could set off a worldwide pandemic in which many thousands might die. The appeal of such an attack to all but the most apocalyptic of terrorist thinkers is questionable. In contrast, a biological attack on agriculture would bring economic woe.

The ultimate fear remains that of a nuclear weapon in the hands of terrorists. Analysts in the 1970s extrapolated hostage-taking by terrorists to nuclear-armed terrorists holding a city hostage. In the era of purely destructive, as opposed to coercive, terrorism, we anticipate no negotiations, just a blinding blast. Some put the odds of a terrorist nuclear attack at better than 50/50.

These are the scenarios of tomorrow's terrorism, not wildly different from those of past terrorism, and no more predictable. It is in this atmosphere of uncertainty that we turn to the issue of counter-terrorism.

From combating terrorism to a global war

The 9/11 terrorists ushered in a new era of terrorism and provoked the ongoing "global war on terror," in which the United States expanded and reorganized its counter-terrorist efforts and changed the rules by which it would operate. In the understandable rush to learn the lessons of 9/11 and stave off new terrorist attacks, there was a tendency to ignore the previous three decades of experience in combating terrorism. After all, that had not prevented 9/11. Nonetheless, the pre-9/11 experience should not be disregarded.

In 2000, at a meeting of the International Research Group on Political Violence, an informal network of American and British officials and analysts concerned with terrorism, I was asked to summarize lessons learned from 30 years of US efforts to combat terrorism and identify the challenges we faced going forward. Delivered a year before the 9/11 terrorist attacks, this briefing provides a starting point for comparison with the situation today.

The first lesson was that terrorism was a phenomenon, not a foe, a set of tactics used by diverse groups to obtain publicity for themselves and their causes, to generate fear and alarm that would cause people to exaggerate the strength of the groups and the threat they posed, to provoke overreaction by authorities, and to create untenable situations that would demand an international response. These tactics were both ancient (assassination, taking hostages) and modern (using the latest explosives, exploiting contemporary mass communications).

The United States and other nations were concerned that terrorism represented a challenge to decades of diplomacy to advance the rules of war: Irregular fighters, regardless of their cause, should not be permitted to use tactics and attack targets denied to formal combatants. This was, however, a theoretical concern. The real driving force behind US efforts was the spillover of violence into the international arena. Urban guerrillas fighting their own governments in South America was one thing, but kidnapping foreign diplomats to advance their cause was another. Hijacking airliners merely to change their destination was a particular problem for the United States, but holding airline passengers hostage to make political demands raised more-serious issues.

This spillover was defined as "international terrorism," an artificial construct that was dealt with as a separate domain, an invention which was itself a not insignificant diplomatic achievement. Creating a special category of international terrorism facilitated the mobilizing of international cooperation, which was seen as prerequisite to success in outlawing and suppressing the phenomenon. The definition itself, meant to be objective, had significant side effects.

It led to the creation of databases to monitor the volume and evolution of terrorism. The RAND Corporation, with government sponsorship, created the first chronology of international terrorism in 1972. This chronology defined terrorism by the quality of the act, not the identity of the perpetrator or the nature of the cause, and then further defined "international terrorism" to include those events in which terrorists operating on their own territory attacked foreign targets, crossed national frontiers to carry out attacks in other countries, or went after international lines of commerce, e.g. hijacking airliners. Within this domain, the United States was found to be the number one target—a reflection of ideological conflicts, US involvement in many contentious issues, and the ubiquity of Americans. Although this was not its original intent, the chronology reinforced the idea that something had to be done about terrorism.

Successfully countering terrorism, especially in its international form, required international cooperation, which was hard to come by. That, in turn, constrained response. The US Department of State remained the lead agency in combating terrorism because that task was seen as ultimately a matter of diplomacy. Almost all major terrorist attacks on US citizens occurred abroad. In America's current "global war on terror," the center of action has shifted to the intelligence community and the Department of Defense.

The United States and its anti-terrorist allies quickly learned that broad anti-terrorist initiatives would not fly. The diverse historical and military experiences of many nations, their ongoing conflicts, and the ideological divide of the Cold War made it impossible to arrive at a universal definition of terrorism, let alone agree on concrete steps to combat it. Although the Cold War ended long ago and there is greater consensus on the threat posed by terrorists today, there is still no universally accepted definition of terrorism.

Unable to outlaw international terrorism, nations instead adopted an incremental approach, defining specific tactics or categories of targets that could be proscribed, without defining terrorism. Hijacking or sabotage of aircraft, attacks at civilian airports, attacks on diplomats and diplomatic facilities, and the taking of hostages were individual tactics on which agreement could be found. Ultimately, 12 conventions were promulgated that together comprised just about everything terrorists did, thus defining terrorism little by little.

The conventions provided a framework for cooperation among nations that chose to cooperate—they did not guarantee cooperation and did not by themselves suppress the actions they defined. Following 9/11, a more ambitious international project was launched at the United Nations in the form of UN Resolution 1373, which mandated the preparation and submission of counter-terrorist plans by all nations. For the first time, an entity was established to monitor progress. Progress has since faltered.

The basic problem was and remains the fact that not all nations are equally affected by terrorism. Most nations are rarely victims of international terrorist attacks, and many therefore view it as a distant problem. Those nations that confront armed challenges at home are inclined to manipulate perceptions of the threat to gain international support in suppressing domestic foes. As a result, cooperation is patchy and pragmatic, less concerned with broad principles or the unique problems of the most-targeted nations.

Meanwhile, counter-terrorism policy in the United States remained just that—counter-terrorism. It did not concern itself with attacking root causes or fundamentally altering political structures. Until the very late 1990s, terrorism was a source of periodic crises and episodic concern; rarely did it remain a paramount policy issue. That changed with 9/11, following which the "global war on terror" was elevated to become the dominant US national security issue—the new Cold War.

But beyond ensuring coordination of US efforts and seeking international cooperation, the United States had no counter-terrorist strategy. Again, the United States, unlike Israel, the United Kingdom or Italy, was not dealing with specific terrorist foes, but rather with the phenomenon of terrorism. This inhibited the creation of a specific strategy to guide a specific campaign. What the United States had instead was an accumulation of policy precedents and instruments of coercion that derived from responses to specific terrorist events.

For example, two terrorist attacks in 1972—the Lod Airport attack and the Munich hostage-taking incident—led to the first formal organizational response by the US government, the creation of the Cabinet Committee to Combat Terrorism, which was to ensure the coordination of US counter-terrorist efforts. Policy came later.

When American diplomats were first kidnapped by South American urban guerrillas in 1969 and 1970, the United States expected the local governments to do whatever was necessary to bring about their release, including yielding to the kidnappers' demands, but, as the tactic spread, attitudes gradually hardened. The United States rejected the demands of Palestinian terrorists holding two American officials in Khartoum in 1973, demands that included, among other things, freedom for the convicted assassin of Senator Robert F. Kennedy. When the terrorists then murdered the two diplomats, "no concessions" became official US policy.

The botched rescue attempt at Munich, in which all of the hostages held by the terrorists were killed, galvanized many governments to create special commando units to rescue hostages. The United States, which had earlier contemplated sending paratroopers to force the release of hundreds of passengers held by hijackers at Dawson Field in Jordan in 1970, moved cautiously in this area. (The use of paratroopers probably would not have worked anyway, and the idea was dropped.) The United States did not create the Delta Force until after Israeli commandos successfully rescued hostages held by terrorists at Entebbe Airport in Uganda in 1976 and German commandos rescued hostages aboard a hijacked airliner in Mogadishu in 1977. In 1980, the Delta Force was used in an ill-fated attempt to rescue Americans held hostage at the American embassy in Tehran. The attempt failed, but it established the precedent that military force would be considered a legitimate option in hostage situations anywhere in the world.

In 1983, a suicide bomber blew up a building in Beirut that housed US Marines, killing 240 Americans. A subsequent commission of inquiry headed by Admiral Long concluded that American officers, from the local commander to the top of the chain of command, had not paid adequate attention to the threat of terrorist attack. The finding made force protection a paramount concern and also made it clear to the Pentagon that combating terrorism was a military matter. (At the time, I wrote an essay declaring that, with the Long Commission report, combating terrorism became a war; but while the term "war" was often used rhetorically, the actual concept of war against terrorists did not become a reality until after 9/11.)

In 1985, terrorists hijacked the cruise ship *Achille Lauro* and murdered one hostage, an elderly American. Subsequent negotiations promised terrorists their freedom in return for the release of their remaining hostages, and US fighter aircraft forced the airliner carrying the escaping terrorists to land at a US base in Sicily. Italian officials, however, insisted on taking custody of the terrorists, whose leader they subsequently released, but the force-down of a civilian airliner created another precedent.

State sponsorship of terrorism had been a concern since the late 1970s. The United States formalized this concern by identifying and annually updating a list of those nations it considered state sponsors (a Congressional initiative) and imposing diplomatic and economic sanctions. Throughout 1985, the United States grew increasingly frustrated by Libya's support for terrorism, in particular its connection with the Abu Nidal group that was responsible for a series of attacks in Europe in which US citizens were killed. A subsequent terrorist attack on Americans in early 1986 led to military retaliation, which may have included an attempt to assassinate Libyan leader Muammar Qaddafi himself.

In 1993, the United States accused Iraq of involvement in an attempt to assassinate former President Bush and responded with military force. Military force was used again in 1998 in response to terrorist attacks on American embassies in Africa. That response included the bombing of a pharmaceutical plant in Khartoum that was suspected of being used to fabricate biological weapons.

This created another new precedent: The United States would use whatever measures it deemed appropriate, including unilateral preemptive military action, to prevent terrorists from acquiring or employing weapons of mass destruction. This idea had been discussed in policy circles more than 20 years earlier and later was part of the justification for the 2003 invasion of Iraq.

In retrospect, it is remarkable how few terrorist-created crises have occurred. Over a period of more than 35 years, there have been only 25 major foreign and domestic terrorist crises, headline events that kept the lights on past midnight in Washington, that caused significant casualties, and that affected policies and legislation:

- the kidnappings of American diplomats in Latin America from 1968 to 1974 (along with the kidnapping of US airmen in Turkey);
- the coordinated multiple airline hijacking that put three airliners at Dawson Field, Jordan, in 1970;
- the Lod Airport massacre in 1972;
- the seizure of the Saudi Arabian embassy in Khartoum and the murder of two diplomats in 1973;
- the seizure of the American consulate in Kuala Lumpur in 1975;
- the sabotage of an American airliner in the Mediterranean in 1977;
- the Hanafi Muslim hostage siege in Washington in 1977;
- the seizure of the American embassy in Tehran in 1979;
- the seizure of the Dominican Republic's embassy and a protracted hostage crisis in Bogota in 1980;
- the kidnapping of American General Dozier in Italy in 1981;
- the bombing of the American embassy in Beirut in 1983;
- the bombing of the US Marine barracks in Beirut in 1983;
- the bombing of the American embassy in Kuwait in 1983;
- the kidnapping of Americans and a protracted hostage crisis in Lebanon from 1982 to 1991;
- the hijacking of an American airliner and the subsequent hostage crisis in Beirut in 1985;
- the hijacking of the cruise ship *Achille Lauro* in 1985;
- the Libyan-backed terrorist campaign of 1985 to 1986;
- the sabotage of Pan Am flight 103 in 1988;
- the World Trade Center bombing in New York in 1993;
- the attempted assassination of former President Bush in Kuwait in 1993;
- the Oklahoma City bombing in 1995;
- the bombing of Khobar Towers in Saudi Arabia in 1996;
- the bombing of the American embassies in Nairobi and Dar es Salaam in 1998;
- the bombing of the USS *Cole* in 2000;
- the 9/11 attacks.

One might argue for the inclusion of one or a few more incidents, and there are thousands of lesser incidents involving American targets, but this is a basic list. What does it tell us? For one thing, Middle East antagonisms account for 20 of the 25 major terrorist-caused crises. The United States has been attacked because of its continued support for Israel, its support for the Shah of Iran and hostility to the Khomeini revolution, its opposition to Hezbollah in Lebanon, its confrontation with Iraq, and its support for and deployment of troops to Saudi Arabia. In addition, the majority of state sponsors of terrorism identified over the years (Afghanistan, Iran, Iraq, Syria, Sudan, Somalia, Libya, North Korea and Cuba) are in the Middle East or North Africa.

A continuing conflict over four decades would seem to lend weight to a "clash of civilizations" thesis, but we must take into account the relatively small numbers of persons actually involved in these terrorist attacks. Combined, the various terrorist groups comprise no more than several thousand people, and the number actually involved in attacks on the United States may be only in the hundreds. We are at war with tiny armies, handfuls of people whose actions may reflect broader antagonism, but whose tactics are not widely endorsed. Indeed, direct interventions have in some cases radicalized and enlarged the opposition.

This would suggest caution in counter-terrorist efforts aiming at a broader transformation of society, whatever the merits of transformation itself. That may disappoint those who demand that counter-terrorist policies address root causes. It may reassure those who see large-scale American intervention as counterproductive. To me, it indicates that, while counter-terrorism is an ideological contest that must include a political-warfare component, implementation must be focused on reducing the appeal of terrorists' ideology and disrupting their recruiting. Bringing down the Taliban in Afghanistan was justified because the regime in Kabul had become inextricably intertwined with al Qaeda—one had to go with the other.

By the end of the twentieth century, then, while the United States still considered terrorism to be primarily a law-enforcement problem—it certainly had not formally placed terrorism in the context of war—it had evolved a policy that gave wide latitude to the use of military force. And it had done so without formally elaborating an overall strategy or doctrine but, rather, in response to specific provocations.

The situation was not so different after 9/11. The United States took immediate action, destroying al Qaeda's bases and toppling Afghanistan's government. These were strategic moves, but without an overall strategy. Efforts to elaborate a national counter-terrorism strategy came later.

Prior to 9/11, whenever the US used military force in response to terrorist actions, it was to further diplomacy, not to achieve military results. This rationale changed significantly after 9/11. No longer was the specific military objective to disrupt terrorist operations or support for terrorism; it was to defeat the terrorists themselves. Where "combating" terrorism had implied an enduring task, one that did not envision ultimate victory, this objective also changed after 9/11—although it was recognized that "victory" might not look the same as it would in a conventional war. Finally, prior to 9/11, it proved difficult to sustain a military campaign. In each case where military force was used, the United States may have wanted terrorists or their state sponsors to believe that US forces might attack again without further provocation, but that never happened. Terrorists retained the initiative. The deterrent effects were therefore minimal, and they were matched by terrorist attacks or bloody encounters that led to American withdrawal (Lebanon in 1983 and Somalia in 1993). The message was mixed. US counter-terrorist efforts may have affected terrorism calculations only marginally.

Were there alternatives?

Theoretically, there were alternatives to US policy decisions. The government could have tried to pay less attention to terrorism or, at least, to adopt a more phlegmatic approach, lowering the volume of its bellicose rhetoric, casting terrorism in a different light and denying terrorists the official attention they sought.

Indeed, there was often internal opposition to the actions taken. In 1980, the US Secretary of State resigned in protest over the use of military force to rescue hostages in Iran, and the rescue mission was an embarrassing failure. The Pentagon did not welcome the Long Commission report that told it to prepare for terrorism. Some thought the attack on Libya was a mistake; certainly its effects as a deterrent were debated. The evidence indicating the manufacture of biological weapons at the plant in Khartoum was not convincing and eroded US credibility.

Nevertheless, it is difficult to imagine how American officials could have successfully avoided the confrontations, especially as terrorism escalated; 9/11 rendered inaction impossible, although swaggering rhetoric remains a problem.

The United States could have adopted a more flexible policy for dealing with hostage situations. There was, in fact, very little evidence to support the assumptions upon which the no-negotiations, no-concessions policy was based; a relationship between a particular policy and the absence or occurrence of further kidnappings simply could not be demonstrated. Even without concessions, terrorists holding hostages still got publicity, created crises and obtained local leverage. These were sometimes sufficient incentives. Rigid adherence to declared policy inhibited creative responses. And it cannot be denied that there conceivably might be circumstances in which quiet negotiations and indirect pay-offs could be preferable to prolonged, distracting crises.

In fact, US policy was never consistently applied. It did not preclude other governments from making concessions to effect the release of American hostages. It never tried to prevent corporations from paying ransoms in return for the release of executives held abroad. And, of course, in 1986, it was revealed that US officials had secretly made deals to bring about the release of some of the Americans held hostage by terrorists in Lebanon, a revelation that caused great embarrassment. One lesson learned by Presidents Nixon (in the Khartoum incident), Carter (in the Iran hostage crisis) and Reagan (in the Lebanon hostage crisis) was that hostage situations could be politically dangerous.

Assassination, although specifically outlawed by a presidential executive order in 1973, was in fact an option. US officials never officially admitted that Qaddafi himself was a target in the Tripoli attack, although it seemed that he was. After 9/11, "targeted killings" of terrorist leaders became an accepted component of the "global war on terror."

In a speech defending the US war effort in Iraq, President Bush in September 2005 criticized the feeble US responses of the Carter administration to the hostage crisis in Iran, the Reagan administration to the bombing of the US Marine barracks in Lebanon, and the Clinton administration to various terrorist attacks, asserting that the terrorists "saw our response . . . and concluded that we lacked the courage and character to defend ourselves."

While it is true that al Qaeda's leaders have inspired their followers with references to the American withdrawal from Lebanon following the 1983 terrorist bombing and from Somalia following the deaths of 17 American soldiers in a single battle, President Bush's assertions ignore historical reality. Obtaining the release of American hostages held in Tehran would not have been served by military action, let alone by an invasion of Iran. The United States would have found itself in a major war, for which there was little political support only five years after the fall of South Vietnam, and the hostages almost certainly would have been killed. As it was, the attempted military rescue was the first use of military force after the Vietnam War and contributed to a change in public attitudes.

President Reagan entered office warning that military force could be used in response to terrorism, despite strong opposition from American military leadership. Following the bombing of the Marine barracks, President Reagan did authorize an air strike and in 1986 ordered the bombing of Libya. Military occupation of Lebanon, still in the midst of a bloody civil war, or the invasion of Libya were not serious options.

President Clinton ordered the bombing of Iraq in 1993 and of Sudan and Afghanistan in 1998. There was no political or military support for an invasion of Afghanistan.

Perhaps understandably, President Bush omits the presidency of his father from his broadside. Yet it was the previous Bush administration that rejected further military action against Libya despite clear evidence of Libyan involvement in the 1988 sabotage of Pan Am flight 103—the worst terrorist incident suffered by the United States prior to 9/11. Instead, the senior President Bush decided to proceed with individual criminal indictments against the Libyan officials involved, thus keeping the US response to terrorism firmly in the realm of law enforcement. And it was the elder Bush who in 1991 decided not to go beyond the original objective of liberating Kuwait in the first Gulf War, which may have affected Iraqi leader Saddam Hussein's calculations in 2002 and 2003.

Polemic is not the point here. The point is that responses are determined by the situation and the terrain. Military action comparable to that in Iraq would have been just as inappropriate in these earlier cases as the absence of a muscular riposte would have been after 9/11. And while action may create new realities, it does not always do so in ways that are predictable or desirable. The lesson of the Iraq War is yet to be drawn.

Few limits to American policy today

The United States today accepts few limits to its counter-terrorism efforts. It has declared and demonstrated that it will itself apprehend terrorist suspects anywhere in the world, keep them in its custody at offshore prisons or at hidden locations, or deliver them to other governments for incarceration and interrogation. It has, most controversially, altered its rules on allowable methods of interrogation. It has asserted, thus far successfully in US courts, that it can detain US citizens as enemy combatants, holding them indefinitely without charge or access to legal counsel.

The United States has for many years endorsed the use of diplomatic and economic sanctions against states accused of sponsoring terrorism, although the effect of these sanctions is debated. Since 9/11, the United States has, with international cooperation, blocked the financial assets of terrorist organizations and institutions used by them for raising or moving funds.

The United States continues to seek international cooperation and to offer various forms of assistance to allies and friendly nations that are making efforts to combat terrorism, but it will not hesitate to act unilaterally, overtly or covertly, when it deems it appropriate.

The United States will use force to rescue hostages held by terrorists. It has demonstrated its willingness to force foreign commercial aircraft to land at US-controlled facilities to apprehend terrorist leaders as it did in 1985. It will offer huge bounties for the capture or death of terrorist leaders. It will kill known terrorist leaders on foreign soil. Confronting suspected suicide bombers, police departments in the United States will shoot to kill.

The United States will retaliate with military force. It will use military force to preempt terrorist attacks or to prevent terrorists from acquiring or developing chemical, biological or nuclear weapons. It may even employ nuclear weapons when they are deemed necessary to destroy a chemical or biological weapons cache or facility. It has declared and demonstrated that it will topple governments and effect regime changes as part of the "global war on terror," as it has done in Afghanistan and Iraq.

It is, in sum, difficult to define what the United States will not do. Instead, the question is whether these looser rules of engagement are limited to the duration of the current "global war on terror." My own view is that they represent permanent change, because there is no foreseeable formal end to the war. And even if the international environment changes dramatically and the terrorist threat is significantly reduced, having been established as precedents these measures will always be available in the future, depending on the circumstances. This is not to say that the United State will do these things, but rather to assert that having taken an action once puts it on the table.

At the same time, the courts could ultimately impose limits on assertions of executive authority. Public revulsion at abuses could restore adherence to stricter compliance with international conventions on treatment of detainees. Changes in public attitudes could also constrain future action, even if an option were theoretically available. Finally, caution itself will preclude routine employment of extreme measures, although "extreme" has been redefined. But these would be marginal adjustments. Just as 9/11 hurled us into a new era of terrorism, so our responses have permanently redefined the arsenal of counter-terrorism.

Counter-terrorism's future trajectory

Counter-terrorism in the future may follow a different trajectory. Historically, the United States has always sought international cooperation to address an international phenomenon, not because it is politically correct but because it is prerequisite to success. The results have often been disappointing; real cooperation has always been limited to a small number of like-minded governments. However, the magnitude and indiscriminate nature of recent terrorist attacks, along with terrorist targeting of countries that otherwise may have preferred to sit on the fence, has increased the number of governments in the active counter-terrorist camp.

The 9/11 attacks brought cooperation among intelligence and security services to an unprecedented level, resulting in numerous arrests of terrorists worldwide and the foiling of numerous terrorist plots. Despite profound political differences, primarily over Iraq, that cooperation continues. The potential of future terrorist attacks of equal or even greater magnitude than those of 9/11 makes it more imperative that cooperation be institutionalized, intensified and expanded. This cannot be accomplished by mere exhortation or by mandates dictating levels of cooperation beyond those to which governments are prepared to go, especially in the sensitive area of intelligence. These are complex issues, and progress is likely to be slow. It will take many years. One need only look at the decades that have been spent imposing rules of war or institutionalizing free trade. The necessity (and the temptation) for unilateral action will always be there, as will the option, but enhancing true international cooperation in counter-terrorism may offer the only means of countering global terrorist threats over the long run.

12

THE NEW TERRORIST THREAT ENVIRONMENT

Continuity and change in counter-terrorism intelligence

Stephen Sloan

Introduction

Over the past 40 years the world has witnessed a transformation in the capabilities of terrorists to pose an increasingly potent threat to a fragile international order. Further, it appears that this threat will intensify in the next decades as terrorists refine their abilities through more sophisticated technology and strategies. In this protracted conflict, where there may be no definitive outcomes, it is essential that policy makers in the counter-terrorism arena re-evaluate their policies, strategies and doctrines. They must consider the imperatives not only of responding to terrorist attack but, more importantly, of taking the initiative against a form of violence that may be as old as the Assassins but as contemporary as the most technologically sophisticated attack.[1]

Many terrorists display the imagination, creativity, innovation and operational capability to place the authorities in a reactive and defensive mode. Terrorism is an international problem: all states will have to continue to confront a challenge where the "conventional wisdom" no longer works in an environment where terrorists, even if they are motivated by the most traditional grievances, are innovative in their strategies and tactics.

Nowhere is the problem of relying on the conventional wisdom more clearly demonstrated than in the arena of intelligence, where old values, old organizational formats, a culture of secrecy and bureaucratic battles on all levels of government, both domestic and international, have acted as a serious impediment to counter-terrorism efforts. This failure is especially significant owing to the obvious fact that intelligence as information-gathering process and administrative organization is necessarily at the forefront in combating terrorism.

First, however one might seek to identify and eliminate the underlying causes of terrorism, such a task will never be achieved completely. Further, even if one could identify and resolve its primary causes, terrorism has its own dynamic. It is not a clearly defined linear process. There are not only underlying, but precipitating and accelerating, causes of violence that may no longer be related to the underlying causes used as a rationalization for acts of violence.

Second, irrespective of the measures that can be taken to prevent or deter threats, there will never be enough security to counter them completely. The technology of physical security and surveillance has come a long way from gates, keys and basic sensors. But in the final analysis these technologies are all too often a rationale for target displacement to terrorists. Terrorists, unless they are committed to a specific locale, facility or person, have a constellation of targets, and therefore none can be totally secured. Well aware of most countermeasures, terrorists simply shift the focus of their attack to a "softer target" in order to succeed. Further, it is vital to question the value of the trade-offs society must make in order to secure potential targets. Physical security in what one author has called "the Age of Surveillance" could be seen as, in effect, separating government from its people, eroding the civil rights of the individual, and increasingly attacking what could be called our private zone.[2]

If civil liberties are seriously challenged, terrorists could already claim putative victories, since they—through their provocative attacks—have provoked governments to overreact, potentially proceeding down the dangerous path to the development of a security state. These are serious issues that have been the subject of much recent debate, particularly in light of the US Patriot Act.

Third, and this is perhaps the most troublesome of all, is the fact that the terrorists now have and will increasingly be willing to use weapons of mass destruction. If Sun Tzu's dictum was correct, "Kill one person, frighten one thousand," the events of 9/11 cogently affirm and magnify his observation, for now it is "Kill almost three thousand, frighten millions." If one recalls how two snipers, who were not part of a terrorist organization and did not have a political agenda, almost paralyzed Washington, DC and the surrounding area, one need only ponder what would happen if and when a mass attack took place.

Aside from the psychological effects, the physical destruction of future terrorist attacks, particularly in the extreme case of weapons of mass destruction, would cause immense human and property damage to a community. While the United States, for example, has refined its response capabilities a great deal since the first bombing of the World Trade Center, the bombing of the Murrah Federal Building in Oklahoma City, and the attacks on 9/11, the specter of a mass biological, chemical or nuclear attack must force authorities to recognize that, in spite of pre-attack preparation, "crisis management," "incident management" or "emergency management"—however vital—would mainly be a vast effort in human and physical "damage control" and "reconstitution."

For the above reasons it is crucial that individuals and organizations in intelligence have the capability to identify, apprehend or neutralize terrorists before they go tactical, that is before they initiate their movement to the target. In the words of a senior counter-terrorism expert, speaking about his own country's experience, "If there is a terrorist's incident there has been a 90 percent failure in my country's counter-terrorism policies and operations and that failure is in intelligence."[3]

This chapter seeks to identify critical aspects of the transformation of terrorism by identifying the "technological/terrorist interface" and discussing its implications for intelligence operations. It is hoped that an understanding of this "interface" could assist intelligence organizations in addressing the changing threat. The chapter is directed at intelligence professionals, policy makers and political leadership at all levels. A major theme throughout the chapter will be how the "technological/terrorist interface" has transformed contemporary terrorism and will continue to transform future terrorist activities.

The technological/terrorist interface:
non-territorial terrorism

If the Zealots represented the ancient tradition of terrorism and the French Revolution initiated the age of state terrorism, in many ways it was the skyjackings and hostage-taking of the 1960s, culminating in the Munich massacre of 1972, that can be viewed as having inaugurated the modern age of terrorism. A new generation of terrorists emerged, assisted by the revolution in satellite communication and the introduction of commercial jet aircraft. As a result of these developments, the "new terrorists," even if they were motivated by traditional grievances, were "functionally different from those who preceded them," for they were engaging in "non-territorial terrorism—a form of terror that is not confined to a clearly delineated geographical area."[4]

In effect, holding hostages at 30,000 feet and broadcasting their "armed propaganda" to a wider audience than ever before, these modern terrorists were able to ignore the arbitrary physical and legalistic boundaries of nation-states. Moreover, since their operations were not geographically limited, they were in effect engaging in a form of what could be called low-intensity aerospace warfare—using a plane as a weapons delivery system. This form of warfare would unfortunately reach its zenith with the events of 9/11 where the aircraft involved were for all intents and purposes low-intensity intercontinental man-guided missiles. The ability to engage in "global propaganda by the deed" has been enhanced by the development of the internet and what could now be called global terrorist information warfare or cyber-warfare.

Despite these technological innovations, the initial response by the authorities was primarily based on geographical and legal considerations associated with state sovereignty. At the international level the terrorists were able to seize the world's headlines, and dramatize their causes. In contrast nation-states, focused on their own national interests, could not arrive at a definition of terrorism much less consensus on how to deal with it. This inability carried over to the policy arena where debates over what types of approaches should be employed to respond to a transforming threat were characterized by a lack of unity. This disunity often acted as an impediment in the operational arena in bilateral and regional collection and utilization of intelligence. The need for transformation in intelligence to match and surpass the transformation of terrorism will be discussed later in this chapter. While there has been marked, if grudging, progress since the early days of modern terrorism and a wide variety of anti-terrorism conventions and treaties have been incorporated into international law, the application of the law has often been the victim of interpretation colored by the parochial interests of individual states. Moreover, whatever progress has occurred serves to underscore the fact that cooperation remains primarily *reactive* in character and is often only initiated after a particularly violent terrorist act or campaign. While technological change has been effectively employed by terrorists, nations have been slower to take advantage of the opportunities it provides. This is due to the forces of inertia resulting from issues of sovereignty, conflicting national interests, and bureaucratic immobility. These factors have acted as a barrier to the technological and political integration required to combat what Brian Jenkins so many years ago aptly called "a new mode of conflict."[5]

The challenges to counter-terrorism intelligence

At the outset it is unfortunate but true that, while terrorists are increasingly and effectively engaging in non-territorial terrorism, governments, including the military, police and particularly intelligence organizations, suffer from their own "territorial imperative."[6] It is an often unfortunate natural tendency of bureaucracies to define their activities in large part based on a particular geographic and, more commonly, jurisdictional area of operations. Whether the mandate for such activities focuses on geographical or legalistic boundaries, the fact remains that bureaucracies face internal constraints and compete with other bureaucracies in protecting what they regard to be their organizational prerogatives, missions and resources. Competition acts as a barrier to achieving a unified approach, particularly when public attention and, potentially, additional funding become focused on a particular issue as in the case of "the war on drugs."

202

Certainly the ability to go beyond the classic geographical fields of operations has been greatly enhanced through aerospace power and is based on the recognition that new threats have created ever greater demands for regional and global integration. In addition, the diversity created by jurisdictional concerns is often exacerbated within a particular country, especially when a federal as contrasted to a unitary system is in place. The continued tensions between the federal law enforcement community and those on the state and local level within the context of "homeland security" unfortunately illustrate this fact. But it is especially in intelligence where the bureaucratic turf battles often act as a barrier to the necessary geographic and functional integration that is required to meet the threat of international terrorism.

In the US this is partly a result of the fact that there has been a legal barrier—or firewall—between domestic law enforcement and national security intelligence organizations. This "wall," which carefully controls cooperation between agencies, has limited, to a degree, the level of domestic intelligence collection possible. The purpose of this is to avoid potential violations of civil liberties and due process. In addition there is recognition that police intelligence has been traditionally tactical, short-term and reactive in nature while long-term intelligence ideally should have strategic focus and global reach.

The results of these complexities have been addressed in the findings of the 9/11 Commission.[7] The report documents in detail the inability and unwillingness of the intelligence community at the national, state and local levels to share information. Further, in addition to the strained relationships between agencies, there has always been an inherent intra-agency tension between the area specialists, who focus on a geographic region, and their counterpart functional specialists, who specialize in the type of collection used, ranging from TECHINTS (technical intelligence) to HUMINT (human intelligence). In the past there have been attempts to "fuse" these resources through the creation of working groups that focus on specific global issues such as drugs, terrorism and narco-terrorism. It remains, however, to be seen whether fully integrated non-territorial intelligence will become a reality.

The creation of the Department of Homeland Security and the position of National Director of Intelligence are fundamental parts of the largest structural reform of intelligence and national security since the National Security Act of 1947. While reorganization may be a step in the right direction, it is by no means clear whether these reforms will lead to further complexity, new turf battles and jurisdictional squabbles rather than the hoped-for improved organization and clarity of mission.[8]

The technological/terrorist interface: the erosion of the nation-state system and the rise of non-state actors

It is not without irony that one of the primary targets of terrorists since its initial transformation in the 1960s and 1970s has been the multinational corporation. These entities have operated outside the restrictions faced by purely domestic enterprises. The field of operations of this type of corporation is transnational and in a very real sense non-territorial—much like that of the terrorists who view the multinational corporations to be a major adversary. This similarity is reinforced further by the fact that both terrorists and multinational corporations have taken advantage of technological advances, allowing them, to some degree, to transcend the boundaries of the nation-state. According to this argument, both terrorist organizations and multinational corporations are non-state actors that have over the past decades begun to challenge the primacy of the "state-centric" system created by the Treaty of Westphalia in 1648. Other examples of non-state actors include "universal and regional governmental organizations (IGO's), transnational guerrilla and terrorists groups, multinational corporations (MNC's) and a rapidly growing number of nongovernmental organizations in a wide variety of functional areas."[9]

The position of the state as the principal actor in international affairs is also eroding as a result of the impact of technological interdependence and networked communications. The accelerating and often highly disorderly process of technological advance does not mean, however, that the state will lose its preeminent position, at least in the short run, although its primacy will be challenged. With the end of the Cold War, authoritarian governments could no longer control restive ethnic, religious and other subnational or transnational groups through traditional instruments of state coercion. One of the major instruments for maintaining state repression and terrorism, the control and manipulation of information, has been subject to the assault of a global web where knowledge is increasingly available to subnational and transnational organizations pursuing their own agendas.

The challenge to counter-terrorism intelligence

While the non-territorial nature of contemporary terrorism has challenged governments, particularly the intelligence services, to break down the arbitrary geographic and jurisdictional barriers to unity of action, yet another challenge has been created by the role non-state actors are increasingly playing in international affairs. This is particularly the case in the "war against terror." Recognizing this new role is difficult enough for those critical of privatizing functions traditionally the responsibility of the public sector. But it is especially difficult for intelligence organizations, in which the monopoly of secret intelligence often justifies their existence and where such secrecy translates to political power and, ultimately, funding. As noted earlier, the proliferation of non-state actors and the availability of information outside the traditional official realm are additional factors that intelligence organizations must address.

We are entering a world where "open-source intelligence" is not only readily available but may increasingly supplant the need for secret intelligence.[10] The sources of critical intelligence will increasingly come from non-state actors such as large multinational corporations, small businesses and non-governmental organizations. Moreover, those who threaten national and international security are also refining their own sources and methods of collecting, analyzing and disseminating information. Unfortunately, this is especially the case for some terrorist groups, which have done so very effectively because of their flexibility and small size. While traditional intelligence services are swamped with data as they seek to separate "the noise from the signal" through information handling, non-state actors not only can have a smaller focus for their requirements but can also readily communicate with each other without the barriers to communication that characterize a large bureaucratic organization.

A positive development after the attacks on 9/11 has been that the intelligence community began to acquire the necessary funding to recruit more collectors and analysts. Unfortunately, at least in the short term, these new members of an expanding community may, in their own right, have generated "more noise than signal." Further, in the current phase of reorganization it is by no means clear—despite the calls for centralization—whether such reorganization could potentially create more rather than fewer barriers between agencies, not only on the domestic level but also in the international arena. In contrast, we have seen the explosive growth in private sector intelligence entities, which increasingly rival governmental intelligence. The reasons for this are varied, but three specifically can be noted. First, these entities are motivated by profit and therefore do not have to address and quantify the services to meet national security requirements. How does one, for example, measure success in counter-terrorism intelligence, when success should in part be measured on the basis that an event did not occur? Second, new, non-state intelligence entities, particularly if they are international in scope, do not have to address the bureaucratic turf battles that often characterize the official intelligence community. Third, "private sector" intelligence entities may be held accountable to their stockholders, clients and, to a degree, governments, but they do not face the level of accountability expected in a democratic society. A classic conundrum in democracies has always been the reconciliation of secret intelligence with the expectations of an open society.

Given these realities, there will have to be a more effective effort to achieve a meaningful level of cooperation between unofficial and official intelligence organizations. This will not be an easy task for corporate, competitive intelligence services, which want to protect proprietary and client information, while official intelligence services must protect state secrets in the name of security. Yet the need for meaningful cooperation is vital; contemporary terrorists increasingly have the capability to attack personnel and facilities in the corporate and governmental sectors as well as in international and non-governmental organizations.

The technological/terrorist interface: the erosion of the state monopoly on coercive power and the privatization of public violence

The penetration of the nation-state system is particularly manifest in the erosion of the state monopoly on force. This has resulted in the emergence of the "gray areas—immense regions where control has shifted from legitimate governments to new half-political, half-criminal powers."[11]

In what, in some states, is near-anarchy, governments have themselves sought to hire mercenaries to control the emergent threats discussed in this chapter. Further, some multinational non-governmental and international organizations in these countries have sought to protect their personnel and assets by hiring private security firms. This trend extends itself even to major powers, which have begun increasingly to rely on private contractors—now called "civilians in the battlefield" (COBs)—to conduct military operations and even wage war. Equally significant is the fact that the public perception—even in stable post-industrial societies—is that the police and security agencies cannot deal effectively with the threat of crime, particularly terrorism. This has led to the massive growth of private security firms and services ranging from traditional "rent a cop" to highly professional and technologically proficient services which now represent the emergence of "the corporate warrior."[12] As a result there are a whole host of new players in the security arena. This has been termed by some as "the privatization of public violence." Along with this privatization comes the need for the collection, analysis and dissemination of information that traditionally fell under the control of state intelligence systems.[13]

The challenge to counter-terrorism intelligence

The emergence of significant new actors requires further cooperation between the public and private sectors regarding the systematic sharing of intelligence. There will clearly be challenges, especially concerning particularly sensitive information directly related to national security. This problem can, however, be resolved to a degree by sanitizing information and not revealing sources and methods. At the same time, another solution could be that more US citizens in the private security sector could be vetted and receive the necessary classifications to acquire pertinent information.

It must be noted that, just as the official intelligence community might not want to share certain types of information, it might also be necessary to recognize that corporations would also be reluctant to share proprietary information, as it could hurt their competitive edge. Nevertheless there is room for negotiation. The problem of sharing information with non-governmental organizations is even more problematic given the fact that in many instances, especially with humanitarian organizations, they do not wish to be associated with any particular government or policy. But even then there could be a degree of cooperation based on informal arrangements and mutual needs. An NGO might want to have information to provide better security in its area of operation and in exchange share its insights about this area. By the same token, the intelligence community is also able to create its own proprietary organizations, as it has demonstrated in the past. The focus, however, should be on cooperation with independent and corporate intelligence entities. The task will not be easy but there is a real need for both sides to reach out. Counter-terrorism intelligence is not a monopoly of the intelligence community.

The technological/terrorist interface: redefining warfare?

While conventional warfare will continue and the danger of a nuclear one will unfortunately hang over the globe like the "sword of Damocles," the international system has entered a new version of an old form of warfare that goes back to traditional guerrilla warfare, partisan warfare and insurgencies. What makes this development different is that it is all of those but enhanced by technological muscle. Contemporary war has increasingly taken on the form of asymmetrical warfare. "In broad terms it [asymmetrical warfare] simply means warfare that seeks to avoid an opponent's strength; it is an approach that tries to focus whatever may be one side's comparative advantages against its enemies relative weaknesses ... in the modern context, asymmetrical warfare emphasizes what are popularly perceived as unconventional or nontraditional methodologies."[14]

The most current application of asymmetrical warfare is clearly terrorism, which has unfortunately aptly illustrated how small, highly dedicated groups can not only strike at the core of the symbols of the economic and political power of the one remaining superpower but also inflict massive casualties such as in the attacks of September 11. Such actions, even on a lesser scale, have influenced the foreign policy of superpowers, as in the case of the 1981 bombing of the Marine barracks in Lebanon, which made Washington disengage from that crucial country.

This type of warfare is also a manifestation of another change within the international system. The very forces of globalization that have fostered interdependence have also led to the reassertion of traditional loyalties by those who have the willingness and increased capability to engage in "small wars" against far more powerful adversaries.[15] This increased enhancement is not simply based on the availability of a wide variety of both conventional and unconventional weapons, but is a result of another even more profound technological innovation—the development of the worldwide web and the internet. Small groups have increased their range of operations while being able to enhance their security in the vacuum of cyberspace by practicing netwarfare . . . an emerging mode of conflict (and crime) at societal levels, short of traditional military warfare, in which protagonists use network forms of organization and related doctrines, strategies and technologies attuned to the information age. These protagonists are likely to consist of dispersed organizations, small groups and individuals who communicate, coordinate and conduct their campaigns in an internetted manner, often without a precise central command.[16]

The organizational implications and challenges to intelligence services to meet these requirements "attuned to the information age" will, as we shall see, require political will and imagination by leaders, policy makers and the public.

The challenges to counter-terrorism intelligence

Perhaps the redefinition of warfare—particularly in the context of the emergence of netwar—represents the most immediate and demanding challenge to counter-terrorism intelligence. The challenges of this increasingly sophisticated threat will require a change in mindset, organizational culture and structure. Perhaps even more important, this threat will demand a degree of imagination and innovation which is at odds with the inertia of a large-scale organization and compounded if that organization has a cloak of secrecy around it. Along these lines, there has been a major emphasis on centralization under the Secretary of the Department of Homeland Security and a new National Director of National Intelligence.

While the process is a difficult one, and the battles among the various agencies continue in a changing bureaucratic environment, the fact remains that such a reorganization may be counterproductive in combating terrorism. When one considers that the September 11 attacks only involved 19 direct participants, it is quite clear that, through use of the internet and planning in small cells, the perpetrators engaged in what the military calls "force multiplication" with the most profound impact on the US and the global economic and political order.

The answer to the challenge will not take place if the response continues to be in the true American tradition of "throwing money at the problem" and "reorganizing." Rather, any meaningful change will require a movement away from classic ladder hierarchies, top-down command and control and micro-management at the operational level. Clearly, modern communication in this case can be a blessing and a curse. On one hand, remote sensors in the form of drones and other delivery systems have given senior decision-makers a real-time view of what is happening on the ground, but also the capability of micro-managing the battlefield—be it a road in the Bekaa Valley or a safe house in Berlin.

While there will always be the need for accountability and oversight, especially in a democracy, such oversight should not negate the ability of highly trained analysts and operators to work together and, when necessary, have their own counter-terrorist cadre cells make operational decisions as part of a counter-terrorist network. Such a network will not work if it suffers from its own form of over-institutionalization, or if it only is used by one country's intelligence services. Rather, it must be a network, based on trust and experience, that can enable a counter-terrorism effort that parallels the terrorists, who have refined the capability to be decentralized yet coordinate regional and global attacks with a minimum of personnel and institutional layering. The requirements of a counter-terrorist cadre will not be easy to achieve but this capability is necessary in order to deal with this form of asymmetric warfare.

As noted in an earlier article, the development of such a cadre could also be actively involved in operations including the preemption of terrorism. Such a force would be small, include both analyst and operators, have very clear and uncluttered lines of communication, and include elements from the military—notably the special operations community—and the clandestine services. Further, given the development of "seamless terrorism," elements from the Federal Bureau of Investigation—considering its new counter-terrorism role—would also be helpful.[17] While progress has been made in this area since the article was written in 1986, such progress will be operationally constrained if the proposed counter-terrorist cadre is subject to micro-management by a large and cumbersome bureaucracy.

Conclusion: there are no magic bullets in counter-terrorism intelligence, but creativity can make a difference

In the protracted war against terrorism there will be both victories and defeats. But it will be difficult to achieve a clearly recognizable decisive outcome. Yet the United States and its allies and others who share a mutual interest in combating terrorism can increasingly contain and preempt threats and acts of terrorism. In pursuing this goal, two factors should be considered—especially in these days of the availability of and willingness to use weapons of mass destruction (WMD). First, the role of counter-terrorism very much should be on the forefront of such an effort. But secondly, counter-terrorism intelligence will not be effective if there is not resolve on the part of the political leadership to meet the long-term threat as well as the will to engage in meaningful intelligence reform. It is still too early to judge how successful the first stages of reform have been. However, such reform focuses more on bureaucratic change than on substance, and, if jurisdictional and other turf battles continue, the intelligence community will not have incorporated the tragic lessons from the attacks on September 11.

If in the reform process the community does not adjust to the realities of terrorism as a form of asymmetric warfare and continues to emphasize classic hierarchy coupled with micro-management, the past errors could be replicated. In order to lessen, if not realistically totally eliminate, the threat, there is the need to think creatively, in an environment where, while "all terrorism is local," its impact is international in scope and magnitude. We must be creative and have the imagination to out-think those who declared their own war against the international order. This is no easy task for, as Craig R. Whitney noted in his Introduction to *The 9/11 Investigation*:

> The most difficult thing to get a bureaucracy or political leadership in any system to do after something goes terribly wrong is to acknowledge responsibility for failure. The natural bureaucratic response is to be defensive. Officials hide behind the veil of secrecy, national security or executive privilege. They fear embarrassment, personal or institutional. Yet demanding accountability from the elected and appointed officials of government, and insisting on revealing and correcting their shortcomings are the basic rights and duties of citizens in a democracy.[18]

We cannot afford to continue to engage in incrimination, indignation and, sadly until the next major incident, public indifference. The adversary is determined to combine ancient hatreds, modern grievances and developing technology to continue a protracted assault on the international order.

Notes

1 Sean Kendall Anderson and Stephen Sloan, *Terrorism: Assassins to Zealots* (Lanham, MD: Scarecrow Press, 2003).
2 Frank J. Donner, *Age of Surveillance: The Aims and Methods of the American Intelligence System* (New York: Vintage Books, 1981).
3 Remarks by a senior official at a meeting attended by the author.
4 Stephen Sloan, *The Anatomy of Non-Territorial Terrorism, Clandestine Tactics and Technology Series* (Gaithersburg, MD: International Association of Chiefs of Police, 1978), p. 3.
5 Brian Jenkins, *International Terrorism: A New Mode of Conflict*, Research Paper No. 48, California Seminar on Arms Control and Foreign Policy (Los Angeles: Crescent Publications, 1974).
6 Robert Ardrey, *The Territorial Imperative: A Personal Inquiry into Animal Origins of Nation and Property* (New York: Atheneum, 1966).
7 Steven Strasser (ed.) with an Introduction by Craig R. Whitney, *The 9/11 Investigation: Staff Report of the 9/11 Commission* (New York: Pubic Affairs, 2004).
8 For an excellent study of how the events of 9/11 have impacted, and will impact, on governance and administration, as well as homeland security, see Larry D. Terry and Camilla Stivers (eds.), *Public Administration Review*, vol. 62, Special Issue, Sept. 2002.
9 Maurice A. East, "The International Systems Perspective and Foreign Policy," in

Why Nations Act: Theoretical Perspectives for Foreign Policy Studies, ed. Maurice A. East *et al.* (Beverly Hills, CA: Sage Publications, 1978), p. 145.

10 Robert D. Steele has been a pioneer in this area. See his home page, OSS.Net/ Global Intelligence Partnership Network, www.oss.net

11 Xavier Raufer, "Gray Areas a New Security Threat," *Political Warfare*, 20, Spring 1992, p. 1.

12 P.W. Singer, *Corporate Warriors: The Rise of the Privatized Military Industry* (Ithaca, NY: Cornell University Press, 2003).

13 See Stephen Sloan, "Technology and Terrorism: Privatizing Public Violence," *IEEE Technology and Society Magazine*, vol. 10, no. 2, Summer 1992.

14 Charles J. Dunlap Jr., "Preliminary Observation: Asymmetrical Warfare and the Western Mindset," in *Challenging the United States Symmetrically and Asymmetrically: Can American Be Defeated?*, ed. Lloyd J. Matthews (Carlisle Barracks, PA: Strategic Studies Institute, US Army War College, 1998), p. 1.

15 For a classic on insurgency that still provides insightful guides to the current challenges of unconventional warfare, see the US Marines classic study, *Small Wars Manual* (Washington, DC: US Government Printing Office, 1940).

16 John Arquilla and David Ronfeldt, *Networks and Netwars* (Santa Monica, CA: National Defense Research Institute, RAND, 2001), p. 6.

17 Stephen Sloan, *Beating International Terrorism: An Action Strategy for Preemption and Punishment*, rev. edn. (Maxwell Air Force Base, AL: Air University Press, 2000).

18 Strasser, *9/11 Investigation*, p. xi.

13

ACTIONABLE INTELLIGENCE IN SUPPORT OF HOMELAND SECURITY OPERATIONS

Annette Sobel

Terrorism is an amorphous beast, and the methodology to defeat the beast is also amorphous and highly adaptive.

Background: lessons learned from 9/11

The tragedies of 9/11 taught us that an integrated intelligence mechanism at the federal, state and local levels is essential to early warning and, potentially, to prevention of attacks.[1] Although overt advance warnings are unlikely, continued situation awareness to anomalous events and behaviors may be useful to averting a disaster, and ensuring a metered, appropriate and timely response. We learned that human targets, to include first responders, may be the primary objectives of the attack. In addition, chronic hazardous and environmental effects need to be considered, countered and mitigated, if necessary. Scalability of intelligence and response are essential to smooth transition to post-attack recovery and continuity of operations and government. We also learned that the media are an important element of open-source information gathering, dissemination, recovery, and denial or fulfillment of terrorist objectives. The public reaction to disasters of all types and the ensuing chaos which characterizes such events is, at best, unpredictable and, at worst, a victim to the aggregate psychological objectives of terror.

This chapter will describe the significance, process and objectives of deriving focused, actionable intelligence: intelligence products that support and enable the full spectrum of tactical/strategic homeland security (HLS) operations. The spectrum of operational environments includes: pre- and trans-threat senior decision-making for policy and resource allocation, law enforcement, incident command, preventive health measures spanning vaccination and epidemiologic early warning, and countering transnational threats *prior* to impact on the homeland.

The true measure of value of actionable intelligence is its ability to influence outcome (either positively or negatively) early enough in the pre-, trans- or post-attack process. The term "outcome" specifically refers to the action of mitigating or minimizing destructive effects (direct or indirect) associated with the event. This is far from a linear process, and therefore, in order to understand the complexities of the intelligence process and best take advantage of opportunities to influence outcome, it is important to understand the overall process.

The description of the intelligence process in this chapter will begin to provide the reader with an appreciation of the types and value of the intelligence process and relevant intelligence products, an understanding of the power of intelligence fusion to support and counter threats of terrorism and, finally, a series of recommended approaches to open-source intelligence (OSINT) collation, validation, analysis, fusion, and support to operations. An emphasis will be placed on the generation of a set of baseline information requirements to support homeland security operations. This process is most effective if driven by the end-user of the information, whether policy maker or senior decision-maker or first responder.

The intelligence process and its significance in homeland security

Traditionally, intelligence consists of five steps. These steps are: requirements generation, collection, processing and exploitation, analysis and production, and dissemination. In addition, consumption and feedback are critical to assessing the usefulness of this process. These steps may occur in parallel or sequentially. This process best serves the customer's needs when establishing redundancy at critical junctures in order to ensure continuity of operations during a disaster. For the novice, it is important to understand the difference between raw information and intelligence. Raw information is generally unvalidated, unprocessed and of variable certainty regarding sources and methods. In contrast, final intelligence products (validated and post-analysis) have a measurable level of certainty regarding sources and methods employed in collection.

A systems approach to defining requirements

Operations security (OPSEC) is essential to assuring operational intelligence capability and information surety. The five principles of OPSEC are: identification of critical information; threat analysis; vulnerability analysis; risk assessment; and countermeasures development and implementation. These principles should be embedded throughout intelligence operations. Consider every component of homeland security intelligence operations as an element of analysis for strategic warning.[2] Strategic warning includes identification of significant indicators and warnings that, if detected or tracked by an adversary, would enable the adversary to know what we intend to do prior to that action being taken. Effective homeland security operations rely on overt deterrence and the element of surprise afforded by actionable intelligence.

Identification of critical information

Achieving the information advantage of time or location requires a heavy reliance on measures well beyond traditional security measures. Anticipation or prediction of events impacting national security is a complex business, requiring constant analysis, assimilation and pattern recognition, in order to push the envelope of early warning.

Essentially, to achieve the element of surprise requires development of a systems approach to threat recognition. Intent, motivation and capability are the general constituents of threat analysis. These categories define collection strategies to defeat or counter an adversary through effective targeting. Disruption of the adversary's strategy or denial of opportunities is the ultimate end-goal of OPSEC. One approach to this process seeks to actively anticipate actions and critical information ("red teaming"), also referred to as a process of getting "inside" the adversary's OODA (observe–orient–decide–act) loop. Ultimately, this process adds specificity and sensitivity to actionable intelligence products.

The system elements necessary to achieving enhanced intelligence capability include assessment of: target (asset), vulnerability, consequences, threat, and human processes (behavior, motivation, etc.). Targets may be hard or soft, and include infrastructure, assets, persons, facilities and cyber-systems. Targets may have innate economic value (e.g. gas pipelines) or be symbolic (e.g. schools and churches).

A process for threat agent, e.g. weapons of mass destruction or mass effect, characterization is essential to defining requirements for intelligence gathering. Threat types may be conventional or unconventional in the form of explosives, biological, chemical, radiological or nuclear. During the requirements definition phase, it is important to understand the highly interactive and dynamic interface of threats, vulnerabilities and (un)intended consequences of terrorist actions describing the operational environment. Timelines should consider immediacy and magnitude of effects, observable effects to include unintentional and secondary effects, and chronic effects such as contamination, psychological effects, and community recovery. Comprehensive risk management strategies include baseline or "pre-event" intelligence collection, and are based on access to intelligence, opportunities to implement courses of action, and acceptable level of risk.

Requirements provide a framework for intelligence gathering and assist in ensuring operational relevance. The categories of requirements may be as generic as the who, what, when, where, how and why of an activity or as specific as the characteristics of a target. Establishing criteria for verification and validation of information is an important step in this process which enables the end-user to judge the utility and confidence measure of final analytic products.

One method of generating requirements is to perform a baseline assessment of customer needs and definition of the relevant operational environment(s). In the HLS environment, potential customers and operational constraints are myriad. However, often a generic set of requirements has wide utility. In the phase of intelligence preparation (often referred to as intelligence preparation of the battlefield or IPB), the critical functions of establishing the threats, vulnerabilities and time–space order of battle are performed. This is a highly dynamic process requiring the continuous re-evaluation and assessment which are critical to IPB or situational awareness. The verification and validation (V & V) process of information must be an integral part of this process, and necessarily must be reported to the customer as a "level of certainty" in the intelligence estimate. When not reported, conclusions should be subject to a high degree of scrutiny.

Collection taxonomy

OSINT or open-source intelligence is probably the most valuable source of intelligence for HLS operations. Areas of collection include the worldwide web, media, documents, books, newsletters, conference materials, etc. The challenge to open-source intelligence is establishing the V & V of such intelligence, as previously discussed. In addition, the difference between open-source *intelligence* preparation, implying assessment, and raw data or *information* presentation must be recognized by the homeland security professional and decision-maker.

The diversity of sources for open-source intelligence leads to a condition of information overload. Overload may be managed by a number of approaches to include filtering and data extraction.[3] Filters may be customized to meet the end-user's requirements by employing such items as key words or topics. A prime example of this application includes list-serve and controlled-access chat rooms. The two approaches are opposite: the former is a push of information and the latter is a pull.

Despite the use of automated tools, it is often still difficult to extract the most timely and relevant information. To this end, text analysis software can be a very effective tool to support analysis. In addition, techniques such as clustering can be very helpful in categorizing and place information in "bins" for current or future analysis. Clustering is extremely helpful in trend and link analysis, particularly when a known pattern does not exist or subtle trends are suspected.

The diversity of information which must be mined to develop early indications and warnings of terrorism activities naturally lends itself to open-source analysis. Subsequently, open-source material may be incorporated into more comprehensive multi-source products or, alternatively, used as "tip-offs" or cues to explore more comprehensive and higher-fidelity sources of information.

HUMINT or human intelligence

HUMINT is one of the oldest and most time-proven intelligence collection methodologies. This information is derived from or collected by human sources and may include surveillance, social engineering or technical means. Soft factors such as human behavior, observable changes in patterns of activities or interactions may provide critical early warning to an impending significant event. For example, the recent tragedy of Beslan, Russia had some clear early warning measures which were inadequately appreciated and derived from observation of changes in human behavior noted by residents.[4] Specifically, a number of cohorts of the terrorists vacated the town the day prior to the onset of the tragic series of hostage-taking events. As a result, 330 deaths ensued.

IMINT or imagery intelligence

Imaging techniques range from hand-held cameras to overhead imaging systems used to collect high-resolution images of particular targets or assets of interest. Targets may be of particular interest owing to their activities, programs or proximity to or association with other high-interest targets.

SIGINT or signals intelligence

SIGINT is derived from signals intercepted either individually or in combination with communications, electronic, or foreign instrumentation. There are three sub-categories, namely COMINT (communications intelligence), ELINT (electronic intelligence) and FISINT (foreign instrumentation signals intelligence).

COMINT targets voice and teleprinter traffic, video, Morse code or facsimile. ELINT is derived from interception and analysis of non-communications transmissions, such as radar, and identifies the source of the emitter. FISINT is intercepted from telemetry from an adversary's weapons systems being tested.

MASINT, or measurements and signature intelligence, is information derived from technical sensors. Characterization of features associated with the source emitter, for example, facilitates the identification and measurement process.

The link between terrorism risk management, actionable intelligence and homeland security operations

Social systems are the fabric of America and, as such, represent prime terrorist targets, either independently or collectively.[5] Hence, any credible discussion of actionable intelligence must include a discussion of the underlying vulnerability and threat assessment processes determining risk management strategies. Ultimately, the objective of actionable intelligence is provision of time-sensitive, validated information in a format which influences outcome and minimizes vulnerabilities.

Any sustainable intelligence capability to support operations must include an assessment of critical social systems. These systems may be defined as those essential to maintenance of critical services and functions of a society. Explicitly, the critical social systems for consideration include: agriculture, water, energy, transportation, communication, information, health, banking and finance, security, and education.

To summarize, the ability to effectively link terrorism risk management, actionable intelligence and homeland security operations requires an in-depth pre-event baseline assessment of vulnerabilities and critical nodes necessary to ensure continuity of operations and government. To this end, actionable intelligence provides indicators and warnings of the most vulnerable assets of targeting which, if not systematically addressed, reflect the most deadly of terrorist motives.

Notes

1 9/11 Commission Report.
2 Cynthia M. Grabo, "Anticipating Surprise: Analysis for Strategic Warning," Center for Strategic Intelligence Research, Joint Military Intelligence Research, December 2002.
3 Donald Pearson, "Effective Use of Commercial Databases," *Journal of Counterterrorism and Security International*, vol. 6, no. 3, p. 7.
4 *Economist*, September 2004.
5 New Mexico Surety Task Force, "Terrorism Risk Management Process Guide," September 2004.

14

THE TERRORIST WAR ON THE MARKET-STATE

A plague in a time of feast

Philip Bobbitt

The coming market-state

A constitutional order may be described by the unique claim it makes on legitimate power. Thus the order of princely states, which flourished in the sixteenth century, demanded power on the basis on the legitimacy of the princes with which it was associated. Give us power, the State said, and we will better protect the person and the possessions of the prince. The constitutional order within which most states lived for most of the twentieth century can also be characterized in a unique way. Nation-states, that is states that existed to serve national groups, asked for legitimacy on the basis of a characteristic claim: give us power, the State said, and we will improve your material well-being. The record of economic and material progress during the twentieth century amply justified this claim. Nevertheless, in the past decade, there has been an increasing recognition that we are entering the transition from one constitutional order to another.[1]

The State is not declining, nor is the nation dying, but the relationship between the two is changing and the particular version of the State that has dominated the developed world for more than a hundred years is undergoing a profound change.

Now, at the moment of its greatest triumph, the parliamentary nation-state is increasingly unable to fulfill its legitimating premise: states are finding it more and more difficult to assure their publics, that is their nations, of increasing equality, security and community. In its place there is emerging a new constitutional order, that of the market-state, that seeks not equality but diversity, not security but opportunity, and not community but conscience.

Five developments are driving this change.

First, human rights are being recognized as norms that require adherence within all states, regardless of their internal laws. Nation-states cannot determine the laws to be applied within their borders, because these are being superseded by an international system of human rights. The reason Slobodan Milosevic is in the dock today is not because he failed to obey the laws of Serbia or because he was not democratically elected by the Serbian nation. Rather he ran afoul of a set of norms, some not codified, that Serbian national institutions had not endorsed or had given him authority to defy.

Second, nation-states find it increasingly difficult to protect, much less improve, the cohesion and influence of national cultures. These are strained by immigration, without which the demographically challenged developed world cannot sustain its material well-being, and by the electronic media— e.g. TV and the internet—that make it impossible for nation-states to manage their cultural lives. In China, 60 percent of the educated population gets its news from abroad, despite strenuous efforts on the part of the government to block these channels.

Third, the development of nuclear and other weapons of mass destruction renders ineffectual the defense of state borders as a way to protect the nation. The development of delivery systems (like the ballistic missile) and weapons (like biological and nuclear warheads) radically shifts the balance between offense and defense, because no nation-state can protect itself by simply fortifying its national borders or by increasing the size of its armies. The United States now has about a million and a half men and women under arms; its defense budget is larger than that of the next 14 countries combined and will, by 2006, be larger than the total aggregate of all the other defense budgets in the world. Yet the United States is probably in greater danger today, and will be in increasingly greater danger tomorrow, than it has been at any time in the past century. For a state that claims power on the basis of steadily improving people's material well-being, this is not an encouraging development.

Fourth, national states are unable to govern the value of their currencies because of a global system of trade and finance from which no state can withdraw without plunging itself into falling living standards. The commodification of money in finance must rank with the relativity of time in physics as one of the great breakthroughs of the century in which what were hitherto fixed measurements suddenly became varying objects of value themselves. One consequence of an international market in capital was to remove the control over national currencies from states and give it over to pitiless market forces.

Fifth, global and transnational threats have proliferated, such as AIDS or SARS, climate change, drug trafficking, and terrorism, that no nation-state can hide from nor control within its borders. It is reasonable to expect that sometime in the early twenty-first century a mutated, drug-resistant strain of influenza will strike the human population with a terrible ferocity, quickened by international travel and urbanization. All the finger-pointing among nation-states after the collapse of the Kyoto agreement only emphasizes how poorly placed this kind of state is at coping with transnational problems. It is hard to know whom to blame more for the failure of policy for climate change: those states that insisted on industrial reforms they knew were unachievable or those states—like the United States—that denounced the agreement and then proposed nothing of substance to solve the problem.

How did all this happen? These developments occurred, ironically, as a consequence of the greatest success of the society of nation-states—the end of the wars (World Wars I and II and the Cold War) that constitute the Long War of the twentieth century. They are a consequence of the triumph of market-based democracies over competing forms of the nation-state, communism and fascism. It was our success in building an international system of trade and finance, winning acknowledgement for human rights norms, bringing rapid industrial development to virtually every northern-tier and many southern-tier states, achieving higher living standards and reproductive control, creating international communications, and inventing and deploying weapons of mass destruction, that defeated our competitors and discredited their systems. The very tactics, technologies and strategies that brought us success in war have now brought us new challenges, challenges that cannot be met by the currently prevailing constitutional order.

It would be a mistake, however, to conclude that the decay of the nation-state as a constitutional order portends the withering away of the State itself. This conclusion is a tempting one if one believes, as many do, that the development of the nation-state is synonymous with the development of the State, that the nation-state originated in the Peace of Westphalia in 1648 and has been with us ever since. But if one sees that several constitutional orders have existed since Westphalia (and at least one before) then it is not hard to imagine that the State will, as in past eras, undergo reform in order to accommodate changed circumstances.

A new constitutional order that reflects these five developments and hails them as challenges that only it can meet will eventually replace the nation-state. Indeed this is already happening. As the literature suggests,[2] a new constitutional order will resemble that of the twenty-first-century multi-national corporation rather than that of the twentieth-century state. The new order, like a corporation, will outsource many functions to the private sector, rely less on law and regulation and more on market incentives, and respond to ever-changing consumer demand rather than to voter preferences expressed in relatively rare elections. This new constitutional order, the "market-state," has not arrived, but one can already see evidence of its approach.

When states move from raising armies by conscription to all-volunteer forces; when they introduce vouchers into the allocation of educational funds; when they deregulate not only vast areas of enterprise by repealing industrial statutes but also deregulate the reproduction of our species by striking down anti-abortion and anti-contraception laws; when states replace relatively generous unemployment compensation with retraining programs designed to prepare the unemployed for re-entry into the labor market; and when they rely on NGOs and private companies as outsourced adjuncts to traditional government operations—when all these factors apply they reflect the emerging market-state. When states permit their officials to be removed through ad hoc recall votes and permit their laws to be replaced by voter initiatives and referenda—when all these developments occur we are witnessing the characteristics of the emerging market-state.

This transition will occur over many decades, and there are many forms that the market-state might take. If the past is any guide, the transition will not be complete without violent conflict. In the past, decades-long epochal wars brought about transitions at this scale. It may be that the war on terrorism is the first engagement of this new conflict.

In every era of the State, throughout the evolution of its constitutional orders, societies have confronted the problem of finding the proper relationship between strategy and law. The State seeks to be free of coercion from outside its boundaries; this is strategy. Inside its boundaries, the State seeks to monopolize violence; this is law.[3] In the twentieth century, national liberation and ethnic secessionist groups used terror to gain or keep the power of nation-states. In the twentieth century, terrorists did not ordinarily challenge the idea or the inevitability of the system of sovereign nation-states; rather, they used violence to keep or to acquire power within that system. Indeed, terrorism in the period immediately past typically represented national and nationalist ambitions, pitting established powers against nascent ones to control or create states.

Twenty-first-century terrorism will present an entirely different face, as it does with al Qaeda. It is global not territorial; it is decentralized and networked in its operations like a mutant non-governmental organization (NGO) or a multinational corporation; it outsources its operations; it does not resemble or seek to become a centralized and hierarchical bureaucracy like that of a nation-state. Terrorism in its new guise has no national focus or nationalist agenda; it operates in the international marketplace of weapons, targets, personnel, information, media influence and ideological persuasion, not in the national arena of state formation or reformation.

The nations of the West, particularly the United States but also the United Kingdom, are not winning the war against terrorism. We are not winning because we are neglecting to reconcile strategy and law, in part because separating them in the twentieth century was so successful. As to strategy, we too often rely on habits of mind associated with strategies by which nations in the past have protected themselves from each other. These fail against the threat terrorists pose because it cannot be located in another state. As to the rule of law, our habits of mind associate it with institutions that command power within but not between states. Law must now extend internationally to constrain terrorism. Within the Alliance, the Americans tend to practice strategy to the exclusion of law; the Europeans tend to emphasize law without an accompanying strategy.

But to combat twenty-first-century terrorism successfully we must conceive and create an interdependent relation between strategy and law. Strategy (the role of the State in defending itself from violence from other states) and law (the role of the State in monopolizing violence within its own borders) begin to coincide. Terrorism in its new form makes it essential for societies to recognize and strengthen the crucial and symbiotic relationship between strategy and law.

The terrorist war

A revolution in military affairs, itself the product of complex global change, is transforming the idea and function of the State. In the Renaissance, the first modern states formed in response to the need—dictated by the military revolutions of that time—to raise huge armies on which military success and thus strategy relied. By the twentieth century, the State offered its citizens welfare as well as security in exchange for military service and civil obedience. As the military has become more professionalized, more dependent on skill and equipment than on numbers of soldiers, and more centered in technology, the State has become less dependent on raising vast armies. Weapons of mass destruction make civilian populations vulnerable no matter how many soldiers they can send to the front. At the same time, citizens have come to understand that the State is no longer the ultimate source of welfare; people must seek their welfare and opportunities within a global economy. Thus the State has come to promise not so much welfare as opportunity, not so much equality as diversity, not so much ideology as freedom of conscience. The transformation of military and economic conditions has led to a transformation in political organization. The market-state is supplanting the nation-state.

Terrorism in its new form also undermines the legitimacy of the nation-state and supports its transition to the market-state. The market-state, in turn, encourages the transformation of twentieth-century, nationalistic terrorism into twenty-first-century, global, networked terrorism. In a system of market- rather than nation-states, strategy has to change as well—in other words, states have to change the way they think about securing their borders against violence and coercion from abroad. States have to rely on the rule of law not only within but also beyond their territories to control external threats—particularly the threat posed by international terrorism but also by the drug trade, the spread of disease and other such forces. Because we have neglected the relationship between strategy and law, we in the West have delegitimated our own efforts in international and domestic law— particularly by the Great Power stalemate in the UN, and American behavior at Abu Ghraib and Guantanamo—and we have abandoned strategic initiative to our enemies, for example on September 11, 2001, in Baghdad in the winter of 2003, and in Madrid in the spring of 2004.

Since September 11, the United States has declared war; and has received the unprecedented invocation of Article Five of the North Atlantic Treaty by its allies on its behalf.[4] The US Congress and the British Parliament have passed various statutes aimed at making the prosecution and detection of terrorists easier. The United States has reorganized its bureaucracy and authorized vast new funding for fighting terrorism. US/UK-led coalitions have invaded and conquered Iraq in a lightning campaign, and the UN has sanctioned, for the first time, the invasion of a member state, Afghanistan, in order to suppress terrorism. Coalition forces have for the first time acknowledged targeted assassinations against terrorists. Much of the senior leadership of al Qaeda has been killed or detained, with all the fruitful possibilities for interrogation. What remains—the senior figures of bin Laden and Zawahri—is in desperate flight.

And yet, at the same time, al Qaeda has continued to strike; indeed there has been a drumbeat of violence and, far from abating since the invasion or Iraq, it has picked up momentum. The deadliest year of terrorist violence in 20 years occurred in 2003, and if we exclude terrorism waged by traditional states it was the deadliest year on record. Every year since has surpassed the preceding totals. In Bali, Kenya, Pakistan, Tunisia, Afghanistan, Iraq, Israel and Morocco, as many people have been killed and wounded in terrorist attacks since September 11 as died on that day, itself the most deadly terrorist attack in history. Virtually every week, US soldiers are killed in terrorist attacks. Arab television networks and al Qaeda websites show the beheading of innocent civilians, a grotesque *coup de théâtre* never depicted before on television. US and British citizens, and non-citizens who are in US or British custody, have seen their rights diminish. As Americans experience countless alerts, color-coded to indicate threat levels, they can reasonably conclude that they are less safe than before, and some believe perhaps less safe than ever. There is a widespread sense in the West of the inevitability of further major terrorist attacks on the scale of the 9/11 atrocity. And many professionals expect that terrorists will acquire and use weapons of mass destruction.

We must step back and ask the most basic questions about winning the war on terrorism. Do we know what we are doing, in the way that we knew what we had to do to defeat Germany and Japan in World War II? Are we developing new strategic doctrines of the kind we had to develop to confront the Soviet Union in the context of mutual deterrence in the Cold War? Are we writing new international law and creating new institutions to cope with global problems in the twenty-first century in the way we did when we faced similar global challenges in the twentieth century? I think answers to all these questions are evident. Our legal and strategic habits, which are enshrined in international institutions like the UN and NATO, in military plans that contemplate invasion and conquest, and in intelligence operations that are geared toward renditions and prosecutions, are not appropriate to the decentralized operations of a mutated market-state like al Qaeda, which finds lucrative targets in the emerging market-states of the West.

Secretary of Defense Donald Rumsfeld put the situation well in a memo-
randum of October 2003 that asked, "Does the US need to fashion a broad,
integrated plan to stop the next generation of terrorists? The US is putting
relatively little effort into a long-range plan, but we are putting a great deal
of effort into trying to stop terrorists. The cost–benefit ratio is against us.
Our cost is billions against the terrorist's cost of millions."

This is the right question, and it applies to both the United States and the
United Kingdom. One may ask, however, whether the US administration of
which the Secretary was a member understood the extent to which the
September 11 attacks were a violent reaction to America's preeminence.
Recent American behavior on the international stage suggests an arrogant
neglect of the importance of the way people in other countries perceive
American hegemony—a disregard of the critical importance of public perceptions in the emerging world of media-centric market-states. Writing in
2004, the International Institute for Strategic Studies in London concluded
that:

> the manner of US preponderance had to change . . . the appearance
> of American unilateralism needed to be tempered. Strategic ends
> had to be more adeptly coordinated with tactical means. The neces-
> sary tools included more nuanced public diplomacy, which could
> portray a less parochial and chauvinistic society while emphasizing
> religious pluralism; less doctrinaire political and economic condi-
> tionalities attached to foreign assistance . . . and an approach to
> international law that—after more than two years delay—openly
> admitted that the older standards of intervention and the laws of
> war that applied to state-based security problems and standing
> armies did not easily fit new security problems and that these
> required systematic, collegial reconsideration on a multilateral
> basis.[5]

With equal validity, one might also have concluded that the manner of
much European behavior in collective institutions like NATO and the UN
has to change. The appearance of strategic neglect must be tempered. Legal
and diplomatic means have to be better coordinated with the strategic goals
of the Alliance, which include the defeat of terrorists and the anti-Western
states that support them. At times French, German, Belgian and more
recently Spanish policies appear to have few other strategic goals than
thwarting wherever possible the efforts of the US and the UK. Instead of
strategic goals, some European nations appear more interested in tactical
maneuvering; they neglect the role of force, while they insist on an international law that plainly does not fit current circumstances.

Rebukes of this sort may be wholesome, but they require, to lead to
reform, an examination of the basic questions involved in waging war

against terrorism. The problem is not simply a matter of the difference in transatlantic styles. The problem arises from the perceived hegemony—or will toward hegemony—of the United States and from the inability of many nations to cope constructively with this preponderance of power.

It is clear that the US, owing to its great economic and military presence, is the chief cause of twenty-first-century, networked terrorism. That is not a matter for blame, anymore than one can blame urbanization for the Black Plague in the fourteenth century. It is equally clear that terrorists believe they can win the war in part because they perceive that other states, including historical American allies, lack a sincere interest in collaborating with the US. This too calls more for reflection on basic issues than for tirades against the leadership of any state. All technologically advanced states will ultimately be threatened by twenty-first-century terrorism. It is understandable but regrettable that some of these states would want to protect their publics by disassociating themselves from the United States and thus from the most prominent political target in the West.

The unwritten script

Here is a brief characterization of some of the basic questions states of the West must answer and of how these questions might be addressed.

First, American policy at present defines the problem of winning the war against terrorism in a way that makes winning impossible. The ways we conventionally understand "winning" (a victory with an armistice agreement followed by occupation and a peace treaty), "war" (a conflict between nations over issues of statehood and sovereignty) and "terrorism" (a criminal act by the disenfranchised or the psychopathic) mean that we simply cannot defeat terrorism. "The war on terrorism" becomes little more than a metaphor for propaganda purposes; we can no more win such a war than we can win a war against disease or disillusionment. We must reconceptualize our key ideas to bring them into accord with the changing nature of war, terrorism and victory.

We are going to have to understand terrorism from the supply side, not simply the demand side. By that I mean we must change our exclusive focus on who is the terrorist and what troubles him—what creates the demand for terrorism—to the vulnerabilities we have created and how to reduce the supply of opportunities to the terrorists. This will become ever more urgent, regardless of what happens to al Qaeda, as we enter a period in which it will be increasingly difficult to determine precisely who is striking at us and from what remove. At some point, we must have a strategy for coping with attacks without knowing who is attacking us.

We shall have to abandon the nation-state's dichotomy of crime and war—the inner and outer dimensions of state violence—and replace this with a world view that admits a free flow between these two dimensions. The

225

IRA were criminals who hungered to be treated as soldiers of Ireland as they conceived it; the Waffen SS were soldiers who behaved like criminals, and deserved to be treated as such. But the atrocities of September 11, though crimes of historic proportion, were not committed by criminals who acted as soldiers or by soldiers who acted as criminals. The terrorists plotted their actions with military precision, against military and political targets, and they were willing to sacrifice themselves for purely political goals. Yet, they served no nation-state or its goals; they were not soldiers. On the other hand, their actions were entirely motivated by political or religious ends rather than by personal interests or viciousness; in this sense it is hard to regard them simply as common criminals.

Second, we must be clear about what we are fighting for, and what that fight requires of us, lest terrorists effectively defeat us through our own misguided attempts to protect ourselves. The question "What are we fighting to defend?" is not nearly as easy to answer as it is made to look by those civil libertarians who alarm us by claiming that any diminution in our rights means "the terrorists win" or by those bureaucrats and politicians who soothingly reassure us that all the necessary measures can be taken without compromising our civil freedom of action.

But what measures are appropriate for a state, within its territory, to prosecute the war outside when inside and outside have lost their clear boundaries? Governments must explore the changing relationship between the intelligence agencies (as they become more dependent on open sources) and the media (as they become more powerful purveyors of secrets); between the political parties who seem to have shunned the traditional bipartisanship of governance during war; and between federal unions and their constituent parts (both in the US and in the EU) where intelligence, in the case of the US, is not shared by the central union and, in the case of the EU, is not shared with the central union owing to national distrust. Governments will have to learn how to find and work with private sector collaborators, partly because they own most of the critical infrastructure that has to be protected or made less vulnerable, and partly because they are market-oriented and global, thus arcing some of the gaps between the nation-state and its terrorist adversaries. Governments must rethink ideas like "homeland security" when the threats to security cannot be neatly cabined as in or out of the homeland. Similarly, the American and British governments must revisit the issues of cooperation between the CIA and the FBI and between MI5 and MI6— issues that arise owing to jurisdictional divisions between domestic and international operations, largely because these agencies are so completely defined by the wars of the twentieth century,[6] which were based in the idea of a territorially demarcated nation-state.[7]

The rights of the People and the powers of the State are like the shoreline and the sea, constantly shifting but generally staying within high and low tides, the movement of one line matching exactly the retreat or advance of the other. This two-dimensional way of looking at private rights and public authority can be misleading, however, because it omits the role of alternative, possible worlds that allow us to compare not rights against powers but rights in one context of state power versus rights in another context. The relation between private rights and public power becomes more complex in periods of great change, such as the present one. We should not make the mistake of measuring our liberties against those which we once enjoyed; the threat of terrorism requires us to think beyond analogies with the past. Instead we must ask whether our rights, in the future, will be greater or lesser if a particular security policy is pursued, in the present. A state that fails to protect its people's security in order to keep their traditional liberties intact will end up with a society that possesses neither security nor liberty.

States must maintain their legitimacy to keep their power; it is essential, then, that governments consider the impact on their legitimacy of their tactical and strategic policies. If the United States were to abandon its Executive Order prohibiting assassinations, what would be the cost to its legitimacy as a State that follows the rule of law, one principle of which is that no criminal penalty can be levied without a fair and open trial? On the other hand, if the US is at war, is the Executive Order even relevant? The domestic environment of states can steadily be militarized, however, if the "war" against terrorism is fought at home as well as abroad. Similarly, do the goals we are fighting for possess intrinsic merit and legitimacy or is "one man's terrorist another man's freedom fighter"? We do not apply murder statutes to soldiers in battle, even enemy soldiers. Soldiers are permitted to maim and kill civilians if that is not their aim, while we condemn the terrorist whose objective is to kill civilians. By such means, however, the foreign environment can be degraded into a sea of "collateral damage." Put all this together and the war on terror can make our soldiers into organized vigilantes, using the methods of warfare against civilians, domestic and foreign.

If our governments engage in torture, perhaps by turning over prisoners to less squeamish national intelligence services, are they substantiating the charges made against them by those who say ours are the true rogue states,[8] and that the state terror of the US and its allies, including Israel, is every bit as much a threat to mankind as the terrorism of al Qaeda? These are essentially constitutional issues, yet they are not matters of civil liberties, but rather of the self-definition societies achieve through their constitutional development. They are matters of constitutional legitimacy because they are matters of self-respect. States must have clear answers to these questions in order to decide to what extent they will sacrifice the legitimacy of the rule of law to meet the threat posed by terrorism. If the legitimacy of the State is too deeply compromised the State will seed its own terrorists who in revulsion will take up arms against it.

If the US and the UK ally themselves with undemocratic autocracies who share our fear of al Qaeda but with whom we have little else in common, are we simply borrowing against a future in which those peoples whom those autocracies suppress will rise up and blame us—much as we have been blamed for collaborating with dictators in the Third World to fight communism (though we are seldom blamed for the equally awful collaboration with communism to defeat fascism)? Is there a realistic choice? If it is true that full and fair elections in a dozen Islamic states would bring bin Laden to power, does the international community dare to risk such an election? And if it does not, does this make us hypocrites 1) to claim that the sovereignty of other states, like Iraq, is forfeited owing to its undemocratic practices and then 2) to turn a blind eye to the legitimacy of regimes that are allied to ours but that deny their citizens basic human rights? Or does the possibility that free and fair elections would bring to power those whom we regard as terrorists mean that our commitment to globalize the systems of democracy itself—or what we mistake for the pluralistic system we have evolved and called "democracy"—must be rethought? Answering urgent strategic questions about terrorism will also require us to give some thought to larger constitutional issues about sovereignty, democracy and the claims of empire because if we ignore these questions we will find we have answered them, inadvertently, in the unthinking acts of crisis that ultimately proved determinative.

Third, while the United States must play a leading role in the war against terrorism, winning will require the collaboration of many states, including some states that fear and even loathe American hegemony. The risks of leadership are twofold: if the US is out in front, it becomes the target for every terrorist group that simply wants a free hand for its various predations while, at the same time, America becomes the focal point of charges by other states that it is seeking an empire. Some of those who make the latter charge believe simply that power corrupts and overwhelming power necessarily leads to empire, indeed that overwhelming power is itself the definition of empire.

The United States is very powerful, economically and militarily. It has the world's largest economy, greater than those of all the other members of G8 combined, and it is growing at a faster rate than they are. The US is the only state that can settle its debts in its own currency. It is, militarily, the only remaining superpower, owing to the collapse of the Soviet Union and US defense budgets that approach half a trillion dollars. Yet we should not be misled by these figures; like the much-cited increase in the gap between high- and low-income earners, these statistics conceal an equally important truth—that the development gap between high and low is closing. This means that, while the US has a large army equipped with infinitely superior weaponry and communications, the harm that can be done to the nation is growing more quickly (as technology disperses and becomes cheaper) than its lead is growing. In other words poor states—or rich terrorist groups—who could not begin to mount a challenge by invading across a contested plain, can hope to do enough damage to dissuade the US or any other powerful state from attempting to coerce them. This paradox—the increasingly greater power and greater vulnerability of the US—means that America is the indispensable leader of the war on terrorism (because it alone has the resources) and that it has a vital interest in actually being the leader (because it is also very vulnerable).

American leadership so far, alas, has invited disarray and non-cooperation. The former French Foreign Minister, Hubert Vedrine, spoke for many when he said, "We cannot accept a politically unipolar world."[9] This conclusion is shared by many outside the US. It is, sorry to say, actually true that, when in the midst of ongoing hostilities in Iraq the French Foreign Minister Dominique de Villepin was asked at the International Institute for Strategic Studies which side he wanted to see win that war, he simply declined to answer after a long and irritated pause.

Indeed there are many who see the war on terrorism as a kind of stalking horse for the creation of an American empire. One research center has provided a list of what it takes to be America's true intentions in the war against Saddam Hussein's regime. The war, the center maintains, was undertaken in order to:

- instigate a "clash of civilizations" that will provide the US with an excuse to reorganize the world under the tutelage of an American empire;
- secure control of the oil- and gas-rich lands of Central Asia and the Middle East;
- undermine the political and economic development and integration of the Eurasian landmass;
- maintain economic power during the course of the current financial crisis by using US taxpayer money (and lives) to force on the world that which a truly free market would not have otherwise allowed—unchallenged American economic and political supremacy.[10]

One must shudder at the consequences for the world, to say nothing of the war on terrorism, of such attitudes, for they invite an anti-American multipolarity with which the worst and most retrograde forces can tacitly combine. Multipolarity is not simply a condition of mutually affecting forces but of mutually opposed forces. How many persons who have called for a European army in order to achieve multipolarity to "balance" the Americans have actually thought through what such an army would do to achieve the objective of thwarting US unilateralism? If that army were to join American expeditions then it might well have influence on allied policy. But this is not what the opponents of US hegemony have in mind. Indeed they have frantically (and successfully) tried to keep NATO forces out of Iraq. If, however, the objective is to prevent US forces from intervening in Serbia or Afghanistan or Iraq or Sudan, then such an army must be used to threaten the use of force. What other role could it possibly play in achieving such an objective? That was how multipolarity checked US polices before 1989 when the Soviet army stood ready to oppose any allied attempts to liberate Eastern Europe. Is it possible that any sane person would want to recreate the conditions for such an armed confrontation in the twenty-first century?

If neither unilateralism nor multipolarity is acceptable, what about multilateralism? Should the war against terrorism be prosecuted under the auspices of a multinational organization such as the United Nations, or perhaps NATO or the EU? Or should what have come to be called "coalitions of the willing" become an acceptable means of fighting this war? There is, at present, no more important question before the world because failure to resolve the problem of legitimate cooperation will frustrate our efforts not only against global terrorism but also against climate change, regional and global epidemics, the international drug trade and other pressing matters.

What constitutional and strategic models can help us reconstitute societies that have been ravaged by conflict and have sheltered terrorists? It may be that we can revive the otherwise outmoded provisions of the UN Charter that create trusteeships for failed states such as Afghanistan or post-war Iraq. Or it may be that we will need new models that are less territorial and exclusive, such as free trade zones for both the US and the EU that embrace areas such as Palestine, Kashmir, the Koreas, Iran and Iraq, creating incentives rather than trying to coerce these societies toward humane constitutional development. The EU's market in sovereignty is one such model, admission to which is priced in terms of relinquished sovereignty, thus creating incentives to achieve reforms in human rights within nations wishing to join.

Fourth, we must urgently develop legal and strategic parameters for state action in the war on terrorism. This will be a matter, ultimately, of evolving legal concepts of sovereignty and its relationship to lawful, legitimate governance while fighting terrorism. We might start with a pragmatic definition of what constitutes terrorism. Perhaps this: "Terrorism is the use of violence to prevent persons from doing what they would otherwise lawfully do when that violence is undertaken for political goals without regard to the protection of non-combatants." Beginning with such a definition, we can then work out what a state is permitted to do in its search for terrorists and in its efforts to suppress them. When a state is acting lawfully—respecting basic human rights, adhering to the rule of law and protecting the institutions of uncoerced consent—violent acts against it amount to terrorism; when the state acts unlawfully on a systematic scale, it may be resisted. In neither case can civilians be targeted. With such a definition we could seek an international convention universally outlawing terrorism as we outlaw piracy. With such a definition we could determine when a group comprises terrorists or "freedom fighters," and when other states may intervene to stop them.

A definition of terrorism would help up to assess the new US National Security Strategy and its call for preemption in spite of the obvious conflict with Article 2(4) of the UN Charter, which prohibits the use of force by any state outside the Charter's carefully circumscribed limits. This article asserts that it is unlawful for a state to use force in the absence of an actual or imminent attack[11] or authorization by the Security Council. Does this mean it is also unlawful—in the absence of a Security Council resolution—for one state to preempt another's war-making capabilities before these are ever put to use? In the era we are entering of disguised attack using terrorist networks, preemption is an absolute necessity where the proliferation of WMD to violent groups is concerned. For once any state, no matter how repugnant, acquires nuclear weapons, a moment that no UN or US monitoring seems capable of predicting with precision, it is too late to compel de-proliferation. The genie cannot be put back in the bottle and will do the bidding of its new master. The chief reason why Saddam Hussein is not in power and Kim Jong Il is—at least of this writing—is because the latter got to the nuclear finish line before he was preempted (despite, it should be noted, UN inspections).

A definition of terrorism could also assist analysis of the so-called "root" causes of terrorism. The developed world might reasonably suppose that it will substantially reduce the threat of terrorism by aiding the peoples of less developed states—to improve their health and longevity, their per capita incomes and education, their human rights and political liberties, and the like. But, in fact, the tie between causes and effects is too tenuous. A "supply-side" approach to terrorism better fits the global, anonymous networks we shall have to face in the twenty-first century than the "demand-side" approaches that were relevant to the national liberation movements of the twentieth. The search for root causes, moreover, leads to an unexpected conclusion: it is the United States' position as world leader, economically, politically and militarily, that is the principal driver of twenty-first-century terrorism. This uncomfortable conclusion makes even more problematic American state-sponsored acts of assassination and torture. If the assassinations and torture by allied states are countenanced, indeed financed, by the United States, either because the US supports their war aims or because they are US proxies, then, it is argued, the US is rightly subject to the same accusations of terrorism it would hurl at any other state that employed such methods. Can the United States persuade its citizens and its allies that these tactics are the only effective means of protecting a society at war with those who can easily infiltrate it and whose operations prefigure the tactics the United States will itself be forced to adopt? If the US does adopt these methods, are they more like the strategic bombing of World War II, which relied on an *in terrorem* effect to achieve its military goals (as at Hiroshima and Nagasaki), or more like the bombing of civilian populations that we now condemn as war crimes (like the blitz against London or the Allied bombing of Dresden)?

We must develop new rules of international law that may be used to determine when it is permissible for one state to intervene in another's affairs in order to protect itself or it allies from terrorism (*jus ad bellum*). Similarly these rules would govern the ways states may lawfully treat prisoners during the war on terrorism (*jus in bello*). Obviously we need to amend the Geneva Conventions to deal with the question posed at Guantanamo: What treatment is to be accorded prisoners of the war on terrorism? They are not combatants in uniform, with a publicly acknowledged chain of command. But they are not spies or partisans either. As soldiers, even if unlawful ones, who are captured on the field of battle they can be held in prisons until the end of the conflict without trial or arraignment. This scarcely makes sense, however, when there is no nation-state with which to agree to end the conflict or to make arrangements for prisoner exchanges—when, that is, these prisoners may be held forever because the field of battle is everywhere and the conflict is perpetual.

Fifth, we must confront the possibility that we will not extinguish global terrorism because we and the rest of the international community will be unable to transform the way we think about strategy and law successfully. We must, that is, consider the question: If winning the war against terrorism is not losing, what constitutes losing? Much important work remains to be done on the question of losing the war on terrorism. The use of global scenarios—a technique pioneered by Royal Dutch Shell, and eloquently recommended by Joseph Nye when he was head of the US National Intelligence Council and imaginatively implemented by his successor, Robert Hutchings—is an appropriate but at present underutilized means of anticipating such failures and coping with or even preventing them.

If the United States loses it strategic hegemony and its legal legitimacy as the foremost leader of the West—and if the West therefore finds that its strategic hegemony as an alliance and its legitimacy as the chief formulator and adherent to the rule of law within the international community are significantly weakened over what they would have been had the United States done nothing to oppose terrorism after September 11—then the West has lost. This could happen in various ways that scenario planning can illuminate.

Plague in the time of feast

There have been no more than a half-dozen constitutional orders in the last 500 years. For more than a century at a time, the constitutional orders of the leading states typically have remained stable. It happens, however, that we are entering one of those rare periods of seismic change—from the nation-state of the twentieth century to the market-state of the twenty-first. The nation-state is being challenged in a number of fundamental ways, such as the five described above. By using market techniques—such as outsourcing, market incentives and the like—to supplement or even replace legal regulations, nation-states are gradually moving toward a market-state model. They will need to devolve power and build decentralized institutions, adopting looser structures in the international context that will, perhaps paradoxically, ultimately strengthen the State. One consequence of this devolution, however, is the radical increase in the influence of ad hoc, direct democracy reflected in the growth of referenda, voter initiatives and recall movements at local scales. This devolution gives far greater power to publics, which is why the theatrical elements of twenty-first-century terrorism are bound to increase. Precisely because publics—and of course the media that inform and persuade them—are becoming so much more powerful, political movements of many kinds will seek to influence them directly by terrorism, making terrorism the extension of diplomacy by violent means.

The emergence of market-states will bring greater wealth to mankind than it has ever known. This new constitutional order will make life more abundant by increasing productivity, more spacious by increasing accessibility to more varied environments, and more connected by means of a global network of telecommunications. Although it will bring important cultural challenges to societies, as an engine of wealth creation the market-state surpasses all of its predecessors. Its *raison d'être* is to maximize opportunity. On this basis it lays claim to power. The legitimating premise of the market-state is: Give us power and we will maximize your opportunity. It has the potential to bring a kind of perpetual feast to the developed world.

The market-state is also responsible for creating the conditions for twenty-first-century terrorism. The dramatic growth in wealth and productivity the market-state harnesses occasions a parallel increase in vulnerability. Market-states provide the model for global, networked and outsourced terrorism. Market-states enable the commodification of weapons of mass destruction. The global presence of the United States, the first and most dynamic of the emerging market-states, constitutes the principal target as well as the main precipitating factor of twenty-first-century terrorism. American military and economic power, American empathy and ideals, and American ubiquity have brought forth both American hegemony and al Qaeda, and will bring forth other global, networked terrorists in the future. The appearance of mutated market-states like al Qaeda represents the emergence of a form of plagues, propagated by the very conditions that brought us feasts.

We are not winning the war against terrorism because we don't understand its deep connections to historic changes in the nature of the State that are currently under way—principally the transition from the constitutional order of the nation-state to that of the market-state. Terrorism will be waged by state proxies and entities that are not controlled by conventional states, which seek to influence the politics of states by theatrical killings and atrocities. Strategy and law, which were carefully separated in the twentieth century of nation-states, will have to be reintegrated in the twenty-first century of market-states. Neglecting this task is the reason we are not winning this war, but we have not lost the war either. It is time for a serious rethinking of first premises.

Notes

1 Mark Tushnet, *The New Constitutional Order* (Princeton, NJ: Princeton University Press, 2003).
2 Philip Bobbitt, *The Shield of Achilles: War, Peace and the Course of History* (New York: Knopf, 2002); see also Rowan Williams, Archbishop of Canterbury, Richard Dimbleby Lecture 2002, Dec. 19.
3 Why only "legitimate" violence, as in original?
4 Article Five provides that the members of NATO will treat an armed attack upon any member state as an attack against all.
5 International Institute for Strategic Studies, 2004 Strategic Survey, at 18.
6 The Long War (1914–1990) was an epochal war fought ultimately to determine whether the nineteenth-century imperial constitutional order would be replaced by nation-states governed by communism, fascism or parliamentarianism. See Bobbitt, *Shield of Achilles*, pp. 21–64.
7 See, for example, the provisions forbidding the CIA to investigate persons within the US.
8 A survey by Eurobarometer conducted for the European Commission of the European Union asked respondents which of 14 countries presented a threat to peace in the world. Among EU respondents, North Korea and Iran were tied with the United States. See http://europa.eu.int/comm/public_opinion/flash/fl151)iraq_full_report.pdf
9 R.W. Apple, "Power: As the American Century Extends Its Run," *New York Times*, Jan. 1, 2000.
10 Center for Cooperative Research, http://cooperativeresearch.org/index.jsp.
11 But see also Article 51, providing for the right of self-defense.

15

COUNTER-TERRORISM IN CYBERSPACE

Opportunities and hurdles

Neal Pollard

Introduction

In the Cold War, the enemy of the US threatened the homeland with a strategic triad of hardware platforms: intercontinental ballistic missiles, strategic bombers, and ballistic missile submarines. Today, religious extremists threaten the homeland, exploiting a new strategic triad of terrorism, failed states, and proliferation of high technology and weapons of mass destruction.[1] However, the capabilities provided by this strategic triad are not based in hardware—they are platforms of international networks, enabled by the engines of globalization. Terrorists exploit globalization to form these networks and wield disruptive power.

Terrorist networks flourish in cyberspace, and cyberspace is a front in the global war on terrorism. Terrorists have proved adept at exploiting information technology (IT) as a tool or target. As a tool, terrorists have used cyberspace—namely, the internet and the worldwide web—to recruit and transmit propaganda, disseminate training manuals and doctrine, conduct transactions, move and bury their finances, communicate and coordinate with operational cells, discover vulnerabilities and conduct surveillance against potential targets, and develop and disseminate operational plans. Terrorists may even be trying to replicate in cyberspace the training safe haven they enjoyed in Afghanistan before September 11, 2001. Terrorists have also considered IT a target: terrorists have accompanied "conventional" campaigns with propaganda and electronic attacks, and there is concern and evidence that terrorists will someday target and attack the information systems that they rely on today.

Fighting in cyberspace is not one-sided. The US has opportunities to use IT strategically, to compete with terrorists in cyberspace, deny terrorists the use of IT as a tool, and defend cyberspace from attack. New technologies such as data mining, aggregation, and pattern recognition tools, the resources of corporate infrastructure providers, and the cyber-defense assets of the US government represent a formidable array of potential capabilities to prevent terrorists from exploiting or attacking cyberspace.

The US and its allies can use strategic IT to build an effective global counter-terrorism network in cyberspace, denying the benefits of globalization to terrorists. However, the US will not be able to form these networks and secure cyberspace until it resolves uncertainties in public policy and law. Lines must be drawn in policy and law, to resolve these uncertainties and clarify authorities of government and expectations of the people. These areas of policy and law include: data mining and privacy, corporate roles and responsibilities in critical infrastructure protection, and roles and authorities in responding to cyber-attack.

Globalization and its bastards

Modern terrorism was born on July 22, 1968.[2] On that date, terrorists from the Popular Front for the Liberation of Palestine (PFLP) hijacked an El Al flight from Rome to Tel Aviv, and demanded the release of comrades-in-arms. It certainly was not the first hijacking, but it was unprecedented because of a number of reasons: its purpose of trading hostages for prisoners, the specific targeting of an Israeli-flagged airliner, forcing Israel to communicate with a *persona non grata* terrorist group, and the specific aim of creating an international media event. Zehdi Labib Terzi, the Palestine Liberation Organization's chief observer at the UN, said in a 1976 interview, "The first several hijackings aroused the consciousness of the world and awakened the media and world opinion much more—and more effectively—than 20 years of pleading at the United Nations."[3] Terrorists found a powerful means of expressing themselves globally, at the convergence of two recent technological developments: cheap intercontinental travel using airplanes, and a pervasive global media using international communications technology. Not coincidentally, cheap intercontinental transportation and global information communication are the two main engines of globalization.

Modern globalization resulted in modern terrorism. Some have argued that the terrorism of September 11, 2001 was the death knell for globalization. Others have argued that terrorists have hijacked globalization, as though terrorism were a force external to globalization. Neither argument is totally accurate: terrorism *is* globalization or, at least, its illegitimate, rebellious, highly agitated offspring. The strategic use of information technology continues to be a key competition between liberal democracies and extremist groups, as both benefit from globalization.

Globalization—and the illegitimate networks that threaten the US—finds its power at the convergence of four trends driven by IT. First, globalization has removed political, economic and technological divisions as a hindrance to commerce, communication and movement. Globalization has created worldwide networks of interdependence, transforming political and economic power.[4]

Second, virtually every individual or business is now connected in cyberspace,[5] the system of electronic networks that includes the internet and the worldwide web. Even if an individual avoids cyberspace, those from whom he receives products or services rely on the internet.[6]

Third, there is the multiplicity of databases of information on individuals and entities. When a private or business actor connects to cyberspace, that connection leaves data, collected and stored for later usage by a variety of entities: government agencies, statisticians, insurance and credit companies, marketing professionals, etc.[7] Most data is provided knowingly and voluntarily, but what happens to the data afterward is more opaque, as illustrated by recent cases with commercial data aggregators ChoicePoint and LexisNexis, in which sensitive personal information, collected legally, was mistakenly sold to fraudsters.[8]

Fourth, there is the increasing power of IT to access disparate databases, mine and aggregate data, and identify patterns.[9] ChoicePoint alone collects 4,000 data sources per month and houses over 100 terabytes of data.[10] These sources of data can be combined with data mining technology to identify patterns, locate non-obvious links among individuals or entities, suggest hypotheses for analysis, and automatically determine which hypotheses are best supported by data. These uses have direct benefits for countering terrorism.[11]

"Cyberspace" is the forum where these trends converge and connect to the "real" world of society, commerce, culture and politics. Cyberspace hosts the communications, finance, information, infrastructures and even capital that support modern society. Even ten years ago, over 60 percent of the US workforce was engaged in activities related to information management; the value of most wealth-producing resources depends on "knowledge capital," rather than financial assets or labor resources.[12]

These trends present both opportunity and danger. They present an opportunity, in that they are enablers of globalization and its benefits to liberty, productivity and progress. The engines of globalization—the information revolution, cheap and open intercontinental transportation, global 24-hour media, electronic finance infrastructure, increasing participation in international organizations, and liberalized trade and investment—increase productivity, business and trade efficiency, cost-effectiveness of human and capital migration, scientific collaboration and development, and the ability of the media and individual communications to overcome government oppression, abuse and corruption. These are arguably boons to humanity, and they are the foundation of effective global networks.

The potential dangers of globalization and cyberspace are twofold. Firstly, cyberspace is an equalizer, and benefits legitimate and illegitimate enterprises alike.[13] Terrorists exploit the same information systems and engines of globalization that foster productivity and open exchange among individuals, businesses and societies.[14] Table 15.1[15] illustrates the nexus between globalization and strategic technology, and how that nexus can empower both legitimate and illegitimate actors and networks.

The second danger of globalization and cyberspace is the concern that terrorists will eventually target cyberspace. The danger is that terrorists will identify the extent to which the US relies on cyberspace and critical infrastructure protection, and attack in cyberspace, exploiting software vulnerabilities or government outsourcing, combining with physical attacks to disrupt critical government services, or even disrupting the infrastructure across a large regional or functional scale, bringing modern society (and safety) to a halt.[16]

The US has opportunities to mitigate both of these dangers, relying on its technological base for resources. However, these opportunities are jeopardized, owing to political and legal uncertainties in data mining and privacy, critical infrastructure protection and public/private cooperation, and government response to cyber-attack.[17]

Opportunities and hurdles

The US has the opportunity to build its own strategic triad of networked platforms to counter-terrorism in cyberspace: networked information fusion and analysis, networked public/private infrastructure partnerships, and networked international coalitions of allies and economic partners for cyber-defense.

Table 15.1 International actors in the globalization age

Type of actor	Legal actor	Illegal actor
State	United Nations member.	"Rogue state."
Profit	Multinational corporation.	Transnational organized crime syndicate.
Policy	Non-governmental organization.	Transnational terrorist group.
Technology/ globalization nexus	Information revolution. Global media. Scientific collaboration and development. Electronic finance. Increased foreign investment. Cheap intercontinental travel and transportation of goods. Greater productivity.	Cyber-terror/crime. Propaganda, al-Jazeera exploitation. Proliferation. Money laundering. Terror sponsorship and fundraising. Smuggling. September 11, 2001. Russian organized crime.

However, the US government will not be able to exploit strategic IT to build these networks, nor adequately regulate its use, until it resolves legal and policy issues raised by these technologies. These issues are challenges of both substance and process.

Intelligence reform, data mining and privacy

Strategic use of information technology can help prevent, detect and mitigate terrorist attacks.[18] "Information awareness" involves promoting a broad knowledge of critical information among law enforcement and intelligence agencies, to identify terrorist activities and important patterns of behavior in cyberspace as indicators of terrorist attack:[19]

> Advances in information fusion, which is the aggregation of data from multiple sources of data for the purposes of discovering some insight, may be able to uncover terrorists or their plans in time to prevent attacks. In addition to prevention and detection, [this technology] may also help rapidly and accurately identify the nature of an attack and aid in responding to it more effectively.[20]

Terrorists use cyberspace as a tool, and leave "footprints" in cyberspace: "This low-intensity/low-density form of warfare has an information signature, albeit not one that our intelligence infrastructure and other government agencies are optimized to detect. In all cases, terrorists have left detectable clues that are generally found after an attack."[21] These clues include data on operational planning and execution, specific acts of surveillance and reconnaissance, transactions, practice runs, and increases in communications (e.g. "chatter").[22] These clues indicate what terrorists are planning, what they are targeting, how they communicate and provide resources, and even how there networks are formed. These clues exist especially for those activities that terrorists have always conducted before an attack (i.e. physical surveillance of a target, communications, and practice runs).[23] This is the demonstrated premise underlying data mining:

> The research into data search and pattern recognition technologies is based on the idea that terrorist planning activities or a likely terrorist attack could be uncovered by searching for indications of terrorist activities in vast quantities of transaction data. Terrorists must engage in certain transactions to coordinate and conduct attacks against Americans, and these transactions form patterns that may be detectable.[24]

The hypothesis underlying the application of data mining, aggregation and pattern recognition for intelligence reform is that these transactional patterns, indicative of terrorist activity, can be identified and interdicted before an attack.

Detecting and identifying an attacker in cyberspace is a significant problem, to which data mining and aggregation technologies contribute solutions. Technological solutions include data mining to extract data and patterns from massive amounts of data spread across disparate databases, "evidence combination" to merge different sources of data and support analysts' reasoning and testing of hypotheses, "natural language" technologies for translating and extracting information from spoken-word transmissions, image and video processing, and data visualization technologies to portray massive amounts of data in intuitive visual formats.[25]

Technology for data mining and aggregation offers opportunities to identify and track patterns of terrorist activity in cyberspace, including plans and preparations, resources and logistics, communications, finance, recruitment, training, and propaganda techniques.[26] Technology for pattern recognition and structured reasoning offers the opportunity for intelligence reform, to compete hypotheses, merge intelligence data and indications of surveillance or attack across various public and private levels, identify hidden links among actors in a terrorist network, and provide insight into linkages between terrorist activity in the "real" world and in cyberspace:[27] "Linkages between hackers, terrorists, and terrorist-sponsoring nations may be difficult to confirm, but cyber terror activity may possibly be detected through careful monitoring of network chat areas where hackers sometimes meet anonymously to exchange information. [These technologies] are intended to help investigators discover covert linkages among people, places, things, and events related to possible terrorist activity."[28]

Data mining and aggregation technologies provide capability beyond simply analyzing information. They provide a solution to the requirement outlined by the Silberman–Robb "WMD" Commission, calling for better tools to develop and compete hypotheses among intelligence analysts. These technologies offer potential for intelligence reform for counter-terrorism, because they:

> support collaborative work by cross-organizational teams of intelligence and policy analysts and operators as they develop models and simulations to aid in understanding the terrorist threat, generate a complete set of plausible alternative futures, and produce options to deal proactively with these threats and scenarios. The challenges such teams face include the need to work faster, overcome human cognitive limitations and biases when attempting to understand complicated, complex, and uncertain situations, deal with deliberate deception, create explanations and options that are persuasive for the decision maker, break down the information and procedural stovepipes that existing organizations have built, harness diversity as

a tool to deal with complexity and uncertainty, and automate that which can effectively be accomplished by machines so that people have more time for analysis and thinking . . . [These technologies] aid the human intellect as teams collaborate to build models of existing threats, generate a rich set of threat scenarios, perform formal risk analysis, and develop options to counter them.[29]

Privacy issues pose a significant hurdle to government implementation of these technologies.[30] Public policy on privacy was not developed with these technologies in mind.[31] Thus, there is the concern that, as pressure mounts for government (and industry) to collect information, the intrusion on privacy will be intolerable to the American people, absent measures to oversee collection and prevent unreasonable intrusion.[32] "Given the limited applicability of current privacy laws to the modern digital environment, resolving this conflict will require the adoption of new policies for collection, access, use, disclosure and retention of information, and for redress and oversight."[33]

Consequently, promising government technology programs have been canceled out of unarticulated privacy concerns.[34] Meanwhile, commercial data aggregators expose consumers to potential identity theft, and perhaps terrorist planners, without regulation or oversight. Federal policy and law regarding new information technologies cannot adequately balance opportunities, risks and vulnerabilities among competing policy objectives (such as privacy and security), without drawing some key substantive lines in law and policy, with respect to expectations of privacy in a world with the above-described trends.

The government has had difficulty in balancing the utility of these technologies with privacy concerns. An example is the cancellation in 2004 of the Terrorism Information Awareness (TIA) program, under development by the Defense Department. The Technology and Privacy Advisory Committee (TAPAC)—a board established at the end of the TIA program to review privacy implications—characterized the program: "TIA was a flawed effort to achieve worthwhile ends . . . It was flawed by its perceived insensitivity to critical privacy issues, the manner in which it was presented to the public, and the lack of clarity and consistency with which it was described."[35] Nevertheless, even TAPAC recognized the benefit—and continuing demand —for these technologies.[36] Expanding its analysis to other data mining programs—government and commercial—TAPAC called for "clear rules and policy guidance, adopted through an open and credible political process, supplemented with education and technological tools, developed as an integral part of the technologies that threaten privacy . . ."[37]

Substantively, the fundamental principles of US privacy law and policy are outdated, and cannot accommodate modern technology. Many of the basic principles of privacy law, vis-à-vis government surveillance and Fourth

Amendment protections, were articulated during the 1960s and 1970s, codified into such statutes as the Omnibus Crime Control and Safe Streets Act of
1968 (specifying procedure for criminal investigation wiretaps), the Privacy
Act of 1974, and the Foreign Intelligence Surveillance Act of 1978.

These statutes, and their underlying principles, come from a different time
in US history. The US faced a different threat. Technology was different,
and modern globalization was beginning to revolutionize society. Even the
government abuses were different that gave rise to such bodies as the Church
Commission. During that period, the concern was that federal investigators
or the intelligence community would gain surreptitious access to information
to which they had no right. Today, the risk of abuse is not so much illegal
government access to information, but the uncertainty of what the government ought to do with information to which it has legal access, and the
efficiencies afforded to government by technology to make more effective use
of massive amounts of legally obtained data.

Reliance on government inefficiency for privacy protection is not good policy. Rather, public policy should be updated to reflect the technological and
social realities of the globalization age. An update begins with new law on use
of public and private databases, based on an extensive and balanced debate
(i.e. Congressional hearings) that answers fundamental questions of privacy
in light of modern trends and expectations. Justice Brandeis's maxim still
holds true, that the right to privacy is the "right to be left alone." But this is
not the same thing as a right to anonymity, to which citizens have no Constitutional right. The court has articulated that privacy is guaranteed around
those personal areas where one has a reasonable expectation of privacy that
society is prepared to accept. But what is a reasonable expectation of privacy,
with respect to information that people freely give to third parties (e.g.
websites, rental car companies, airlines, etc.)? What is the material difference
between privacy and anonymity? How much data must be aggregated before
privacy is violated, and what is the actual harm? Can the potentially intrusive
effect of data mining and aggregation be mitigated by data anonymization or
other technology? Are there duties and liabilities that we as a society want to
impose on those private sector entities that collect and maintain data?

If the answers to these questions preclude data mining, then in a balanced
debate that preclusion will have been based in the informed and balanced
interest of the public, rather than on hunch and hyperbole. In the meantime,
as current and future technology programs go forward, they should include a
meaningful investment in privacy protection technologies and innovative
concepts to protect privacy with technology. The government should foster
such programs as demonstrate a balance investment, in case the market does
not provide adequate incentives for privacy protection investments. If these
technologies are developed and deployed in the private sector or by foreign
governments, the US constituency for privacy protection technology may be
rendered impotent. Without federal investment, the Congress minimizes its

ability to influence development and implementation of technology, and thus the ability for oversight.

Critical infrastructure protection and private responsibility

Critical infrastructure includes basic processes and information systems that support banking and finance, telecommunications, energy, water and food, and government services.[38] Corporations own and operate up to 90 percent of these infrastructures. As described above, terrorists move through, and leave footprints in, global infrastructures. Government processes, including security functions, also flow across these infrastructures. The vulnerability of a single system is compounded by its interdependence upon other infra-structure,[39] and all systems are dependent upon IT.[40] Not surprisingly, terrorists have sought to exploit and target critical infrastructures,[41] and there is a growing body of experience relating to infrastructure attacks by cyber-techniques that can add to the cyber-terrorist's attack calculus.[42]

There are opportunities in bringing the private sector into critical infra-structure protection efforts. Given the extent of control and visibility into infrastructures that their corporate owners and operators maintain, corporate technologies and resources are central to protecting critical infrastructures from attack and exploitation. Yet there is a gap between government and corporate security for infrastructures, and the threshold between corporate and national responsibility for infrastructure protection has not been well defined in policy or regulation.[43] Shareholder imperatives militate against spending corporate resources on information security beyond that for which the corporation is responsible on a daily basis (e.g. common crime and hacking, as opposed to cyber-attack). Privacy issues and liabilities prevent the use of corporate resources in identifying terrorist activities (purchases, movements, finances, etc.) within the "noise" of legitimate transactions ordinarily occurring daily across those corporations' infrastructures. As a consequence, terrorist activities or attacks might slip through the "gap" between corporate and national responsibility.

Strategic use of IT in protecting our critical infrastructures requires the government to strike a public/private partnership that brings to bear the full resources of corporate information technology and security to meet national goals of securing the infrastructure without eroding its social or economic viability. The participation and cooperation of the private sector is not only critical in balancing civil liberties with civil security,[44] but central to infra-structure protection: "if these businesses do not offer their full cooperation, the government is in no position to protect them relying on its own resources alone."[45] Yet much of the private sector still does not take critical infra-structure seriously even within its own corporate responsibilities. Almost half of roughly 100 companies in one survey had not increased annual spending on security after 9/11.[46] Furthermore, nearly 40 percent of the corporate executives surveyed said security was an expense that should be minimized. A quarter of the companies surveyed said their chief executives had not met in the last year with their security chiefs. Experts say the reasons for the faltering revolution are varied, ranging from a concern about costs to simply a lack of sustained focus and vigilance.[47] "There were even some companies that buried their heads in the sand," said William Daly, a former FBI counter-intelligence investigator who directs the New York office of Control Risks Group. "Security is kind of incident-driven, and continues to be that way."[48]

Before corporate resources can be used to fight terrorism, corporate infra-structure providers must be convinced that it is both a public and a private problem. Critical infrastructure protection is about trading risks between the government and the private sector. Homeland security is the ultimate responsibility of the government, even while the interconnectivity and effi-ciencies of critical infrastructures are largely driven by corporate-owned information technology. The threshold between homeland security and cor-porate responsibilities has not been well defined, and there is a policy gap between homeland security interests and shareholder interests. As a result, private sector security investments and processes might not be sufficient to meet national needs for critical infrastructure protection.

This creates a hurdle to efficient, reasonable or even practical assignment of appropriate risks to be borne respectively by the government and share-holders. How then to enlist the private sector for public goals? How do we secure these infrastructures without degrading their efficiency? How do we deny terrorists the use of these infrastructures without hindering social and economic commerce? What tools can the government use to reconcile private commercial objectives with public policy objectives? What national responsibilities and obligations can be reasonably and efficiently borne by shareholders? Can these responsibilities overlap, rather than fall victim to a gap?

At least since the Cuban Missile Crisis, the US government has system-atically tried to answer these questions. The most promising opportunities to protect our infrastructures are oriented around the strategic use of corpor-ate IT, but this use will be limited to local patches in system vulnerabilities, until key balances are struck that will enable the strategic use of IT to protect critical infrastructures while maximizing their efficiencies. This will come about only after the private sector infrastructure providers are full partners with the government and international community, marshaling the full extent of corporate resources and technologies in securing infrastructures.

To date, regulation has been one of the most popular tools the government has used in the interest of public welfare. As any corporate manager will tell you, regulation usually costs shareholders money. But not always and, even when regulation does cost shareholders money, it is not in the same manner across all infrastructures. Regulation is an unwieldy tool to use evenly across all infrastructure sectors, because different infrastructure sectors respond differently (in economic and operational terms) to regulatory measures. For example, the nuclear power infrastructure was heavily regulated from its inception. Corporate burdens imposed on nuclear power providers are neither unexpected nor disruptive to commerce. On the other hand, information and telecommunications systems—especially the internet—are notoriously hard to regulate, and ham-handed regulation frequently imposes great corporate costs, without necessarily resolving the public safety issue the regulation was crafted to solve. This has been apparent from anti-trust regulation to internet taxation regimes to regulations supporting subpoena powers against internet service providers.

Regulation has in the past had unintended consequences contrary to national security interests, despite successful fulfillment of regulatory goals. Both anti-trust policy and homeland security policy have at their heart the economic interests of the United States—one through prevention of predatory competitive practices, the other through assurance of the economic viability of critical infrastructures. The tools wielded in pursuit of these policies are not always as complementary. The 1984 break-up of AT&T was a victory for anti-trust policy, but many national security officials at the time, including Defense Secretary Caspar Weinberger in court testimony, opposed the break-up of AT&T on grounds of national security policy. Weinberger and others argued that a dominant and robust telecommunications provider like AT&T, and its support of the national security telecommunications system developed in response to the Cuban Missile Crisis, was critical to the national security. One concern was that several telecommunications providers, in competition, would result in lack of interoperability and unity of effort in crisis, and thus jeopardize critical communications in times of national emergency. Thus, there were potentially bad unintended consequences for homeland security goals, even though equally valid anti-trust goals of public policy were achieved.

Governments must develop tools in law and policy more imaginative than regulation, to bring the corporate sector into the homeland security mission. These tools might include liability limitation and "safe harbors," subsidies or tax breaks to cover the gap between corporate and homeland security, perhaps even quasi-socialization of specific components or processes of infrastructure as "public goods," or a combination of one or more of these tailored to specific infrastructure sectors, their respective needs and interests, and gaps or overlaps between their specific interests and broader national interests. Different sectors and different countries have had different experiences and lessons in all of these approaches. Furthermore, America's vulnerability in cyberspace does not stop at its shores, but rather is enmeshed in the interconnectivity with other nations' critical infrastructures, the cyber-fabric of globalization. Thus, international conventions and agreements should also be considered as useful tools to balance international commerce, domestic economics and homeland security.

Innovative tools and approaches for infrastructure protection must reconcile, and seek common ground between, business interests and homeland security interests, and balances must be tailored to specific sectors and industries. For example, *Business Week* recently described some innovative tools:

> a bill now pending, sponsored by Senators John D. Rockefeller IV (D-W.Va.) and Olympia J. Snowe (R-Me.), [will] let companies expense equipment costs when they build networks of at least 20 megabits a second. A U.S.-backed bond program would encourage municipalities to build their own fiber networks and then lease them to upstarts. And government can attract broadband to sparsely populated regions without tax dollars by creating pools of local buyers—a measure Canada has adopted to reach its vast rural expanses.[49]

These tools speak to corporate interests of tax incentives and market interests of greater broadband. Greater broadband also benefits homeland security, in that it provides more graceful degradation if parts of the national network fail. Such commonalities of interest will vary from sector to sector, and will require different mixes of tools to reach common ground. It is up to the government to investigate into best practices and lessons learned in developing these tools, and derive either new policy or legal tools or innovative ways to combine and apply existing tools, define the varying and dynamic thresholds between corporate and government responsibility, reconcile interests, and deploy these tools to exploit technology and protect our infrastructures without degrading them.

Cyber-conflict, defense and international law

A cyber-attack on the US critical infrastructure is the ultimate doomsday scenario posited when discussing critical infrastructure protection. The strategic use of IT in cyber-conflict is, simply put, waging war in cyberspace— using IT in support of computer network attacks (CNA) on terrorist infrastructure or processes in cyberspace, while defending against terrorist electronic attack against our physical or electronic systems. Technology opportunities for cyber-conflict extend well beyond hacking, viruses and corporate espionage, to include analytical tools, and models to predict the effects of alternative approaches to network disruption. They include technologies that provide the ability to target, access and sustain disruption of critical electronic systems, enabling the US to track, intercept or shut down communications on terrorist-affiliated websites and internet accounts, divert terrorist funding, shut down regional electronic infrastructures on which terrorists rely, and even wreak havoc on the infrastructures of terrorist sponsors.

These technologies can also be used to defend against enemy cyber-attack. An article in *Wired* characterizes the technological potential of the US military underlying these capabilities, operated by US Strategic Command's Joint Functional Component Command for Network Warfare.[50] However, as the article points out, despite a 2002 presidential directive to develop guidelines for offensive cyber-warfare, the rules of engagement and supporting legal and policy structures are emerging more slowly than the technological capabilities. Similarly, the private sector has the technological tools to detect, report and mitigate computer penetrations and attacks on its own systems, which might possibly be the opening salvos of a larger cyber-attack. However, marshaling the technologies and resources of the private sector to form such a national cyber-defense capability is a policy challenge of public/private cooperation. Equally important, the US government must coordinate its policies and strategies with its allies and economic partners. A cyber-conflict—especially one that includes defensive responses from the US in cyberspace—could easily wreak unintended consequences on the infrastructures and economies of allies and economic partners, without proper analytical preparation. IT tools for data mining and analysis provide opportunities to forecast and track the effects of disruption in cyberspace, in the event of a "cyber-exchange" with terrorists or enemy nations.

Cyber-conflict and computer network attack (CNA) are poorly understood in the context of war: in terms of both its usage, and responding to it with conventional military force. What kind of a cyber-attack would be considered an act of war? Can we use CNA to disrupt al Qaeda's presence in cyberspace without due process or covertly, if it means trampling on infrastructures of other Western economies, which are interconnected with our own? Would such a tactic be an act of war? During wartime, what is a legal use of cyber-weaponry? Are there any "civilian" targets that would be prohibited by the law of armed conflict? How does one restrict the effects of a cyber-attack to a specific enemy or region, and prevent the effects from cascading throughout the global information infrastructure? What is the role of commercial infrastructure providers—whose electronic systems will likely serve as part of the cyber-battlefield—in detecting, mitigating and responding to attacks?

Most problematic are questions of how to respond to cyber-attacks. It is not clear what magnitude of cyber-attack would rise to the threshold of an act of war (thus justifying a conventional military response in self-defense). Confidence in attribution will challenge the rationale and type of response. How confident must a defending state be before it responds, and does that confidence vary depending on the severity of either attack or response? International law requires that actions taken in self-defense must be necessary and proportionate. How does one measure necessity and proportionality, especially if one is responding to a cyber-attack with conventional military force, with varying levels of confidence about who the enemy is?

It is undisputed that the United States has (along with perhaps many other nations) the information weapons to attack and defend in cyberspace. Using those weapons will be extremely risky, without clear policies and legal authorities—supported by international consensus—on the conduct of warfare in cyberspace, cooperation with the private sector in detecting, mitigating and responding to attacks in electronic infrastructures, and the technological capabilities—rooted in data mining and analytical tools—to forecast and track the effects of cyber-attack and responses. The technology already rests in the hands of the public and private sector—it now requires a modern legal and policy framework to guide its application.

Conclusion: process problems

The substantive legal and policy questions posed above can be answered only through a modernized policy process that integrates the co-development of technology, policy and law. The US government has reached a point where technology development cannot flourish in a policy vacuum. Similarly, law and policy development is ineffective unless it recognizes the possibilities, policy implications and market demands for technology.

Technology developers must consider the policy ramifications of their technology, at the onset of development. This is especially true for IT, where the pace of development and rush to implementation are so quick. Policy and legal objectives and options ought to be included alongside technical objectives in program plans. For example, if a technology program focuses on data mining and aggregation, it ought to identify potential policy ramifications (for example, on privacy policy) and identify options—in policy and technology development—for addressing any potential negative ramifications. If a data mining program identifies privacy as a possible policy concern, it ought to take steps to articulate policy objectives as well as technical objectives to mitigate this concern, and provide possible policy and technology options, such as anonymization technology and policy processes that prevent unwarranted de-anonymization. A tool to facilitate this might be a "legal impact statement," analogous to the environmental impact statement required by the Environmental Protection Agency.

The government's policy process must also inform policy makers and lawmakers of the opportunities and implications of emerging technology. Policy makers and legislators need to understand when general technology trends, or "technology push," challenge current policy or law, or even suggest the need for change. This calls for a more robust, proactive role for policy offices such as the Office of Science and Technology Policy, as well as independent bodies such as the national academies, to provide policy requirements and guidelines to technology initiatives, as well as guidance and advice to policy makers about technology trends and implications.

Finally, competing with terrorists requires cooperation from allies, especially in cyberspace. This means engaging international organizations, and perhaps constructing international agreements or other diplomatic and commercial instruments, setting international standards for data access, sharing or pooling, reconciling and finding common ground among different countries' privacy laws and policies, and their respective use of commercial data providers, formulating common expectations of corporate responsibility in securing and protecting infrastructure, identifying opportunities and lessons learned in implementing policy and legal tools for infrastructure protection, and forging partnerships to track terrorist activity in cyberspace, deny terrorists cyber-safe havens, and respond quickly to cyber-attack as its effects and perpetrators emerge.

A global network to counter-terrorism can derive great capability from strategic information technology and the benefits of globalization. However, this network must exist in the real world as well as cyberspace. The US and its allies—particularly the Council of Europe—have the opportunity for leadership in the United Nations and other bodies such as the World Trade Organization, for developing a foundation for crafting common standards, expectations, and limitations for competing with terrorists in cyberspace. This is an opportunity that democratic nations ought to seize upon, while we have the luxury to modernize privacy expectations and corporate obligations, and before the threat of cyber-conflict truly manifests as a destructive mode of warfare. Democratic nations have two significant advantages over terrorism: the capabilities and resources of advanced technology, and the values and resiliency of liberal democracy. One advantage need not erode the other.

Notes

1 This notion of a strategic triad was proposed by Dr. Robert Popp, Deputy Director, Information Exploitation Office, Defense Advanced Research Projects Agency.
2 Bruce Hoffman, *Inside Terrorism* (London: Victor Gollancz, 1998), p. 67.
3 Ibid., p. 68.
4 See, for example, Joseph Nye, *The Paradox of American Power* (Oxford: Oxford University Press, 2002), p. 81; Defense Science Board, *Final Report of the Defense Science Board Task Force on Globalization and Security* (Washington, DC: US Defense Department, 1999), p. 5; A.T. Kearney, "Measuring Globalization: Economic Reversals, Forward Momentum," *Foreign Policy*, vol. 54, Mar./Apr. 2004, p. 141.
5 Simson Garfinkel, *Database Nation: The Death of Privacy in the 21st Century* (New York: O'Reilly, 2001).
6 See., for example, "Crowned at Last: A Survey of Consumer Power," *Economist*, Apr. 2, 2005, pp. 1–16.

7 Garfinkel, *Database Nation*; Robert O'Harrow Jr., *No Place to Hide* (New York: Free Press, 2005).

8 Jonathan Kim and Robert O'Harrow Jr., "Data under Siege: ID Thieves Breach Lexis-Nexis, Obtain Information on 32,000," *Washington Post*, Mar. 10, 2005, E01; Evan Hendricks, "When Your Identity Is Their Commodity," *Washington Post*, Mar. 6, 2005, B01; Robert O'Harrow Jr., "ChoicePoint Data Cache Became a Powder Keg," *Washington Post*, Mar. 5, 2005, A01.

9 Mary DeRosa, *Data Mining and Data Analysis for Counterterrorism* (Washington, DC: Center for Strategic and International Studies, March 2004); O'Harrow, *No Place to Hide*.

10 DeRosa, *Data Mining and Data Analysis for Counterterrorism*, p. 11.

11 National Research Council, *Information Technology for Counterterrorism: Immediate Actions and Future Possibilities* (Washington, DC: National Academies Press, 2003), p. 12.

12 Defense Science Board, *Final Report of the Defense Science Board Task Force on Information Warfare—Defense* (Washington, DC: US Defense Department, 1996), p. 2–1.

13 Jacques S. Gansler and William Lucyshyn, "Trends in Vulnerabilities, Threats, and Technologies," in *Information Assurance: Trends in Vulnerabilities, Threats, and Technologies*, ed. Jacques S. Gansler and Hans Binnendijk (Washington, DC: National Defense University Press, 2004), p. 15.

14 Neal A. Pollard, "Globalization's Bastards: Illegitimate Non-State Actors in International Law," *Law Enforcement and Low Intensity Conflict*, vol. 11, no. 2/3, Winter 2004, pp. 215–16; Moisés Naím, "The Five Wars of Globalization," *Foreign Policy*, Jan.–Feb. 2003, pp. 29–33; Michael A. Vatis, "Trends in Cyber Vulnerabilities, Threats, and Countermeasures," *Information Assurance: Trends in Vulnerabilities, Threats, and Technologies*, ed. Jacques S. Gansler and Hans Binnendijk (Washington, DC: National Defense University Press, 2004), p. 104; Dorothy Denning, *Information Warfare and Security* (Reading, MA: ACM Press, 1999), p. 68; President's Commission on Critical Infrastructure Protection (PCCIP), *Critical Foundations: Protecting America's Infrastructure* (Washington, DC: Government Printing Office, 1997), p. 8.

15 This table was originally published by the author in Neal A. Pollard, "Globalization's Bastards," pp. 215–16.

16 Lars Nicander and Magnus Ranstorp (eds.), *Terrorism in the Information Age—New Frontiers?* (Stockholm: Swedish National Defense College Press, 2004); Vatis, "Trends in Cyber Vulnerabilities, Threats, and Countermeasures"; Greg Rattray, "The Cyberterrorism Threat," in *The Terrorism Threat and U.S. Government Response: Operational and Organizational Factors*, ed. James M. Smith and William C. Thomas (Colorado Springs, CO: US Air Force Academy Press, 2001), pp. 79–102; Denning, *Information Warfare and Security*, pp. 68–76; Matthew G. Devost, Brian K. Houghton and Neal Allen Pollard, "Information Terrorism: Political Violence in the Information Age," *Terrorism and Political Violence*, vol. 9, no. 1, Spring 1997, pp. 72–83; Defense Science Board, *Final Report of the Defense Science Board Task Force on Information Warfare—Defense*.

17 Clay Wilson, *Computer Attack and Cyber Terrorism: Vulnerabilities and Policy Issues for Congress* (Washington, DC: Congressional Research Service, 2003).

18 National Research Council, *Information Technology for Counterterrorism*, p. 12.

19 Ibid.

20 Ibid.

21 Defense Advanced Research Projects Agency, Broad Agency Announcement 02–08, Information Awareness Proposer Information Pamphlet 2 (2002), http://www.eps.gov/EPSData/ODA/Synopses/4965/BAA02–08/IAPIP.doc

22 Neal A. Pollard, "Indications and Warning of Infrastructure Attack," in *Terrorism in the Information Age: New Frontiers?*, ed. Lars Nicander and Magnus Ranstorp (Stockholm: Swedish National Defense College Press, 2004), p. 51.

23 Ibid.

24 Defense Advanced Research Projects Agency, *Defense Advanced Research Projects Agency's Information Awareness Office and Total Information Awareness Project*, http://www.darpa.mil/iao/iaotia.pdf (visited Jan. 28, 2003).

25 National Research Council, *Information Technology for Counterterrorism*, pp. 63–71.

26 James Jay Carafano and Paul Rosenzweig, *Winning the Long War: Lessons from the Cold War for Defeating Terrorism and Preserving Freedom* (Washington, DC: Heritage Books, 2005), pp. 113–16; DeRosa, *Data Mining and Data Analysis for Counterterrorism*, pp. 3–12; Wilson, *Computer Attack and Cyber Terrorism*, pp. 14–19; Markle Foundation, *Protecting America's Freedom in the Information Age* (New York: Markle Foundation, 2002), pp. 2, 12–15, 48–9, 53–64.

27 Wilson, *Computer Attack and Cyber Terrorism*, pp. 14–19; Laurence H. Silberman and Charles S. Robb, *Final Report of Commission on the Intelligence Capabilities of the United States Regarding Weapons of Mass Destruction* (Washington, DC: Government Printing Office, 2005); Richard Shelby, "September 11th and the Imperative of Reform in the U.S. Intelligence Community: Additional Views of Senator Richard C. Shelby, Vice Chairman," Senate Select Committee on Intelligence, Dec. 10, 2002.

28 Wilson, *Computer Attack and Cyber Terrorism*, pp. 14–15.

29 Defense Advanced Research Projects Agency, Broad Agency Announcement 02–08, pp. 22–3.

30 National Research Council, *Information Technology for Counterterrorism*, p. 71.

31 Mary DeRosa, "Privacy in the Age of Terror," *Washington Quarterly*, vol. 26, no. 3, Summer 2003, pp. 27–41; Office of the Inspector General of the Department of Defense, "Information Technology Management: Terrorism Information Awareness Program," Report to Congress, Dec. 12, 2003.

32 National Research Council, *Information Technology for Counterterrorism*, p. 71.

33 Carafano and Rosenzweig, *Winning the Long War*, p. 113.

34 O'Harrow, *No Place to Hide*, pp. 190–213.

35 Technology and Privacy Advisory Committee (TAPAC), *Safeguarding Privacy in the Fight against Terrorism* (Washington, DC: US Government Printing Office, 2004), p. 43.

36 Ibid., p. 8.

37 Ibid.

38 PCCIP, *Critical Foundations*. See also Michael J. O'Neil and James X. Dempsey, "Critical Infrastructure Protection: Threats to Privacy and Other Civil Liberties and Concerns with Government Mandates on Industry," 12 DePaul Bus. L. J. 97, 125 (2000).

39 Stephen J. Lukasik, Seymour E. Goodman and David W. Longhurst, *Protecting Critical Infrastructures against Cyber-Attack*, Adelphi Paper 359 (London: International Institute for Security Studies, 2003), p. 7.

40 National Research Council, *Information Technology for Counterterrorism*, p. 11.

41 Vatis, "Trends in Cyber Vulnerabilities, Threats, and Countermeasures," p. 100.

42 Lukasik *et al.*, *Protecting Critical Infrastructures*, p. 9.

43 Ibid.

44 Ibid.

45 Ibid.

46 Benjamin Weiser and Claudia H. Deutsch, "Many Offices Holding the Line on post-9/11 Security Outlays," *New York Times*, Aug. 16, 2004, Internet source http://www.nytimes.com/2004/08/16/nyregion/16security.html (accessed Aug. 17, 2004).

47 Ibid.

48 Ibid.

49 Catherine Yang *et al.*, "Behind in Broadband," *Business Week*, Sept. 6, 2004, http://www.businessweek.com/magazine/content/04_36/b3898111_mz063.htm (accessed Apr. 22, 2005).

50 "U.S. Military's Elite Hacker Crew," *Wired*, http://wired-vig.wired.com/news/privacy/0,1848,67223,00.html?tw=wn_tophead_1 (accessed Apr. 21, 2005).

16

THE GLOBAL CHALLENGE OF OPERATIONAL INTELLIGENCE FOR COUNTER-TERRORISM

Gregory F. Treverton and Jeremy M. Wilson

The United States and its major global partners in the war on terrorism all face the same challenge: bringing intelligence and law enforcement agencies much closer together than they had been, and doing so in ways that do not run high risks of infringing on privacy and civil liberties. In the United States, the "wall" between intelligence and law enforcement was built rather consciously, from the time of forming the CIA in the late 1940s, and it was reinforced by the Congressional investigations of intelligence abuses in the 1970s. It extended not just across agencies but also inside them, especially the FBI. It sharpened what were not just bureaucratic differences but also deep differences in purpose, method, time horizon and standard between intelligence and law enforcement. Now, that wall is being dismantled, and law enforcement and intelligence are being pushed together.

At the same time, both intelligence and law enforcement are being challenged, in many countries, to work with state and local officials (or equivalents) in new ways. The process is often called "information sharing," but that label is misleading, for it presumes not only that the process is technical but also that it is the federal level that has the information to be shared. In fact, the challenge is changing the way intelligence and law enforcement do their business, a challenge that runs from how information is classified to how the levels of government reach a division of labor. For instance, local officials have neither time nor manpower to conduct special intelligence collection in the war on terrorism; nor do they have much of any capacity to analyze information. What they do have is "eyes and ears" on the street *if* they have some sense of what to look for and some confidence that what they report will be processed in ways ultimately useful to them. This chapter concludes with suggestions about the division of labor between levels of government in the intelligence of counter-terrorism.

Constructing the "wall"

In an important sense, it should not be surprising that cooperation between the CIA and the FBI before September 11 was ragged at best.[1] Americans wanted it that way. Out of concern for civil liberties, they decided the two agencies should not be too close. The FBI and CIA sit astride the fundamental distinctions of the Cold War—distinctions between intelligence and law enforcement, between foreign and domestic and between public and private. The distinctions run very deep.

Those distinctions were deepened, for the United States, by the nation's first ever investigations of intelligence in the wake of the Watergate scandal. Those investigations uncovered abuses of the rights of Americans, especially in a curious mixing of intelligence, or counter-intelligence, and law enforcement at the FBI during J. Edgar Hoover's long tenure as director.[2] The justification and ostensible target of these "counter-intelligence programs," COINTELPRO in Bureau acronym, was the operations of hostile foreign intelligence services.[3] But most of COINTELPRO's specific targets were American citizens, in civil rights and anti-war groups. People like Rev. Martin Luther King were not only put under surveillance but harassed, and worse.

In reaction to the revelations, the domestic intelligence activities of the FBI were sharply restrained, and the Chinese wall separating intelligence from law enforcement was built higher. A compromise between presidential discretion and civil liberties resulted in the creation of the Foreign Intelligence and Surveillance Act (FISA) of 1978, which created the Foreign Intelligence and Surveillance Court (FISC), a court operating in secret to grant covert wiretap and other surveillance authority for intelligence—as opposed to law enforcement—purposes. Before FISC, presidents had claimed the right of searches for national security purposes with no warrants whatsoever.

Quite apart from the investigations, law enforcement and intelligence are very different worlds, with different missions, operating codes and standards. Intelligence, what John Le Carré refers to as "pure intelligence," is oriented toward the future and toward policy—that is, it seeks to inform the making of policy.[4] Living in a blizzard of uncertainty where the "truth" will never been known for certain, it seeks to understand new information in light of its existing understanding of complex situations. Thus, its standard is "good enough for government work." Because intelligence strives above all to protect sources and methods, its officials want desperately to stay out of the chain of evidence so they will not have to testify in court.

By contrast, law enforcement is oriented toward response. It is after the fact. Its business is not policy but prosecution, and its method is cases. It strives to put bad guys in jail. Its standard is high, good enough for a court of law. And law enforcement knows that, if it is to make a case, it must be prepared to reveal something of how it knows what it knows; at least it is aware that it will face that choice. It has no real history of analysis; indeed, the meaning of the word "intelligence" is different for law enforcement, where it means "tips" to finding and convicting evil-doers more than looking for patterns to frame future decisions. Law enforcement and policing also traditionally have been defined in geographical units. These definitions are more and more mismatched to threats, like terrorism, that respect no geographical boundaries.

A second distinction, that between foreign and domestic, magnifies the intelligence–law enforcement disconnect. American institutions and practices both during and prior to the Cold War drew a sharp distinction between home and abroad. The FBI had conducted wartime espionage and counter-espionage in Latin America, and in December 1944 Hoover had proposed that the FBI run worldwide intelligence operations on the lines of its Latin American operations.[5] The proposal had some support outside the FBI, at the State Department in particular. But President Harry Truman worried openly that giving the intelligence mandate to the FBI would risk creating a "Gestapo-like" organization, and so foreign operations went first to the Central Intelligence Group, the CIA's predecessor, and then to the CIA. Both, however, were barred from law enforcement and domestic operations.

Relations between the two agencies were ragged from the start and, by the 1970s, it was literally true that the directors of the CIA and the FBI didn't speak to one another. The National Security Act of 1947 was clear in pro-scribing the police function for the CIA. The National Security Agency (NSA), created later, was and is also barred from law enforcement and from domestic spying, so if the trail of conversations or signals it is monitoring becomes "domestic"—that is, involves a US person, corporation or even resident alien—then the trail must end. The FBI was required to provide information to the Director of Central Intelligence (DCI) but only if that information was "essential to the national security," and only "upon the written request" of the DCI.[6] The FBI also was responsible for protecting material before federal grand juries and, while sharing was possible, in prac-tice information came to be shared only with a court order. Both these sets of provisions were an invitation for the FBI to hoard information.

A third distinction is public versus private. During the Cold War, national security was a government—federal government—monopoly. To be sure, private companies and citizens played a role but, for most citizens, fighting the Cold War simply meant paying their taxes. That does not seem likely to be so for the campaign against terrorism and for homeland security. Civilians' lives will be affected—ranging from the inconvenience of waiting in long lines at airports, to harder questions about how much security will make use of pre-screening, national databanks and biometrics. Across the country, there are three times as many "police" in the private sector as in governments.

All three of these distinctions were all too vividly on display before

September 11. At the federal level, according to the joint Senate–House investigation of September 11, the CIA's procedures for informing other agencies—FBI, State, NSA and the Immigration and Naturalization Service (INS)—of suspected terrorists were both restricted and haphazard.[7] The number of names the CIA put on the watchlist soared after September 11, from 1,761 during the three months before September 11 to 4,251 in the three months afterwards.

So, too, the ragged cooperation between the CIA and the FBI was visible in their misdealings over the al Qaeda-affiliated terrorists Khalid al-Mihdhar and Nawaf al-Hazmi.[8] They were monitored by the CIA while attending a terrorist meeting in Malaysia in January 2000, and NSA had independent information that linked al-Hazmi to al Qaeda. Yet their names were not put on the main watchlist until August 2001, shortly before the attacks and after they had been training as pilots and living under their real names in San Diego. Al-Mihdhar had applied for and received a new visa earlier that summer—since his name was not on the watchlist, neither State nor INS had strong grounds for suspicion. No agency told the Federal Aviation Administration (FAA) to be on the lookout for the two men, apparently because it was not in the law enforcement business. And the airlines were not informed because they were private, not public.

Dismantling the wall

Almost literally overnight, the "wall" was perceived to need to become a window. September 11 overturned all the presumptions on which the distinctions had been based. The threat was not just "over there"; it was both here and there. It was not a matter for intelligence *or* law enforcement but, rather, for both working together. And the war on terror was neither purely a federal one nor purely a government one. It extended not only to local police on the beat but also to private citizens.

For the United States, as for other countries, dismantling the wall has come about through both legislative and administrative changes. Given the US constitutional structure, some of the changes raised vexing constitutional issues. The most immediate of these were the detentions without charges or trial of several American citizens after September 11.[9] The most enduring has been what to do about the hundreds of foreign citizens still held at Guantanamo Bay, Cuba.[10] For these purposes, though, the most important issues were raised by the USA Patriot Act, passed in the immediate aftermath of September 11, and by the accompanying changes in procedures within the FBI.[11]

Some of the Act's provisions simply corrected oversights in statutory language or updated the law to match new technology. For instance, FISA wiretaps were designed for an era of analog telephones, and the Act authorized the use of "roving" or "multi-point" wiretaps, which allow the monitoring of all devices a suspect might use—a practice of long standing in criminal investigations. Other parts of the Act were more controversial. FISA taps always were permitted to be longer than law enforcement counterparts—90 days rather than 30, with extensions easier to obtain. The Patriot Act extended them further, to 120 days, and it doubled, from 45 to 90 days, the period in which foreign agents, including US citizens, can be subject to clandestine physical searches.

Perhaps of greater concern, the Act made an apparently small change that it is feared will have large consequences. Before September 11, obtaining foreign intelligence information had to be "*the* purpose" of FISA surveillance.[12] If evidence of crime was uncovered in the course of the tap, that evidence was admissible in court, but the foreign intelligence purpose was paramount. The Patriot Act loosened the requirement to "*a significant* purpose."[13] Because FISA taps do not require probable cause of a crime, and are longer, more flexible and less controlled by judges than are law enforcement taps, there is concern that FISA taps will be used to troll for law enforcement purposes.

The expansion of FISA also led to tensions between the FBI and the FISC over who can approve the sharing of FISA data with FBI law enforcement agents. However, in November 2002 a federal court ruling upheld more sharing of intelligence across the intelligence–law enforcement wall within the Bureau, and in October 2003 new guidelines went to the field offices confirming the change.[14] Before the Patriot Act, the Bureau would have had to open separate wiretaps—a criminal one based on a court order and a FISA one for intelligence purposes—and would have been sharply constrained in sharing information between the two. Under the new guidelines, it could open, for example, a single FISA surveillance looking both at whether a suspect was part of a terrorist organization, an intelligence purpose, and whether he planned to buy explosives, a law enforcement one. Agents working on the two aspects of the case could cooperate closely.

So far, the United States has differed from its main partners in dismantling the wall but not constructing a separate domestic intelligence agency. Indeed, while there were calls for such an agency immediately after September 11 and while the logic of the 9/11 Commission's diagnosis points toward one, neither the Commission nor the drafters of the December 2004 intelligence reform legislation recommended a new agency.[15] Both were impressed by the FBI's determination to transform its mission from law enforcement to prevention and intelligence by creating a serious intelligence function and embedding it throughout the organization. In the circumstances of dismantling the existing wall *within* the FBI, creating a new one *between* it and a new agency did not seem a good idea.

The arguments for a separate domestic intelligence agency are two. The first is that the FBI is likely to remain—and perhaps should remain—primarily a case-based law enforcement organization. It is good at that. Yet pursuing cases the way the FBI does simply is contrary to building a comprehensive intelligence picture. If the FBI identified a suspected terrorist in connection with a Hamas investigation, for example, the suspect would be labeled a Hamas terrorist with relevant information kept in a separate Hamas file that would be easily accessible to and routinely used only by Hamas-focused FBI investigators and analysts. The Osama bin Laden unit would be unlikely to know about the FBI's interest in that individual. In the case of Zaccarias Moussaoui—the so-called twentieth hijacker—when agents from the local field office began, in August 2001, looking into his flying lessons at a Norman, Oklahoma school, they did so in ignorance that the same field office had been interested in the same flight school two years earlier because a man thought to be bin Laden's pilot had trained there.

Second, while domestic intelligence services in other countries have been willfully misused for political purposes—Italy and Peru are two cases in point—the lesson of COINTELPRO is that dangers to democracy can arise from mixing domestic intelligence with law enforcement. For similar reasons, Canada took its Royal Canadian Mounted Police (RCMP) out of the domestic intelligence business, replacing it with a separate service, the Canadian Security Intelligence Service (CSIS). Other states have been successful in creating domestic intelligence bodies that have operated effectively within the constraints of liberal democracy, including Britain (Security Service, MI5), France (Direction de la Surveillance du Territoire, DST), Germany (Bundesamt für Verfassungsschutz, BfV) and Australia (Australian Security Intelligence Organization, ASIO).[16]

In all of these democracies, the intelligence function remains subject to legislative oversight and supervision yet retains the latitude to aid government crisis decision-making through covert and, often, unorthodox means. They, along with the COINTELPRO history, suggest that domestic intelligence might be both better and safer for democracy if it is separate, not the tail of a law enforcement dog. The experiences of other countries also can provide useful ideas about how relationships among federal, state and local law enforcement agencies can be strengthened. In Canada, for example, CSIS has established a network of regional liaison officers, who help facilitate the flow of information between local and provincial police agencies and the federal authorities.

Yet the downsides of a new agency are also apparent. Purely practically, it would have all the teething pains of any new agency—pains on vivid view at the Department of Homeland Security (DHS)—and would, to boot, need to duplicate the range of offices and infrastructure that the FBI now has. Moreover, a new agency is hardly a panacea; in Britain, MI5 and Scotland Yard were for years locked in a turf battle over who had primary responsibility for counter-terrorism in Britain outside Northern Ireland. Indeed, Britain is planning to consolidate its fragmented anti-crime efforts into a "British FBI."[17] Finally, the idea of a domestic intelligence service completely unhitched from cases, and perhaps from investigation as well, does raise civil liberties concerns. A more modest version would underscore the transition the FBI is already trying to make, by creating distinct career tracks for counter-terrorism and intelligence within the Bureau—a kind of MI5 within the FBI.[18]

Sharing intelligence—including with ourselves

In sharing, too, September 11 itself provided a powerful impetus to change. Before the attacks, the very different cultures compounded the effect of the wall between intelligence and law enforcement. For instance, FBI agents have Top Secret clearances, but few are cleared into the Special Compartmentalized Information (SCI) that is the woof and warp of intelligence.[19] So, when faced by unfamiliar FBI counterparts in meetings, CIA officers might be sincerely uncertain how much they could say, and vice versa for FBI agents, who feared that inadvertent disclosures might jeopardize prosecutions. The safest course was to say nothing. If the conversation turned to matters domestic, then the CIA officials would also be uncertain how much they should *hear*.

For the United States, at the federal level, the simple fact of September 11 has been a powerful impetus to moving information within and across agencies, and to working together, though there is a long way to go.[20] New institutional creations, like the National Counterterrorism Center and the Director of National Intelligence, can help.

The harder challenge, though, is sharing with ourselves, *across* levels of government and *across* the public–private distinction. The conclusions of a recent Markle Foundation task force on the DHS apply to the government as a whole: "DHS has yet to articulate a vision of how it will link federal, state, and local agencies in a communications and sharing network, or what its role will be with respect to the TTIC [Terrorist Threat Integration Center, now the National Counterterrorism Center] and other federal agencies."[21] There is no gainsaying the difficulty of the task. DHS and FBI share the responsibility. Not only is infrastructure for moving information lacking, but much of the relevant information is classified. To state and local officials, however, the classification problems often look like a smokescreen covering an attitude on the part of federal officials that the war on terrorism is a federal responsibility.

This challenge of sharing with ourselves is common to the United States and its major global partners, but the US federal structure poses special obstacles. It means that there are some 18,000 authorities at the state and local level to coordinate, of widely varying size, capacity and vulnerability. As a starting point, it is worth observing that, while most discussion of information sharing in the war on terrorism still has concentrated on the federal government, state and local law enforcement agencies (LEAs) are the nation's eyes and ears in that war.[22] So it seemed useful to several of us RAND colleagues to look at intelligence from the bottom up, rather than the top down.[23] How widespread is counter-terrorism intelligence activity among state and local LEAs? What are those state and local authorities doing differently now, after September 11, in collecting and processing information? What are courts and other oversight bodies doing to guide that process? And, ultimately, what might an "ideal" division of labor among the various levels of government look like?

LEAs' involvement in intelligence activities designed to counter-terrorist actions ranges from investigation, including electronic surveillance, of possible criminal acts (typically those authorized by Title III) to collecting information in the normal course of policing activities but data that is not related to any specific criminal case.[24] That information would typically be handed over "the wall" to the FBI for its continued investigation and assessment. These activities may occur in collaboration with other agencies. Other relevant state and local LEA intelligence activity occurs in more direct partnership with, or supervision by, federal authorities; the preeminent of these is the FBI-led Joint Terrorism Task Forces (JTTFs), which now exist in all of the 56 FBI field offices in the United States and in many other cities as well.[25] The number of JTTFs increased from 36 in 2001 to 84 in 2003.[26]

Of the 18,000 LEAs across the United States, approximately 1,000 have 100 or more full-time sworn officers. Not surprisingly, for many of those, especially the smaller ones, terrorism is not a major issue. As Table 16.1 indicates, substantial majorities of state law enforcement agencies indicated knowledge of terrorist groups within their state, but only a fifth of local law enforcement agencies indicated knowledge of such groups operating in their jurisdiction. Most local law enforcement agencies (88 percent) indicated that no incidents attributed to a terrorist group had occurred within their jurisdiction within the past five years.

Table 16.1 Reported terrorist groups located within jurisdiction

Type of group	Percentage of all state law enforcement agencies reporting group in their jurisdiction	Percentage of all local law enforcement agencies reporting group in their jurisdiction
Right-wing	85	17
Race/ethnicity/hate-related	82	19
Religious groups utilizing violence	38	3
Single issue/special interests	74	24
Millennial/doomsday cults	8	3
Other	15	7

Similarly, state entities have greater experience with incident management and response, incident investigations, and hoaxes. About 16 percent of local LEAs have a specialized terrorism unit, while three out of four states report such a unit. Local LEA terror units typically have a more limited mission (primarily information sharing); state LEA terror units are more likely to take on more expansive roles such as training.

Most LEAs at both the state and local level have conducted terrorism threat assessments. Local LEAs were more likely to have conducted one only after 9/11; about half of the states had done theirs prior to September 11. Not surprisingly, there is a correlation between the size of the LEA and threat assessment activity: the larger the local LEA, the more likely they are to do a threat assessment. As Table 16.2 indicates, only about one out of three local LEAs collaborate with an FBI JTTF. Again, the larger the local LEA, the more likely that it will participate in a JTTF. For local authorities, participating in JTTFs typically means sharing information and receiving training. In contrast, nearly all state LEAs collaborate with JTTFs for the same reasons as well as for more expansive reasons, such as assisting with investigations.

Most states and close to a majority of local LEAs report needing more and better threat information, and most states and one-third of local authorities register requirements for more manpower (see Table 16.3). The need for better threat information was confirmed in a 2003 survey, which found that

Table 16.2 Participation in terrorism-related task forces

Liaise with or member of?	Percentage of all state law enforcement agencies	Percentage of all local law enforcement agencies
Yes	90	42
Of those that do, with which task force(s)?	Percentage of state law enforcement agencies that liaised with or were a member	Percentage of local law enforcement agencies that liaised with or were a member
FBI's Joint Terrorism Task Forces (JTTFs)	89	36
State Attorney General's Anti-Terrorism Task Force (ATTF)	77	44
State Homeland Security Office Task Force	77	23
City/county task forces	20	42
Other task force(s)	17	10

Table 16.3 Intelligence and information-related support needs

To improve response capabilities	Percentage of all state law enforcement agencies	Percentage of all local law enforcement agencies
More/better intelligence information on threats and terrorist activity in region	64	42
More manpower dedicated to response planning and/or to counter-terrorism activities	87	35

To improve assessment capabilities	Of those organizations that indicated a need for some type of support:	
	Percentage of state law enforcement agencies	Percent of local law enforcement agencies
To inform assessment activities, better intelligence on terrorist threat/capability from federal government	47	17

both state and local organizations were looking to the DHS for intelligence information and information about the terrorist threat within their jurisdiction. Sixty-two percent of local LEAs wanted more such information.[27]

Despite a desire for more detailed intelligence information, few local LEAs were in a position to receive it. Only 7 percent of local agencies indicated having applied for security clearances for their personnel after September 11, and, of those that had applied, only half indicated that all of their personnel who had applied had received the clearances. Indeed, state offices of emergency management and state public health departments were more likely than LEAs to have sought security clearances for their personnel after the September 11 attacks.

Detailed interviews with eight local LEAs confirmed the survey finding that local police generally have not created separate units for counter-terrorism intelligence. Counter-terrorism intelligence gathering and analysis tend to occur as part of a larger criminal intelligence unit. Nor has the terrorist threat led to large-scale changes in the organizational structure of most local police departments. In general, what local police have done is to increase their commitment of human resources to counter-terrorism efforts, which usually has come at the expense of other policing areas.

In general, too, the mandate of the counter-terrorism function is informal and set by the chain of command. Local police departments rely on federal guidelines in shaping their intelligence function, but the terrorist threat has raised awareness about what should and can be done in intelligence gathering, analysis, retention and dissemination. This has led some departments to adopt or refine their own guidelines. So, too, oversight of counter-terrorism intelligence is usually provided internally, through the chain of command in most agencies. Some jurisdictions have some oversight by an external body; a civilian committee, for instance, approves the Los Angeles Police Department's undercover operations.

Those departments have very little capacity to analyze the information they collect or receive, and, while federal grants have been available, most of that has gone for equipment and consequence management, not analysis and training. The September 11 attacks have led to a sharp increase in the amount of counter-terrorism information that is shared within and among local police and their federal counterparts. Paradoxically, though, the sheer number of cooperating agencies sometimes inhibits progress in responding to the terrorist threat.

One issue that did not emerge in the cases, but for which there is anecdotal evidence, is the extent of local opposition to forming intelligence groups. Recently, the Portland, Oregon City Council, acting on a recommendation from the mayor, withdrew the police from the FBI JTTF.[28] The precipitating cause was the refusal of the FBI to grant the mayor a security clearance, an issue that is a problem for the JTTFs. Understandably, many elected officials and senior local police officers chafe at having to be "cleared" by the FBI. Similarly, several states quit the Multistate Antiterrorism Information Exchange, known as MATRIX, out of concerns that included privacy and the social impact of interstate data sharing.[29] They did so even though MATRIX did not have intelligence-gathering functions, but rather focused on enabling information sharing across state lines.

The survey and case studies portray a very varied set of state and, especially, local responses to the threat of terrorism. For many, perhaps most, of the localities surveyed, terrorism is a threat that may come but has not yet. The findings tend to belie the notion that counter-terrorism intelligence is a pervasive function among LEAs. Instead, the survey findings, which reflect heightened awareness associated with the Oklahoma City and September 11 attacks, suggest that the "eyes and ears" capability is concentrated among the larger departments. These are the agencies that are investing in training, response plans, coordination and other preparedness measures. This in turn suggests that the process of shaping and directing state and local LEA involvement in intelligence activities may be a narrower and more focused challenge than is often implied by the "eyes and ears" metaphor.

Overall, however, state and local intelligence gathering has gone up, at least as measured by wiretaps for law enforcement purposes. Not surprisingly, as Table 16.4 indicates, the jump was sharpest from 2000 to 2001. Since 2001, the number of orders has stayed roughly constant, but the number of communications intercepted under each order has gone up sharply, nearly

Table 16.4 Federal, and state and local wiretap orders, 2000–2003

Year	Total federal orders	Average number of communications per order	Total state and local orders	Average number of communications per order	Total federal FISA orders
2000	479	NA	711	NA	1,005
2001	486	2,367	1,005	1,180	932
2002	497	2,354	861	1,335	1,228
2003	578	2,931	864	3,052	1,724

Sources: Administrative Office of the US Courts, *Wiretap Reports*, available at http://www.uscourts.gov/library/wiretap.html (last visited June 14, 2004); for FISA, see http://www.fas.org/irp/agency/doj/fisa/index.html#rept (last visited June 14, 2004). The 2003 report on FISA surveillance from the Justice Department to the Administrative Office of the US Courts is available also at http://www.fas.org/irp/agency/doj/fisa/2003rept.pdf (last visited June 14, 2004).

tripling from 2000 to 2003. Table 16.4 reports the number of intercept orders for law enforcement purposes approved by, respectively, federal, and state and local judges, along with the average number of communications intercepted per order. The sixth column of the table reports the number of federal intercept orders granted under the Federal Intelligence and Surveillance Act, or FISA, for national security purposes.

There has been considerable attention to privacy and civil liberties considerations at the federal level, especially after the Patriot Act, which widened authority not just for FISA but also for investigation and surveillance in other ways. By contrast, there has been much less attention to what is going on, or what might be authorized, at the state and local level, and virtually no research on law and practice at those levels.[30]

The numbers in Table 16.4 should be read with some caution. First, the state and local numbers probably understate the facts, for several reasons. In 2001, for instance, 46 states had laws permitting interceptions, but only 25 reported using that authority. And if the states under-report to the federal government, so, too, localities may under-report to the states. Second, the purpose of the interceptions is not evident because terrorism is a problem for both intelligence and law enforcement. Thus, while by definition the FISA taps were for intelligence, as opposed to law enforcement, purposes, they might have generated leads or other information relevant to criminal prosecution. More to the point, while many states are in the process of broadening their authority to intercept communications, in most cases in most places the purpose is law enforcement. If the wiretaps generate information that is useful in the war on terrorism but *not* germane to any ongoing criminal investigation, that information will be a by-product.

From our interviews with local police departments—Las Vegas, for instance—it seemed likely that, if the locals undertook terrorism-related surveillance for intelligence purposes, they almost always did so with federal officials through the JTTFs. If so, the request for surveillance presumably would go through FISA channels, and any subsequent oversight would be through federal courts. The role of state courts in overseeing police investigations will usually come in the form of Fourth Amendment litigation arising from a criminal prosecution. It has been—and probably will continue to be— rare to see state courts ruling on the constitutionality of post-9/11 legislation like the Patriot Act.

In fact, a search turned up but one case of a state court ruling related to a post-9/11 issue. In that case, civil liberties groups sued New Jersey counties that held detainees for the federal (then) Immigration and Naturalization Service (INS) in county jails, seeking disclosure of information on detainees pursuant to state disclosure laws. The New Jersey court rejected the suit, largely on the grounds that federal authority preempted state action.[31] If surveillance is done through FISA, federal officials will be responsible. State courts will rarely have an opportunity to rule on the conduct of those federal officials. It can happen; federal officers acting pursuant to federal legislation can obtain evidence that a state later uses in a criminal prosecution. A state court could rule on the constitutionality of the federal officers' conduct (and thus on the federal legislation itself). But this would be rare.

After September 11, many states began to discuss more permissive reforms of their wiretap legislation.[32] Those measures typically expanded what crimes would justify wiretaps; who could grant authority; who could implement taps; and authorization to conduct "roving" taps across broader geographic areas, as well as the devices subject to interception.

The last, expanding authority to new devices, merely brings state laws into line with the prevailing federal statute, the Electronic Communications Privacy Act (ECPA) of 1986, which updated the standards for newer technologies, like cell phones and e-mail.[33] The issues raised are mostly those of whether local officials will get the training needed to operate such taps. Similarly, "roving" taps that permit surveillance of any communications device the target may use, instead of specifying a particular telephone or the like, are mostly a modernization of legislation. Roving taps were permitted under ECPA but not under FISA until the Patriot Act brought the two into harmony. States are moving to modernize their statutes in the same way. This does imply, though, that, just as federal judges can issue orders for the entire nation, some states are permitting judges to issue orders that extend beyond the jurisdictional bounds of the court. Florida, Virginia and Maryland have such provisions.[34] While these provisions recognize the fact that terrorism respects few boundaries, they do raise the prospects of "judge shopping" and of lessened supervision of interceptions performed beyond the originating court's jurisdiction.

Approaching an "ideal" in intelligence relations among levels of government

What does all this flux in procedures amount to? One way to evaluate that "so what?" question is to pose an ideal pattern or division of labor among the levels of government, and evaluate what is going on against that ideal. The evaluation, in turn, suggests steps that might be taken to move closer to the ideal.

Given FISA, federal authorities, and the FBI in particular, would naturally lead in intelligence gathering that is not connected to criminal investigation. The locals have neither money nor capacity for that kind of pure intelligence. So, too, the intelligence gathering would be guided by federal regulations and overseen primarily by federal courts. Here, the current pattern is close to the ideal.

Ideally, the state and local authorities would conduct two kinds of information or intelligence gathering: investigation, including electronic surveillance, of possible criminal acts; and collection that is incidental to the normal activities of LEA officers. The latter is the eyes and ears of the cops on the beat, and the goal is domain awareness—what's going on in the jurisdiction, what's the state of possible targets, and so on. Here, the shortfalls of current practice against an ideal are two, one more doctrinal and the other more practical.

The doctrinal problem is that both kinds of state and local intelligence gathering involve enormous discretion—not an unfamiliar issue in policing. But terrorism compounds the problem because the task is inherently preventing crimes, not enforcing the law after the fact. As states emulate the federal government in relaxing their eavesdropping regulations, the line between intelligence and law enforcement blurs for them too. Yet the range of state reporting, let alone state regulation, of eavesdropping is enormous. And the problem of guidelines runs all the way down the chain of command: we saw from the cases that most of the guidelines for the counter-terrorism mission at the local level are ad hoc and derive from the local chain of command.

The more practical shortfall is that local LEAs get neither much guidance about what to look for nor enough intelligence that is specific enough to shape local operations. There has been considerable attention to information sharing, especially looking from the federal level down, for instance by the national 9/11 Commission. It reflects the by-now common wisdom that the problem is only apparently one of hardware, the "pipes" to actually move information. To be sure, the piping remains a considerable problem, especially for many local departments, which have difficulty enough communicating with one another.

Yet policy and guidelines are the still more formidable obstacles. The 9/11 Commission recommends creating a government-wide "trusted information network" to share information horizontally, on the model suggested by a recent task force organized by the Markle Foundation.[35] Yet, as both the surveys and the cases suggested, the principal information sharing mechanism, the JTTFs, is constrained because it requires getting the state and local participants security clearances at the level of their FBI counterparts. Finding new ways to share information and to share it more widely are imperative. The 9/11 Commission notes that intelligence analysts, like other professionals, want to play at the top of their games, so their reports inevitably begin with the most classified—and thus least shareable—information. The Commission suggests the opposite, starting any report by separating information from sources and writing first at the level that can be most easily shared. If intelligence consumers wanted more, they could query the system under whatever rules were in place, leaving an audit trail of requests. Now many, perhaps most, potential consumers would not even know what to ask for.

Secretary of Defense Donald Rumsfeld has focused attention in the intelligence war on terrorism to the "known unknowns," the things we know we don't know, and especially to the "unknown unknowns," the things we don't know we don't know. Yet much of the 9/11 failure turned on another category, the "unknown knowns," the things we didn't know or had forgotten we knew. One of the striking findings from the surveys and cases is the importance of more analysis across all of Rumsfeld's categories.

That importance derives directly from the nature of the counter-terrorism task. A traditional law enforcement investigation seeks to reconstruct the single trail from crime back to perpetrator. By contrast, the counter-terrorism task, especially prevention, needs to look at a number of paths, assembling enough information about each to know when patterns are changing or something suspicious is afoot along one of the paths. It is not only an intelligence-rich task. It is also a task rich in intelligence analysis.

Ideally, the analysis function would be split among the levels of government. The federal level has a comparative advantage in special sources, especially sources abroad. Its analysis will naturally concentrate on those and on the broad, "connect the dots" function. Sometimes, those sources and that analysis will provide warning specific enough to alert particular local authorities. In other cases, though, it will remain general and will serve mostly to tip off local officials about what they might look for—for example, a string of apparently unrelated crimes involving false identities.

The federal government is struggling, through TTIC and DHS as well as a greatly expanded FBI intelligence function, to do better at its part of the ideal. Yet what is still more striking is how limited the analytic capacity is at the local level. Only the very largest police departments have much of any at all. Yet the local role in the division of analytic labor would be to take the general guidance provided by the feds and relate it to local domain awareness. What does new federal information or analysis add to that understanding of local circumstances?

The ideal and shortfalls against it suggest an agenda for doing better. The obvious first need is more training for more intelligence capacity, especially in analysis, at the state and local level. That would include techniques for increasing domain awareness and for undertaking local threat assessments. Yet, so far, federal assistance programs have tended to emphasize equipment for consequence management, not training for intelligence, though that state of affairs is changing.

Moreover, training might also address the other visible concern—the varied and ad hoc nature of guidelines for counter-terrorism intelligence. The federal government might regulate through training. It could require training to specified standards if a jurisdiction is to receive funding from the DHS. More generally, greater and more explicit federal funding for state and local intelligence agencies would permit the federal authorities a greater regulatory role over what is a fairly loose and ad hoc process at this point. It would encourage local police to develop internal guidelines (including mandate) and external oversight by tying them to funding.

More generally, while law enforcement throughout the United States is fundamentally local in its structure, there is no reason that law enforcement intelligence needs to be. A program like COPS (Community Oriented Policing Services), under which the federal government supported additional law enforcement officers for a specified period of time, could be modified to significantly boost local intelligence capabilities. A federal program on intelligence could operate similarly, with the federal government paying the cost of an intelligence "supervisor" for eligible law enforcement agencies. The supervisor could be selected by national authorities (such as the FBI) and trained to national intelligence standards. This federal/state link would embed federal capacity in state organizations. It might increase the flow of intelligence from the local level to the national level. It might also help standardize the flow of information.

While the DHS has a legal mandate to take the lead in sharing intelligence, as a practical matter the lead in sharing is likely to continue to rest with the FBI through the JTTFs. That would let federal, state and local LEAs jointly develop a definition of terrorism and apply it by requiring that terrorism cases, including surveillance, be run through the JTTF. In any case, it would help if *someone* were in the lead. This is consistent with the recommendations of the DHS and the Justice Department in their recent National Criminal Intelligence Sharing Plan, which lays out a number of recommendations regarding law enforcement's intelligence role in this area and how it could be improved.[36]

Finally, it will be up to the courts, and the federal courts in particular, to continue assessing how the relaxed procedures in the intelligence war on terrorism are striking the balance between privacy and civil liberties, on the one hand, and security on the other. What is hinted at in our survey and cases is much more explicit when talking with federal homeland security intelligence officials. They feel they are without much guidance in deciding, especially, what they do with information they collect that happens to be about Americans. Can they keep it in databases? For how long and on what basis? It will be up to the courts to enforce guidelines when constitutional or statutory standards apply, and to put pressure on the executive branch to issue clear guidelines when such standards do not apply.

Notes

1 These shortcomings were graphically illustrated in the report of the National Commission on Terrorist Attacks upon the United States, *The 9/11 Commission Report* (Washington, DC, 2004), available at http://www.9–11commission.gov/ (last visited Aug. 2, 2004). (Hereafter referred to as "9/11 Commission" and "9/11 Commission Report.")

2 In the spirit of full disclosure, one of the authors, Gregory F. Treverton, was a staff member of the Senate Select Committee, chaired by Sen. Frank Church (D, ID). This was the author's first job in government, a fascinating introduction and the beginning of an abiding interest in intelligence.

3 See *Final Report of the Select Committee to Study Governmental Operations with Respect to Intelligence Activities of the United States Senate*, 94th Congress, 2nd Session, 1976, Book II, *Intelligence Activities and the Rights of Americans*, and Book III, *Supplementary Detailed Staff Reports on Intelligence Activities and the Rights of Americans*. For links to these reports, as well as to a rich range of other documents, both historical and contemporary, see www.icdc.com/~paulwolf/cointelpro/cointel.htm

4 John Le Carré, *The Night Manager* (New York: Knopf, 1993), p. 42.

5 See Gregory F. Treverton, *Reshaping National Intelligence for an Age of Information* (Cambridge: Cambridge University Press, 2001), pp. 139ff.

6 National Security Act of 1947, 50 U.S.C. 403, Sec. 102 (e).

7 The findings of the joint House–Senate investigation of September 11 outline the basic story: *Final Report*, Part I, The Joint Inquiry, The Context, Part I, Findings and Conclusions, Dec. 10, 2002. A fuller account is contained in Senator Richard Shelby's long supplementary document, *September 11 and the Imperative of Reform in the Intelligence Community*, Additional Views, Dec. 10, 2002. Both are available at www.fas.org/irp/congress/2002_rpt/index.html. See, in particular, Shelby's report, pp. 15ff., from which this account is drawn, unless otherwise indicated.

8 The fullest account of these episodes is in the 9/11 Commission Report, Chapter 7–8.

9 Gregory F. Treverton discusses this and other constitutional issues in more detail in "Terrorism, Intelligence and Law Enforcement: Learning the Right Lessons," *Intelligence and National Security*, vol. 18, no. 4, Winter 2003.

10 This issue, too, arises from the changed nature of the threat. For if the international laws of war permit would-be combatants to be held until the war is over, does that principle still apply if the "war" has no proximate end?

11 Uniting and Strengthening America by Providing Appropriate Tools Required to Intercept and Obstruct Terrorism Act (hereafter referred to as Patriot Act), Pub. L. No. 107–56, 115 Stat. 272 (2001).

12 50 U.S.C., section 1804 (a) (7) (b), emphasis added.

13 Ibid., as amended by the Patriot Act, section 218, emphasis added.

14 For background, see "FBI Pairs Criminal and Intelligence Cases," CNN.com,

Dec. 13, 2003, available at http://www.cnn.com/2003/LAW/12/13/fbi.terrorism.ap/ index.html (last visited Dec. 23, 2003).

15 The bill, formally the Intelligence Reform and Terrorism Prevention Act of 2004, is available at www.fas.org/irp/congress/2004_rpt/h108–796.html (last visited Jan. 4, 2005).

16 For an interesting discussion of how other industrial democracies handle domestic intelligence and sharing among intelligence and law enforcement agencies, see Peter Chalk and William Rosenau, *Confronting the Enemy Within*, MG-100-RC (Santa Monica, CA: RAND, 2004).

17 Mark Rice-Oxley, "Plans to Fight Organized Crime with a 'British FBI," '*Christian Scientist Monitor*, Feb. 12, 2004, available at www.csmonitor.com/2004/0212/ p05s01-woeu.html (last visited Mar. 8, 2004).

18 The WMD commission went further and recommended creating not just a directorate of intelligence within the FBI but a national security service, incorporating intelligence plus the FBI's Counterterrorism (CTD) and Counterintelligence Divisions (CD). It feared that even a beefed-up intelligence directorate would lack the ability to task the FBI field offices for information or control the intelligence budget, most of which is spent by CTD and CD. *Final Report of the Commission on the Intelligence Capabilities of the United States Regarding Weapons of Mass Destruction* (Washington, DC, 2005), p. 30, available at http://www.fas.org/irp/ offdocs/wmdcomm.html (last visited Apr. 27, 2005).

19 SCI is the broad category covering most intelligence information. But, as the name implies, that category is then broken down into compartments, for consumers usually by source, with individual users then cleared into particular compartments. Some of those compartments, such as signal intelligence, or SIGINT, may have large numbers of people cleared into them; others are more specialized and smaller. But most FBI agents and almost all state and local law enforcement officers are *not* cleared into SCI at all.

20 For Gregory F. Treverton's assessment, see his "Intelligence Gathering, Sharing and Analysis," in *The Department of Homeland Security's First Year: A Report Card*, ed. Donald Kettl, Century Foundation, Mar. 2004, www.tcf.org/Publications/ HomelandSecurity/2.intelligence.pdf

21 For a recommendation for a fleshed-out structure for sharing, one that resonates with many of the ideas in this chapter, see *Creating a Trusted Network for Homeland Security*, Second Report of the Markle Foundation Task Force, Dec. 2003, available at http://www.markletaskforce.org (last visited Dec. 5, 2003). This quotation is from p. 8.

22 Some examples that address domestic and foreign intelligence issues include *Countering the Changing Threat of Terrorism*, National Commission on Terrorism, June 2000; the first through fifth reports from the Advisory Panel to Assess Domestic Response Capabilities for Terrorism Involving Weapons of Mass Destruction (Gilmore Commission), 1999–2004; and *Final Report of the National Commission on Terrorist Attacks upon the United States* (9/11 Commission Report), July 2004.

23 This discussion is drawn from K. Jack Riley and others, *State and Local Intelligence in the War on Terror*, MG–394, RAND Corporation, 2005, available at http:// rand.org/pubs/monographs/2005/RAND_MG394.pdf. The authors thank their RAND colleagues in that venture, Jack Riley, Lois Davis and Gregory Ridgway. This draws on a 2002 survey of all 50 states and some 200 local authorities, and on detailed interviews with eight local authorities around the country. The 2002 survey results are published in more detail in Lois M. Davis and others, *When Terrorism Hits Home: How Prepared Are State and Local Law Enforcement?*, MG-104-MIPT (Santa Monica, CA: RAND Corporation, 2004).

24 "Title III" refers to Title III of the Omnibus Crime Control and Safe Streets Act of 1968. For a discussion of the Patriot Act's relationship to Title III issues, see Testimony of Robert S. Mueller III, Director, Federal Bureau of Investigation, before the United States Senate Committee on the Judiciary, Sunset Provisions of the USA Patriot Act, Apr. 5, 2005, available at http://www.fbi.gov/congress/congress05/mueller040505.htm (last visited May 23, 2005).

25 The JTTFs vary in size and structure in relation to the terrorist threat dealt with by each FBI field office. On average, 40 to 50 people are assigned full time to the JTTFs; however, some task forces, such as New York City, can have as many as 550 personnel, and a number of part-time personnel can also be assigned to the JTTFs.

26 Office of the Inspector General, US Department of Justice, Dec. 2003.

27 Lois M. Davis and others, "Summary of Selected Survey Results," Appendix D in the report "The Advisory Panel to Assess Domestic Response Capabilities for Terrorism Involving Weapons of Mass Destruction Fifth Annual Report to the President and Congress," V. "Forging America's New Normalcy: Securing Our Homeland, Protecting Our Liberty," Dec. 15, 2003.

28 "Portland Becomes First to Pull Out of FBI-led Anti-Terror Team," *Seattle Post-Intelligencer*, Apr. 29, 2005, available at http://seattlepi.nwsource.com/local/222207_fbi29.html (last visited May 23, 2005).

29 See, for example, "Two More States Withdraw from Database Program," *Information Week*, Mar. 12, 2004, available at http://www.informationweek.com/story/showArticle.jhtml?articleID=18312112 (last visited June 10, 2005).

30 That is the conclusion of the one study of law and practice at the state and local level: Charles H. Kennedy and Peter P. Swire, "State Wiretaps and Electronic Surveillance after September 11," *Hastings Law Journal*, 54, 2003.

31 See 352 N.J. Super.44, 799 A.2d 629.

32 The website of the Constitution Project Initiative on Liberty and Security provides information on the status of each state's wiretap legislation, along with an overview across states. See http://www.ncsl.org/logic.htm?returnpage=http://www.ncsl.org/ programs/lis/CIP/surveillance.htm

33 Berger v. New York, 388 U.S. 42, 54–55. 58–59 (1967); and Electronic Communication Privacy Act of 1986, Pub.L. No. 99–508, 100 Stat. 1848 (1986).

34 Kennedy and Swire, "State Wiretaps and Electronic Surveillance after September 11," p. 982.

35 See *Creating a Trusted Network for Homeland Security*, Second Report of the Markle Foundation Task Force, Dec. 2003, available at http://www.markletaskforce.org (last visited Dec. 5, 2003).

36 The Global Justice Information Sharing Initiative Intelligence Working Group, which is a national criminal intelligence council, developed the National Criminal Intelligence Sharing Plan. This plan is available at http://it.ojp.gov/documents/National_Criminal_Intelligence Sharing_Plan.pdf

17

FUSING TERRORISM SECURITY AND RESPONSE

John P. Sullivan

Global society is in the midst of a significant transition, perhaps on the scale of a vast epochal change. Technological innovation, globalization, the emergence of networks and the evolution of state forms are changing the nature of war, crime and threats to society. The security dynamics emerging during this transition constitute what appears to be a new suite of threats. Foremost in this array of threats is global terrorism. Extremist organizations, exemplified by the self-proclaimed global jihadi movement described as al Qaeda and its affiliates, are complex non-state actors operating as transnational networks within a galaxy or nebula of like-minded networks. These transnational entities pose security threats to nation-states and collective global security. New approaches are needed to meet these circumstances. This chapter explores these dynamics.[1]

The first of these dynamics is the emergence of globalization and the early shift toward global society itself. Increasingly, issues that were once local concerns managed within a nation-state are now at once both local and global in reach. These concerns include global crime, global terrorism and global disease. Fueled by global economic, technological and increasingly political forces, these threats demand the development of a global security regime, structures and approaches.

Three overarching influences are driving the globalization of conflict. These are the emergence of networks, "global cities" and the "market-state."

Networks and conflict

Networks are organizations of people and organizations that are linked for a common purpose. Specifically, social networks drive all levels of social, political, economic and consequently criminal activity. Technology allows networks to expand their reach and impact. Networks are central to the current security situation, but, while many talk about networks, a true recognition of network dynamics is often missing from the discussion. At a mechanical or organizational level, John Arquilla and David Ronfeldt's typology of networks is instructive. It provides a basic overview of network organization and summarizes how individual nodes link to each other:[2]

- *Chain.* The chain or line network, as in a smuggling chain where people or goods or information move along a line of separated contacts, and where end-to-end communication must travel through the intermediate nodes.
- *Star.* The hub, star or wheel network, as in a franchise or a cartel where a set of actors are tied to a central (but not hierarchical) node or actor, and must go through that node to communicate and coordinate with each other.
- *All-channel.* The all-channel or full-matrix network, as in a collaborative network of militant peace groups, where everybody is connected to everybody else.
- *Hybrid.* There may also be hybrids of the three types, with different tasks being organized around different types of networks.

As I noted in my paper "Networked Force Structure and C⁴I,"[3]

> The rise of networks (and hence networked adversaries) results from the migration of power to non-state actors that are able to organize into multi-organizational networks (particularly "all-channel" networks where every node is connected to every other node) more readily than hierarchical, state actors. As a result of this trend, network-based conflict and crime are a growing threat. As Ronfeldt, Arquilla and others have often noted, hierarchies have a difficult time fighting networks. Thus, to combat networks, that is to master counternetwar, the police, military and security services must first understand the nature of the networked threat, and then as described later forge the proper balance between networks and hierarchies to combat these emerging threats.[4]

As a result, networks influence all aspects of the discussion of global threats and response, influencing "global cities" and the emerging "market-state," as well as efforts to negotiate this new security environment. Conflict within the context of networked forms is described as "netwar," which will be further explored later in this chapter.

"Global cities"

As networks and technology influence economic processes, new forms of social, political and criminal opportunity arise. Saskia Sassen, a sociologist at the University of Chicago and London School of Economics, describes the rise of "global cities" in this context.[5] Sassen observes that the flow of capital, labor, goods and people—all cross-border economic processes—have traditionally occurred within the context of an interstate arrangement of nation-states. Yet as the result of privatization, deregulation, transnational firms, and technology enabling the participation of national economic actors in the global market, this has changed. The result is the redistribution of strategic economic territories, where national units are no longer the primary spatial focus of transnational economic transactions. Subnational or non-state actors are the principal beneficiaries of this economic restructuring.[6] Cities, subnational regions, cross-border hubs, and supranational markets and trade blocs are among those impacted[7] (as are transnational criminal and terrorist actors).

In short, the geographic dispersal of economic activity that relies on activities geographically distributed among local, regional and global nodes fuels the need for complex centralized corporate functions. These in turn stimulate the need for outsourcing to specialized firms that derive their talent pool from the deep mix of talent, capability and expertise available in metro-politan areas that provide "an extremely intense and dense information loop."[8] To leverage these capabilities, individual firms develop a network of affiliates and partners. These global links strengthen transnational node-to-node interaction. As a result, cities become nodes in cross-border networks—in Sassen's words, a series of transnational networks of cities.[9]

As can be expected, the transnational network dynamics among "global cities" and regions embrace a range of transactions and influence a broad range of activities: political, cultural, social and criminal. Interaction among immigrant and diaspora communities provides richness and context to communities, but, when expectations to benefit from inclusion in the global community are thwarted or not realized, discontent can resonate globally (as in the case of Muslim youth attracted to the global Salafist jihad). The connectedness that results from cross-border networks increases social and political exchanges, including non-formal, issue-specific transnational politi-cal networks focused on environmental and human rights issues among others. Sassen notes that these new "agoras" or political space are "largely city-to-city cross-border networks," since it is currently easier to capture "the existence and modalities of these networks at the city level. The same can be said for the new cross-border criminal networks."[10]

Disease, itself a significant security threat in the emerging global grid, can also be a key security concern in the transnational city network. Consider the impact of a global pandemic, perhaps triggered by highly pathogenic avian flu of the H5N1 variety, or one of a range of other strains,[11] mutating and spreading via city-to-city connections. Yet the most pressing challenge is likely to involve global crime and terrorism. Generally, transnational organ-ized crime groups, while they exploited the seams between states, benefited from the existence of a stable state order. Traditional criminal enterprises did not seek to challenge the state; rather they exploited corruption and political influence to further their enterprises. This appears to be changing as a new range of transnational gangsters exploit shadow economies, the absence of effective states, and endemic corruption.[12]

According to Louise Shelley, director of the Transnational Crime and Corruption Center at American University, "the newer crime groups most often linked to terrorism have no interest in a secure state."[13] They promote and exploit grievances at local levels and through the globalization of conflict to secure the maneuver room to capture profit. Shelley notes that the embedded nature of network crime structures in local communities and the inability of both domestic and international militaries, as well as law enforcement agencies, to control their activities make them a growing danger.[14] These terrorist–transnational criminal relationships are particularly poignant in "global cities" and subnational or cross-border enclaves or "lawless zones."[15] In the case of "new transnational criminals" these relationships go to the "very heart of the relationship between crime groups and the state."[16]

Terrorists and new transnational criminal organizations are not the only criminal enterprises to benefit from networked dynamics and city-to-city interconnectedness. Street gangs, specifically "third-generation gangs" such as Mara Salvatrucha (MS-13), originally from the MacArthur Park neighborhood in Los Angeles, which now operates throughout North and Central America, have also exploited the new spatial and geographic relationships afforded by globalization. Once the province of inner-city ghettos, some gangs have evolved from a sole interest in turf or localized crime to gain in sophistication, political interest and international reach. Exploiting seams in law enforcement and judicial structures, immigration (often by forced deportation), and technologies that foster communication, third-generation gangs have also become de facto global criminals, threatening local stability and potentially fueling broader networked conflict.[17]

The interactions of technology, networks, "global cities" and non-state actors and enclaves set the stage for a discussion of state transition, specifically the transition to "market-state" structures.

Transnational actors and the rise of the market-state

Foremost among the factors challenging law enforcement and intelligence agencies, and government at large, is the emergence of transnational networked actors on both the domestic and the world stage. This situation is a result of the shift from the "nation-state" as the dominant state form to the "market-state." Philip Bobbitt observes that this change is the latest of a series of changes in state form. He observes that the constitutional foundations of states, and their position in the international order, have evolved— largely as a result of changes in war. Thus he notes that the princely state morphed into the kingly state, then into the territorial state, then into the state-nation and most recently into the nation-state. The Habsburg–Valois wars, Thirty Years War, wars of Louis XIV, wars of the French Revolution and finally the "Long War" respectively influenced each state iteration.

Finally, Bobbitt posits that the result of the Long War (World Wars I and II and the Cold War) set the stage for the transition to the market-state, which appears, at least for now, to be dominated by the global war on terrorism (GWOT).[18] Thus, we are seeing a shift in international order, constitutional foundation(s), war-making, and security structures. National security is morphing into global security—and market actors will play a bigger role.

Failures of governance in the face of this epochal shift lead to state failure and the emergence of lawless zones. This in turn fuels conflict and a reordering of the state system. This creates new spaces for private military actors: private military firms or corporations (PMFs/PMCs), private security and private intelligence groups on the one hand, and rogue or criminal actors on the other.

Bobbitt[19] observes that Osama bin Laden's al Qaeda is a malignant and mutated version of the market-state. As such, al Qaeda and its kin are more than stateless gangs. These new networked adversaries possess standing armies, treasury and revenue sources (even if derived from criminal enterprises), a bureaucracy or "civil" service, intelligence collection and analysis organs, welfare systems, and the ability to make alliances (with state and non-state entities). They also promulgate law and policy, and declare war. As such, the al Qaeda network and others like it are virtual states. These virtual states are non-territorial market-states (although they sometimes hold and control territory) and through insurgency and terrorism seek to hold or influence more.

P.W. Singer observes, "Transnational criminals, economic insurgents, warlords for profit, armies of child soldiers, and brutalized civilians are all found in these zones of conflict and lawlessness."[20]

This rise of the role of non-state entities in violence leads to diminished local capabilities to protect. When combined with the global impact of many terrorist or transnational criminal acts, the challenge to the public monopoly on war and security—both internal and external—to individual states is enhanced. At the least this renders the distinction between foreign and domestic security (and intelligence) increasingly anachronistic. More likely, however, this will require a re-engineering of security and intelligence structures and relationships to embrace networked forms and capabilities to navigate this global shift.

This shift, which is currently accountable for the rise of global terrorism, insurgency and instability, requires the development of disaggregated structures for global security and governance. Within these structures, security and response will necessarily adapt to address global threats. The resulting structures will link law enforcement, criminal intelligence, national security intelligence and operational intelligence (including epidemiological intelligence for disease security and pandemic response) from public and private sectors within and among states and non-state organs to develop a networked approach to global security.

Post-modern terrorism, netwar and counternetwar

These "counternetwar" structures are evolving to address netwar, a situation where non-state actors operating across international borders are becoming the significant actors in war, terrorism and transnational crime. Netwar is characterized as "an emerging mode of conflict (and crime) at societal levels, involving measures short of war, in which the protagonists use—indeed depend on using—network forms of organization, doctrine, strategy, and communication. These protagonists generally consist of dispersed, often small groups who agree to communicate, coordinate, and act in an internetted manner, often without precise leadership or headquarters."[21]

B. Rahman, an analyst and former Indian cabinet secretary for intelligence, describes this shift in terms of government capabilities and perceptions of terrorism. Rahman rightly observes that, "before the 1960s, terrorism was seen largely as a threat to law and order within a society."[22] During this era, and indeed up until the 9/11 attacks, the role of the military and national intelligence agencies was largely subordinate to that of the police in the counter-terrorism arena. Al Qaeda and its jihadi international changed this perception. Now professional actors, governments and the public are beginning to recognize that post-modern terrorism not only threatens their own national security, but places regional and global security at risk.

Michael Hardt and Antonio Negri describe this situation as a state of global war requiring global governance—an "empire" ruled by the "multitude."[23] Within this broader analysis, where they see the current security situation as "a state of war in which network forces of imperial order face network enemies on all sides,"[24] they emphasize the need for networked approaches to counter global insurgency. They note that the "networked form of power is the only one today able to create and maintain order."[25]

Robert Bunker and I examined these dynamics in our paper "Multilateral Counter-Insurgency Networks," noting:

> Existing security structures (domestically the police, and internationally the military and foreign intelligence services) designed to counter state-on-state threats find this new operational environment challenging at best. Preserving global and national security requires traditional organs of national security (the diplomatic, military and intelligence services) to forge new partnerships with police and public safety organizations at the state and local (sub-national) level to

effectively counter these threats. Significant operational, policy and cultural challenges must be overcome to forge an effective global network of public safety, law enforcement and traditional intelligence organizations to understand and anticipate current and evolving terrorist threats.[26]

Effective response to this situation demands a "more robust counter-terrorism approach."[27] Rahman goes further:

> National intelligence agencies, by themselves, however strong and capable, may not be able to deal with this new threat of a trans-national nature. Hence, the need for a regional and international networking of the intelligence and security agencies to counter the trans-national terrorist network. The new terrorism calls for a revamped intelligence apparatus at the national level and a reinforced co-operation mechanism at the regional and international levels.[28]

To accomplish this it is necessary to stimulate growth of a high degree of interoperability among all levels of responders (local, state, federal and global), between a variety of disciplines (law enforcement, intelligence, fire service, public health and medical), and between civil and military agencies. Intelligence is an important element of forging the necessary networked response.

Networked solutions: more than information sharing

As John Arquilla and David Ronfeldt are frequently quoted as stating, "it takes networks to fight networks."[29] But networks are more than just a way of organizing criminal and terrorist conspiracies. Al Qaeda is a classic network, or network of networks, but, as Anne-Marie Slaughter observes, we live in a "world of networks: of corporations, of nongovernmental organizations, of criminals, of government officials."[30] Slaughter observes that networks are decentralized, informal and flexible, relying upon regular interchange among participants. Like Ronfeldt, Arquilla, Bunker and Sullivan, she posits that nations can address networked threats by establishing networks of their own: global or regional networks of financial regulators, prosecutors, criminal investigators, immigration officials, transport officials and customs agents.[31]

Slaughter notes that the emerging state form is likely to be a disaggregated state dominated by a new global landscape of government networks.[32] The state has not disappeared but, similar to Bobbitt's view, it has morphed—or is morphing. This network will include horizontal government networks (characterized by peer-to-peer links with professional counterparts across borders) and less frequently vertical government networks between national

and intelligence sharing, including sharing best practices and bolstering bilateral and trilateral security assistance.[42]

Police cooperation and interaction are both vital and common attributes of current initiatives and future multidimensional approaches. This is equally true for developing capacity to address global crime and terrorism. International criminal enterprises and terrorists engage in drug crime, illegal immigration and human trafficking, financial crimes, arms trafficking, and the potential proliferation of weapons of mass destruction by smuggling radiological and fissile materials. There are two major transnational or international vehicles for formal police interaction: Interpol (the International Criminal Police Organization or ICPO) and Europol, the European police agency. These organizations have both improved their capacity to address terrorism, but have traditionally been restrained in their effectiveness in terrorist activities. This requires enhancements to their capabilities, as well as the development of new structures—including police-to-police interaction at the city-to-city level as well as state-to-state efforts as traditionally practiced.

Recent experience affirms the need for such cooperation, as well as developing linkages among metropolitan police, national police forces, customs agencies, prosecutors and judicial authorities, intelligence and security services and, when appropriate, military and private sector actors. Consider the June 2003 arrest of a Thai national allegedly involved in transporting a radioactive isotope to Thailand from Russia through Laos. In other cases, the Royal Thai Police supported by information from the Central Intelligence Agency were able to arrest three suspects linked to Jemaah Islamiyah who were planning to bomb multiple Bangkok embassies. Italian anti-terrorist and intelligence units supported the UK's investigation of the 7/7 London bombings; and Spain's investigation of the Madrid (Once Eme) bombings, which also included contributions by Interpol's National Central Bureaux in Madrid and Serbia, and Serbian police.[43]

Yet Interpol has traditionally played only a marginal role in addressing terrorism, owing to its constitutional restraints on supporting investigations of "political" activities and related crime. Similarly, until the establishment of Europol, European agencies developed their own counter-terrorism police efforts. These included the inter-ministerial Trevi Group, and an informal operational network—the Police Working Group on Terrorism (PWGT)—as a sub-Trevi practitioner-based effort. The PWGT has proved effective at police-to-police exchanges, but excludes civil security services and police with non-judicial roles; additionally, it has no central analytical capability. In 1999, Europol was established as a formal entity and given a terrorism remit, which has been expanded to include analytical capabilities at its situation center in Brussels. Despite this role, Europol does not have independent authority to collect information, as that role was reserved by member states. Another limit is the requirement that Europol receive and disseminate information exclusively through a government-designated "Europol

National Unit."[44] Often, this can be an agency not directly charged with terrorism enforcement (this is often the same with Interpol National Central Bureaux, and precludes direct interaction of metropolitan police without having to transverse a hierarchical structure).

Practical, legal and bureaucratic factors often complicate transnational police cooperation. This is particularly true when subnational police need to engage their international counterparts. This is further complicated when police and law enforcement officials need to interact with those outside their own bureaucracies and disciplines to develop intelligence and craft responses to threats and the aftermath of an attack or disaster. The technology, economy, threat and need are there, but political, organizational, institutional and legal frameworks are immature. Entrenched obstacles slow progress toward the establishment of formal collaborative networks.

A global civil/security network

Protecting post-modern market-states is undoubtedly a multifaceted, complex endeavor. The relationships among the variety of state and non-state actors involved are continually evolving in organizational and legal terms. Much of the critical infrastructure of modern society is privately held and operated. Private corporations and non-governmental organizations (NGOs) provide functions and a range of services traditionally provided by governments. Counter-terrorism (or counter-insurgency) networks must interoperate with a number of governmental organizations and entities at the local, state and federal levels (for example, within the US, Australia or Canada) or with a number of international or supranational entities (e.g. within the European Union).

These security networks will also have to interact with a number of NGOs, corporate security entities, private military firms and private intelligence providers. In today's wars in failing states and lawless zones, humanitarian NGOs, advocacy groups, international organizations (such as the United Nations and its various arms), military and intelligence services of multiple nation-states, and a growing number of PMCs must interact to assess, anticipate and respond to crises.

These diverse entities play varying roles in understanding and containing crises and war. Peacemaking, peacekeeping, stability and support operations rely on a complex networked interaction among actors to operate. NGOs such as the International Committee of the Red Cross (ICRC) advocate— and aid providers rely upon—the services and cooperation of nation-states and their military organs, private security, and intelligence contractors to provide services. Within this setting, state security services (military and police) increasingly rely upon private contractors (PMCs and corporate intelligence services) to perform their tasks. Because of this, an understanding of contemporary terrorist and insurgent networks and threat areas requires diverse intelligence and security agencies to pool and aggregate information.

Conclusion

Contemporary terrorists frequently operate in networks, or more specifically cells and/or groups within networks, that in turn link with other networks to form networks of networks. These "multi-networks" thus constitute a "galaxy" or "nebula" that confounds traditional governmental responses. Indeed, as network structures and organization become more common, these entities may resemble nebulae or clusters or groups of galaxies of different kinds: criminal, terrorist, insurgent, activist, legitimate political or theological actor, etc.

Networked approaches are essential to addressing these already emerging structures and the global instability, global terrorism and transnational threats that exploit the transition to new approaches to societal organization. Intelligence is an essential element of negotiating this environment. Intelligence, however, has always been a competitive endeavor. Sources and means must be protected, but new cooperative and networked approaches must be quickly and accurately assessed to diffuse a global networked threat.

Co-production of intelligence to counter the evolving terrorist threat requires the development of multilateral structures. Much of the information necessary to understand the dynamics of a threat—indeed, even to recognize that a threat exists—is developed from the bottom up, as well as through horizontal (as opposed to top-down) structures. Multilateral exchanges of information, including indicators of potential attacks and alliances among networked criminal actors, are needed to counter networked adversaries. This requires the development of new analytical tradecraft, processes and policy. Intergovernmental instruments are needed to fully exploit lateral information sharing, along with the development of distributed intelligence processing across organizational and political seams, including the development of mechanisms for sharing information among both intra-national and international nodes.[45]

Important as intelligence may be, it—along with traditional police and enforcement efforts—provides only one element of an effective networked approach. It is essential to recognize that "leveraging the benefits of network organization constitutes a new source of power and a new way of accomplishing global governance."[46]

According to network scholar Paul Hartzog, "[a]s individuals and groups engage each other globally, the locus of global governance shifts from state-centered activities to distributed networks. The cumulative effect of this shift from hierarchies to networks is a system of overlapping spheres of authority and regimes of collective action called 'panarchy.' "[47] Hartzog posits that the activities of these transnational networks "not only influence states, but also serve as a means of social governance that *functions independently of and in parallel to state governing.*"[48] This involves both governmental actors and their criminal challengers. Yet networked approaches when embraced by government can enhance societal security, even if they change the position of state structures as currently configured. As Jamie Metzl observes, "enhanced transnational civil society and issue networks may challenge government authority in some cases. But more often they can serve as invaluable tools for sharing information, developing mutual understanding, and solving problems."[49] This is definitely the understanding necessary to craft the networks of networks—formal, informal, state-to-state, supranational, police-to-police, city-to-city, etc.—needed to develop the cohesive and effective responses required to combat global disease, global crime and global terrorism.

Networked cooperation is essential to solving this problem. Metropolitan police and governments need to leverage their local capabilities in partnership with national, supranational and their extra-jurisdictional local counterparts worldwide to develop, nurture and exploit the multilateral and multidimensional capabilities needed. These capabilities include the organizational frameworks required to perform dynamic sensing (of the threat), ensure co-production of intelligence, and provide an adaptive response. This necessitates both node-to-node and professional peer-to-peer efforts, but most importantly requires security, intelligence and enforcement agencies to place their actions in the broader context of global social governance and global security. Embracing cooperative networks as an organizing framework for fusing security and response is an essential element of controlling and managing current and emerging global threats such as natural pandemics and their virulent human counterparts: terrorism, global insurgency and transnational crime.

Notes

1 An earlier version of this chapter was presented as a paper: John P. Sullivan, "Fusing Terrorism Security and Response for a Global Counterterrorism Network" in the panel "Law Enforcement—Intelligence Interactions: Problems and Opportunities," Intelligence Studies Section, International Studies Association (ISA), 2005 ISA Annual Convention, Honolulu, HI, Mar. 3, 2004.

2 John Arquilla and David Ronfeldt, "The Advent of Netwar (Revisited)," in *Networks and Netwars: The Future of Terror, Crime, and Militancy*, ed. John Arquilla and David Ronfeldt (Santa Monica, CA: RAND, 2001), pp. 7–8.

3 John P. Sullivan, "Networked Force Structure and C⁴I," in *Non-State Threats and Future Wars*, ed. Robert J. Bunker (London: Frank Cass, 2003), pp. 145–6.

4 While it is often postulated that it takes a network to combat a network, Ronfeldt, Arquilla and others note that counternetwar actually relies upon hybrids of hierarchies and networks. Either way, the side that masters networks (and their relationship with other organizational forms) is likely to prevail in the now-and-future conflict.

286

5 See Saskia Sassen, *Denationalization: Territory, Authority and Rights in a Global Digital Age* (Princeton, NJ: Princeton University Press, 2005), for a powerful seminal view of the re-emergence of cities in the global political order.

6 Saskia Sassen, "The Global City: Introducing a Concept," *Brown Journal of World Affairs*, vol. XI, no. 2, Winter/Spring 2005, p. 27.

7 Ibid.

8 Ibid., pp. 28–9.

9 Ibid., p. 29.

10 Ibid., p. 32.

11 See Peter Katona's discussion of disease security in Chapter 1. For a discussion of global avian flu potentials and the impact on global commerce and agriculture, see Mike Davis, *The Monster at Our Door: The Global Threat of Avian Flu* (New York: New Press, 2005).

12 Louise Shelley, "The Unholy Trinity: Transnational Crime, Corruption, and Terrorism," *Brown Journal of World Affairs*, vol. XI, no. 2, Winter/Spring 2005, p. 102.

13 Ibid.

14 Ibid.

15 See John P. Sullivan and Robert J. Bunker, "Drugs Cartels, Street Gangs, and Warlords," in *Non-State Threats and Future Wars*, ed. Robert J. Bunker (London: Frank Cass, 2003), pp. 40–53, for a discussion of lawless zones, such as Ciudad del Este and the Tri-Border Region and their impact on the crime–terrorism nexus and state erosion.

16 Shelley, "Unholy Trinity," p. 104.

17 See especially John P. Sullivan, "Gangs, Hooligans, and Anarchists—The Vanguard of Netwar in the Streets," in *Networks and Netwars: The Future of Terror, Crime, and Militancy*, ed. John Arquilla and David Ronfeldt (Santa Monica, CA: RAND, 2001), pp. 99–128. Also see John P. Sullivan, "Urban Gangs Evolving as Criminal Netwar Actors," *Small Wars and Insurgencies*, vol. 11, no. 1, Spring 2000, pp. 82–96; and Max G. Manwaring, *Street Gangs: The New Urban Insurgency* (Carlisle, PA: US Army War College, Strategic Studies Institute, Mar. 2005) for detailed discussions of the threats posed by third-generation gangs.

18 Philip Bobbitt, *The Shield of Achilles: War, Peace, and the Course of History* (New York: Anchor Books, 2002).

19 Ibid. See especially pp. 820–1.

20 P.W. Singer, *Corporate Warriors: The Rise of the Privatized Military Industry* (Ithaca, NY: Cornell University Press, 2003), p. 51.

21 John Arquilla and David F. Ronfeldt, *The Advent of Netwar* (Santa Monica, CA: RAND, 1996), p. 5.

22 B. Rahman, "Intelligence and Counter-Terrorism," Paper no. 989, South Asia Analysis Group, Apr. 21, 2004 at http://www.saag.org/papers10/paper983.html and *Journal of International Security Affairs*, no. 7, Summer 2004, pp. 91–101.

23 Michael Hardt and Antonio Negri, *Multitude: War and Democracy in the Age of Empire* (New York: Penguin Press, 2004).

24 Ibid., p. 62.

25 Ibid., p. 59.

26 John P. Sullivan and Robert J. Bunker, "Multilateral Counter-Insurgency Networks," *Low Intensity Conflict and Law Enforcement*, vol. 11, no. 2/3, Winter 2002, pp. 353–68.

27 Rahman, "Intelligence and Counter-Terrorism."

28 Ibid.

29 Arquilla and Ronfeldt (eds.), "The Advent of Netwar (Revisited)," p. 15.

30 Anne-Marie Slaughter, "We Can Beat Terror at Its Own Game: Networks Are Both the Problem and the Solution," *Los Angeles Times*, Apr. 25, 2004, at http:www.latimes.com/news/opinion/commentary/la-op-slaughter25apr25, 1,1645678.story?coll=la-news-comment-opinions

31 Ibid.

32 Anne-Marie Slaughter, *A New World Order* (Princeton, NJ: Princeton University Press, 2004), p. 12.

33 Ibid., p. 21, and Robert Cooper, *The Breaking of Nations: Order and Chaos in the Twenty-First Century* (New York: Atlantic Monthly Press, 2004), pp. 30–1.

34 Slaughter, *New World Order*. These are discussed throughout her text.

35 Ibid., p. 54.

36 Ibid., pp. 55–6.

37 Nora Bensahel, *The Counterterrorism Coalitions: Cooperation with Europe, NATO, and the European Union* (Santa Monica, CA: RAND, 2003), pp. 39–41.

38 Rahman, "Intelligence and Counter-Terrorism."

39 Ibid.

40 For additional information on the prospects for the TEW model to serve as a counternetwar structure, see John P. Sullivan, "Networked Force Structure and C⁴I," in *Non-State Threats and Future Wars*, ed. Robert J. Bunker (London: Frank Cass, 2003), pp. 144–58, and John P. Sullivan and Robert J. Bunker, "Multilateral Counter-Insurgency Networks," *Low Intensity Conflict and Law Enforcement*, vol. 11, no. 2/3, Winter 2002, pp. 353–68.

41 Barry Desker and Elena Pavlova, "Comparing the European and South Asian Response to Global Terrorism," *Journal of Conflict Studies*, vol. XXV, no. 1, Summer 2005, p. 9.

42 Ibid.

43 "INTERNATIONAL: Police cooperation," *Oxford Analytica Daily Brief*, Sept. 8, 2005.

44 Paul Swallow, "Counter-Terrorist Police Cooperation in the European Union: Political Ambitions and Practical Realities," Paper presented to the Los Angeles Terrorism Early Warning Group (TEW) Conference: Terrorism, Global Security, and the Law, Santa Monica, CA, June 1–2, 2005.

45 John P. Sullivan, "Terrorism Early Warning and Co-Production of Counterterrorism Intelligence," in Panel 5, "In Pursuit of the Analytical Holy Grail: Part 1, Innovation in Analysis, Warning and Prediction," Canadian Association for Security and Intelligence Studies, CASIS 20th Anniversary International Conference, Montreal, Quebec, Canada, Oct. 21, 2005.

46 Paul B. Hartzog, "Panarchy: Governance in the Network Age," Internet paper found at http://www.panarch=y.com/Members/PaulbHartzog/Papers, p. 2.

47 Ibid.

48 Ibid., p. 3.

49 Jamie F. Metzl, "Network Diplomacy," *Georgetown Journal of International Affairs*, Winter/Spring 2001, as quoted in Hartzog, "Panarchy," p. 33.

SELECT BIBLIOGRAPHY

Alibeck, K., *Biohazard* (New York: Random House, 1999).

Allison, G., *Nuclear Terrorism: The Ultimate Preventable Catastrophe* (New York: Owl Books, 2005).

"Anonymous" (M. Scheuer), *Through Our Enemies' Eyes: Osama Bin Laden, Radical Islam, and the Future of America* (Dulles, VA: Brassey's, 2002).

"Anonymous" (M. Scheuer), *Imperial Hubris: Why the West Is Losing the War on Terror* (Washington, DC: Potomac Books, 2004).

Arquilla, J. and Ronfeldt, D. (eds.), *Networks and Netwars: The Future of Terror, Crime, and Militancy* (Santa Monica, CA: RAND, 2001).

Bobbitt, P., *The Shield of Achilles: War, Peace, and the Course of History* (New York: Anchor Books, 2002).

Bunker, R.J. (ed.), *Non-State Threats and Future Wars* (London: Frank Cass, 2003).

Bunker, R.J. (ed.), *Networks, Terrorism and Global Insurgency* (New York and London: Routledge, 2005).

Carus, W.S., *Bioterrorism and Biocrimes: The Illicit Use of Biological Agents since 1900*, 1998, http://www.ndu.edu/centercounter/Full_Doc.pdf

Christopher, G.W., Cieslak, T.J., Pavlin, J.A. and Eitzen Jr., E.M., "Biological Warfare: A Historical Perspective," *Journal of the American Medical Association*, 278, 1977, pp. 412–17.

Creveld, M. van, *The Transformation of War* (New York: Free Press, 1991).

Creveld, M. van, *The Rise and Decline of the State* (Cambridge: Cambridge University Press, 1999).

Davis, L.M., Riley, K.J., Ridgeway, G.K., Pace, J.E., Cotton, S.K., Steinberg, P.S., Damphousse, K. and Smith, B.L., *When Terrorism Hits Home: How Prepared Are State and Local Law Enforcement?*, MG-104-MIPT (Santa Monica, CA: RAND, 2004).

Ferguson, C.D., Potter, W.C, Sands, A., Spector, L.S. and Wehling, F.L., *The Four Faces of Nuclear Terrorism* (Monterey, CA: Center for Nonproliferation Studies, Monterey Institute, 2004).

Ganor, B. *The Counter Terrorism Puzzle: A Guide to Decision Makers* (New Brunswick and London: Transaction Publishers, 2005).

Gunaratna, R., *Inside Al Qaeda: Global Network of Terror* (New York: Columbia University Press, 2002).

Hoffman, B. *Inside Terrorism* (New York: Columbia University Press, 1999).

Howard, R., Forest, J. and Moore, J. (eds.), *Homeland Security and Terrorism: Readings and Interpretations* (New York: McGraw-Hill, 2006).

Huntington, S., *The Clash of Civilizations and the Remaking of World Order* (New York: Simon & Schuster, 1996).

Jacquard, R., *In the Name of Osama Bin Laden: Global Terrorism and the Bin Laden Brotherhood* (Durham, NC: Duke University Press, 2003).

Juergensmeyer, M., *Terror in the Mind of God: The Global Rise of Religious Violence*, 3rd edn. (Berkeley, CA: University of California Press, 2003).

Kean, T.H. and Hamilton, L.H., *The 9/11 Report: The National Commission on Terrorist Attacks upon the United States* (New York: St. Martin's Press, 2004).

Kepel, G., *Jihad: The Trail of Political Islam* (London: I.B. Tauris, 2003).

Lederberg, J. (ed.), *Biological Weapons: Limiting the Threat* (Cambridge, MA: MIT Press, 1999).

National Criminal Intelligence Sharing Plan: Executive Summary, US Department of Justice, 2003.

Pillar, R., *Terrorism and U.S. Foreign Policy* (Washington, DC: Brookings Institution, 2001).

Rapoport, D., "Modern Terror: The Four Waves," *Current History*, Dec. 2001.

Reich, W. (ed.), *Origins of Terrorism: Psychologies, Ideologies, Theologies, States of Mind* (Washington, DC: Woodrow Wilson International Center for Scholars and Cambridge University Press, 1990).

Sassen, S., *Losing Control? Sovereignty in an Age of Globalization* (New York: Columbia University Press, 1996).

Slaughter, A., *A New World Order* (Princeton, NJ: Princeton University Press, 2004).

Sloan, S., *Terrorism: The Present Threat in Context* (Oxford: Berg, 2006).

INDEX

291